AF541007

ENCYCLOPAEDIA OF STEM CELLS - 2

EMBRYONIC STEM CELLS

By

Dr. Amita Sarkar

Dept. of Zoology

Agra College

Agra (U.P.)

(India)

DISCOVERY PUBLISHING HOUSE PVT. LTD.

NEW DELHI-110 002

First Published-2009

ISBN 978-81-8356-408-3

Published by:

DISCOVERY PUBLISHING HOUSE PVT. LTD.

4831/24, Ansari Road, Prahlad Street,
Darya Ganj, New Delhi-110002 (India)
Phone: 23279245 • Fax: 91-11-23253475
E-mail: dphbooks@rediffmail.com
dphtemp@indiatimes.com
Website: www.discoverypublishinghouse.com

Printed at:

Sachin Printers, Delhi

Preface

The present title *Embryonic Stem Cells* is the amazing advancement of biotechnology. It provides the various fundamental aspects of stem cell technologies to be understood adequately. It has been compilated for graduate and undergraduate students, research scholars, teachers, practising biochemical engineers, biotechnologists, applied and industrial microbiologists, cell biologists and scientists involved in bioprocessing research and development. The basic concepts have been clearly explained and their functions are adequately highlighted. The presentation of the text is simple and systematic. The selection of chapters and topics is according to the specified syllabi of several Indian Universities and are so structured as to enable the student to move easily from the fundamental to the complex. It is our earnest hope that this title will be of great value to all our students.

To make the work more comprehensive and informative, the author has consulted many authoritative books, research journals, abstracts, monographs etc. He is grateful to all those great scholars whose work are cited or substantially reproduced.

There can be no claim to originality except in the manner of treatment and much of the information has been obtained from the books and scientific journals available in the different libraries.

The author expresses his thanks to his friends and colleagues whose continue inspirations have initiated him to bring out this book.

The author expresses his gratitude to Mr. Wasan and staff of M/s Discovery Publishing Pvt. Ltd. for their whole hearted co-operation in the publication of this book.

In the mean time, the author will remain sincerely responsible for any shortcomings of the book and the grateful to the readers for their suggestions and constructive criticism for the continuous betterment of the book. He takes this opportunity to appeal to the readers to send their suggestions straightway to his publisher.

Author

Contents

1

INTRODUCTION

Embryonic stem (ES) cells were first isolated in the 1980s by several independent groups. These investigators recognized the pluripotential nature of ES cells to differentiate into cell types of all three primary germ lineages. Gossler et al. described the ability and advantages of using ES cells to produce transgenic animals. The next year, Thomas and Capecchi reported the ability to alter the genome of the ES cells by homologous recombination. Smithies and colleagues later demonstrated that ES cells, modified by gene targeting when reintroduced into blastocysts, could transmit the genetic modifications through the germline. Today, genetic modification of the murine genome by ES cell technology is a seminal approach to understanding the function of mammalian genes in vivo. ES cells have been reported for other mammalian species (i.e., hamster, rat, mink, pig, and cow), however, only murine ES cells have successfully transmitted the ES cell genome through the germline. Recently, interest in stem cell technology has intensified with the reporting of the isolation of primate and human ES cells.

ES cells are isolated from the *inner cell mass* (ICM) of the blastocyst stage embryo and, if maintained in optimal conditions, will continue to grow indefinitely in an undifferentiated diploid state. ES cells are sensitive to pH changes, overcrowding, and temperature changes, making it imperative to care for these cells daily. ES cells that are not cared for properly will spontaneously differentiate, even in the presence of feeder layers and *leukemia inhibitory factor* (LIF). In addition, healthy cells growing in log phase are critical for optimal transformation efficiency in gene targeting experiments.

Targeted murine ES cells have little value if they lose the ability to transmit the introduced mutations through the germline of the resulting chimeras. Therefore, it is critical that murine ES cells have a normal 40 XY karyotype. It is standard practice in our laboratory to have complete karyotypic analysis of all targeted ES cells prior to the production of chimeras. The criteria used in our laboratory to qualify an ES cell clone for making chimeras is that at least 50% of the chromosome spreads analyzed must be 40 XY. In our experience, our DBA/1LacJ ES cells meet or exceed that criterion at least 86% of the time, whereas our 129 strain of ES cells meet or exceed the criteria 45% of the time.

The many opportunities that exist in stem cell biology today, combined with the need to further explore and develop new technologies, makes it necessary to clearly define the process of developing stem cell lines. Therefore, this chapter will present the methods used in our laboratory to develop murine ES cell lines and maintain them in an undifferentiated state.

MATERIALS

Mice for Blastocyst Stage Embryos and Primary Embryonic Fibroblasts

1. DBA1/LacJ, 129/SvJ, and C57BL/6 inbred mice were obtained from Jackson Laboratories.
2. MTK-neo CD1 transgenic mice were obtained from Dr. Colin Stewart for the production of *primary embryonic fibroblasts* (PEF) for feeder cells.

Tissue Culture Plastic and Glassware

1. 35-mm Petri dish.
2. 4-Well multiwell tissue culture dish.
3. 24-Well multiwell tissue culture dish.
4. 12-Well multiwell tissue culture dish.
5. 6-Well multiwell tissue culture dish.
6. T-25 Flask.
7. 100-mm Tissue culture dishes.
8. 60-mm Tissue culture dishes.
9. 50-mL SteriFlip filter unit.
10. 150-mL Stericup filter unit.
11. 250-mL Stericup filter unit.
12. 500-mL Stericup filter unit.

13. Nalgene controlled-rate freezer.
14. Bright-Line hemacytometer (improved Neubauer counting chamber).

Media and Reagents

1. ES cell qualified light mineral oil.
2. M2 Medium.
3. KSOM.
4. Knockout Dulbecco's Modified Eagle medium (KO-DMEM).
5. ES cell qualified *fetal bovine serum* (FBS).
6. 0.2 m*M* L-Glutamine (100×).
7. 0.1 m*M* MEM *nonessential amino acids* (NEAA).
8. 50 U/ml penicillin/50 μg/mL streptomycin (100X).
9. 1000 μ/mL ESGRO or LIF.
10. 0.1 m*M* 2-Mercaptoethanol (BME).
11. Dulbecco's *phosphate-buffered saline* (PBS).
12. 0.05% Trypsin EDTA.
13. 10 μg/mL Mitomycin C.
14. 10% *Dimethyl sulfoxide* (DMSO).
15. 175–300 μg/mL G418 (Geneticin 50 mg/mL).
16. 2 μM/L Gancyclovir (Ganc).
17. HAT supplement (100X) 10 m*M* sodium hypoxanthine, 40 μM aminopterin, and 1.6 m*M* thymidine.
18. 0.1% Gelatin in sterile water.
19. Mouse Y-ES system.
20. Mycoplasma *Plus* PCR detection primer set.
21. Mycoplasma stain kit.

Methods

Preparation of Media Used for Feeders and ES Cells

1. The list of reagents for the different culture media's used for ES cells and PEFs can be found in Table 1.1. All reagents are combined and filtered through 0.2-μm filter units. ES cells are sensitive to pH change, therefore, when a bottle is about half full, the remaining medium is filtered into a smaller bottle. This practice minimizes the air space in the bottle that causes the pH to raise as air gases and medium reach equilibrium.

Preparation of Feeder Layers from PEF

1. PEFs were isolated from 12–14-d-old transgenic MTK-neo CD1 embryos and frozen as described. Frozen vials of PEF cells are

thawed by agitation in a 37°C water bath until cell suspension becomes a slurry. Transfer the cell suspension into 49 mL DMEM with serum, L-glutamine, and BME (sDMEM) in the 50-mL tube. Pipet up and down gently and transfer 10 mL cell suspension into each of 5 labeled 100-mm dishes (approx 1.5–2.0 × 10^6 cells/dish). Rotate plates back and forth to distribute cells evenly over entire dish.

Table 1.1. Media protocols for ES cells and feeder cells

Reagents	*sDMEM*	*SCML*	*G418/Ganc/ SCML*	*G418/ SCML*	*HAT/ SCML*
KO-DMEM	500 mL	500 mL	500 mL	500 mL	500 mL
FBS	50 mL	90 mL	90 mL	90 mL	90 mL
L-Glutamine	5 mL	6 mL	6 mL	6 mL	6 mL
MEM/NEAA	—	6 mL	6 mL	6 mL	6 mL
BME	4 μL	4 μL	4 μL	4 μL	4 μL
LIF	—	60 μL	60 μL	60 μL	60 μL
Pen/Strept	2.5 mL	3 mL	3 mL	3 mL	3 mL
G418	—	—	2.1–3.6 mL	2.1–3.6 mL	—
Gancyclovir	—	—	2 μM	—	—
HAT	—	—	—	—	6 mL

2. Incubate 2–3 d and examine for confluence. When approx 80% confluent, remove media and replace with 6 mL mitomycin C (10 μg/mL in sDMEM) and incubate 2–5 h. After treatment, remove mitomycin C solution, wash with 10 mL PBS, then add 10 mL sDMEM. Incubate in sDMEM until ready to use.
3. The day before harvesting blastocysts to develop new ES cell lines, remove media from one 100-mm PEF feeder layer, and rinse with 10 mL PBS. Incubate 2–3 min in 2 mL trypsin EDTA. Dislodge the PEF cells by tapping the dish against the palm of your hand. When cells release from the dish, add 24 mL sDMEM to neutralize the trypsin and pipet up and down to produce a single-cell suspension (approx 2.5–3.5 × 10^5 cells/mL). Transfer 1 mL/well of six 4-well dishes. Incubate overnight. The next day, remove media, wash with 1 mL PBS/well, then add 1 mL (SCML). These 4-well dishes are ready to receive embryos.

Preparation of Gelatin-Coated Dishes

1. Warm the 0.1% gelatin solution in a 37°C water bath. Transfer enough gelatin solution to cover the bottom of the dish (i.e. 0.5 mL/well for 4 or 24 wells, 1 mL/well for 12 wells, 2 mL/well

for 6 wells, 3 mL for 60-mm dishes and 6 mL for 100-mm dishes). Let gelatin solution sit at room temperature for 30 min in a tissue culture hood.

2. Remove the excess gelatin solution and use dishes immediately. Do not allow the gelatin to air-dry.

Obtaining Blastocyst Stage Embryos

1. Blastocysts can be obtained from super-ovulated or naturally mated females. However, we believe blastocysts are generally more fit from natural matings.
2. For natural matings, place two females per male on Thursday mornings. Check for copulation plugs daily. This is typically done before 10 AM to ensure the identification of all mated females. Separate plugged females and label for blastocysts embryos 3 d later. Set up 10–15 males and 20–30 females this way.
3. On d 3.5 *post coitus* (p.c.), sacrifice plugged females, and flush blastocyst stage embryos from both uterine horns as described. Transfer the embryos through several M2 drops to wash away uterine fluids and debris. Finally, transfer one washed embryo into a 4-well dish with fresh PEF feeder layer in SCML. PEF feeders may be eliminated if you have 1000 U/mL LIF (ESGRO) in the medium.

Culture of the Blastocyst and Picking of the ICM

1. Observe the embryos daily to monitor fitness, hatching, and attachment to the feeder layer or gelatin-coated plastic. When the embryos have attached, the ICM will become apparent.
2. Using a drawn mouth pipet, tease the ICM away from the rest of the embryo and gently aspirate it into the pipet. Transfer the ICM into one well of a 24-well dish previously prepared with fresh PEF feeders and SCML. If you prefer not to use feeder layers, gelatin coat the wells and proceed in the same manner as with PEF feeders.

Isolation of Putative ES Cells from the ICM

1. The ICM should attach to the feeder layer or gelatin-coated dish overnight. The next day, remove the media and wash the cell layer with 0.5 mL PBS/well. Remove the PBS and add four drops of 0.05% trypsin EDTA. Incubate for 1–2 min. Vigorously tap the dish against the palm of your hand to dislodge the cells into suspension. When fully detached, add 2 mL SCML/well and pipet up and down to dissociate cells into a single-cell suspension. Record

this as S11 p1 (split one to one, passage one) and return the cells to the incubator.

2. Twenty-four hours after splitting, remove the media from each well and replace with 2 mL SCML/well. Examine the cells in each well and record the morphology. Following examination, feed the cells daily by removing the old medium and replacing with 2 mL fresh SCML. Every second or third day, the colonies must be dissociated and the passage number recorded. Never allow colonies to become larger than 400 μm in diameter. If the colonies are less than 100 μm in diameter, wait another day before dissociating. We believe that keeping the colonies small aids in maintaining pluripotency. Large colonies tend to flatten and differentiate.
3. The new ES cells generally remain in the 24-well dish for 2–3 passages. When the colonies appear to be evenly dispersed over the dish, it is time to move the cell population to a larger 12-well dish. Individual colonies should never be allowed to overgrow, forming a monolayer. Follow the same procedure as above, to trypsinize the cells.
4. When the trypsinized cells are in suspension and no longer attached to the dish, they are ready to be moved to the next size dish. Using a 5-mL pipet, aspirate 3 mL SCML into the pipet. Tilt the 24-well dish and express 2 mL SCML into the well, then immediately aspirate the entire contents of the well into the pipet. Quickly transfer 2 mL of the volume into one well of a previously prepared 12-well dish (PEF feeders or gelatin-coated). With the remaining 1 mL SCML in the pipet, go back and wash the well in the 24-well dish to ensure that all cells have been removed. Then add the remaining 1 mL to the 2 mL cell suspension already in the well of the 12-well dish. Pipet up and down to completely dissociate the cells into a single-cell suspension. Repeat this procedure for each well and make sure to record passage number. Note that, at this stage, only a few embryos will move into the 12-well dish, because many will die at this stage.
5. The next day, examine each well, record morphology, and change the media with 2.5 mL fresh SCML/well. Follow the same media change and dissociating procedures as described in this section, with the exception that the 12-well dish will use 0.5 mL trypsin. Generally, there will be only one S1:1 in the 12-well dish.
6. When there are enough colonies to move to the next sized vessel, transfer to one 100-mm dish. At this point, the cells are typically

at passage 5. Prepare a 100-mm dish with 10 mL fresh SCML on a PEF feeder layer or gelatin. Remove the media from the 12-well dish and wash with 1 mL PBS. Remove the PBS and add 0.5 mL trypsin. Incubate for 1–2 min, then dislodge the cells from the dish by tapping the dish against the palm of your hand. Once these cells are dislodged, aspirate 5 mL SCML into a 10-mL pipet. Tilt the 12-well dish and express 2 mL SCML into the well, then quickly aspirate the contents of the well into the pipet. Immediately express 3 mL into the previously prepared 100-mm dish. Return to the 12-well dish and express the remaining 2 mL SCML in the pipet into the well, then quickly aspirate the contents of the well back into the pipet. This is to ensure that you have removed all the cells from the well. Add the last 2 mL to the 100-mm dish and pipet up and down to dissociate the cells into a single-cell suspension. There should be approx $0.5–1.0 \times 10^7$ total cells in the suspension. Incubate overnight.

7. The following day, record morphology and change the media with 15 mL fresh SCML. On the second day after the move into the 100-mm dish, either change the media again or, if the cells are ready, split them 1:2 based on colony size (if colonies are less than 100 μm in diameter, feed that day and wait another day to split).
8. From this point on, the new ES cell population is being expanded and cryopreserved. Therefore, every time the cells are split, part of the cell suspension must be passed for expansion (approx 2×10^6 cells/100-mm dish) and part will be cryopreserved. Pass the cells in a 100-mm dish by removing SCML and washing with 10 mL PBS. Remove the PBS and incubate in 2 mL trypsin for 1–2 min. After incubation, vigorously tap the dish against the palm of your hand to dislodge the ES cells from the dish.
9. Once the cells are completely in suspension, tilt the dish and add 8 mL SCML to wash the cells into a pool at the bottom of the tilted dish. Aspirate the cell suspension into the pipet and transfer into a 15-mL conical tube. In the 15-mL tube, gently aspirate the cells up and down 3–4 times to dissociate into a single-cell suspension. Leave 5 mL of the cell suspension in this tube and transfer the remaining 5 mL cell suspension into another 15-mL tube (one tube is for freezing and one is to maintain cells). Pellet the cells by centrifugation at 110*g* for 5 min.
10. While the cells are in the centrifuge, prepare two 100-mm dishes of fresh PEF feeders by washing the monolayer with PBS and

adding 5 mL SCML. (If using a gelatin-coated dish, just add 5 mL SCML to the dish.) After centrifugation, aspirate the supernatant from both tubes, taking care not to disturb the cell pellet. Resuspend the cell pellet from one tube in 10 mL SCML. Count the cells using a Neubauer counting chamber, then transfer 2×10^6 cells/dish into the previously prepared 100-mm dishes with PEF feeders or gelatin and record the passage number (should be around p6). At this stage, there should be enough cells to plate one or two 100-mm dishes. Resuspend the cell pellet in the other 15-mL tube with enough freezing medium to freeze 4–6 $\times 10^6$ cells/mL for each cryovial. Transfer 1 mL of cells in freezing media into 1.5-mL cryovials labeled with the name of the cell line, with or without feeders, the passage number, freeze number (F1 in this case), and your initials. Place cryovials of cells into a controlled-rate freezer at –80°C overnight.

11. The next day, transfer the cryovial of cells into long-term freezer storage, in either liquid nitrogen or a –150°C freezer. Record location in freezing log. Next, examine the cells that were passed and record morphology. Change the media by removing the old media and replacing with 15 mL SCML.
12. Once the cells are into the 100-mm dish, the new ES cell line is usually established. Continue to carry the cells for expansion of the line to ensure many vials in cryopreservation. The next split should be S1:6 or S1:8. Freeze 3 or 4 vials, respectively. Aim to freeze 4–6 $\times 10^6$ cells/vial in 1 mL freezing medium. We typically accumulate approx 50 vials.

Characterization of Putative ES Cells

It is necessary to characterize the ES cell lines to determine sex, karyotype, pluripotency, and absence of pathogens. It is preferred to have a male cell line, because XY ES cells can sex convert an XX blastocyst in a chimeric embryo development, and these resulting chimeric males can produce more offspring than females. In addition, it is necessary to determine the karyotype of the ES cell lines, because transmission of the ES cell genome through the germline of the chimeras is dependent upon the ES cells having a normal chromosome number. Finally, the ability to differentiate into many cell types and the ability to make healthy chimeras is dependent upon the cells being free of pathogens, such as mycoplasma and murine viruses. Therefore, it is necessary to test for mycoplasma contamination and *murine antibody production* (MAP) testing for antibodies against murine viruses.

Sex determination to identify XY ES cell lines

1. The first step in determining the sex of the novel ES cells is a PCR screen. Pick 6 colonies into individual microfuge tubes that contain 10 μL sterile water. Put the tubes in a –20°C freezer for 10 min. Next, remove the tubes from the freezer, vortex mix for several seconds, and then pulse-spin to collect lysate in the bottom of the tube. Follow the instructions for the Y-ES system to PCR screen for the Y chromosome.
2. The next step is to do a full karyotype of all cell lines determined to be male by PCR. Karyotyping can be done according to published protocols or contracted. We typically contract our ES cell karyotyping. At the time of splitting, 1–1.5 $\times$ 10^6 cells are transferred into a T25 Flask in 10 mL SCML and cultured overnight. The next day, the medium is removed, and the flask's lid, if filled to the brim with SCML, is closed tightly, and the lid and neck are wrapped in parafilm to prevent leakage. The flasks are packed and shipped to Coriell Cell Repository for full karyotyping.

Mycoplasma and murine viral contamination testing

1. To test for mycoplasma contamination, you may do a simple Hoechst stain using the Sigma kit (follow insert instructions) or do a PCR of the supernatant (follow Stratagene insert instructions).
2. To test for murine viral contamination, we send a vial of frozen cells to Charles River Laboratories for MAP testing.

In vitro differentiation (IVD)

1. To remove the ES cells from the PEF feeders, aspirate the media from the dish and wash the cell layer with 10 mL PBS. Remove the PBS and add 2 mL trypsin. Immediately take the dish to the microscope and place on the stage. While observing the cells through the eyepieces of the microscope, tap the dish to dislodge the rounded ES cell colonies. As soon as many of the colonies are floating and the feeder layer is still attached, return the dish to the hood and aspirate the colony suspension and transfer into a 15-mL conical tube. Add 8 mL SCML, pipet up and down to dissociate the colonies, then pellet by centrifugation at 110*g* for 5 min. Resuspend the pelleted cells in 15 mL SCML, plate in a 100-mm tissue culture dish without PEF feeder layer, and incubate overnight. The next day, change the media on the feeder-free ES cells by removing the old media and adding 15 mL SCML.

2. To begin the IVD experiment, change the media and add 15 mL SCML, approx 1–2 h before dissociating the cells. Next, remove the media and wash the cell layer with 10 mL PBS. Remove the PBS, add 2 mL fresh trypsin, and incubate 1–3 min. Check the cells every 30 s for dissociation by tapping the dish against the palm of your hand. When the colonies are completely free-floating, return the dish to the hood, add 8 mL SCML, and pipet up and down until the cells are in a single-cell suspension. Count the cells using a hemocytometer, then pellet the cells by centrifugation at 110*g* for 5 min.
3. After centrifugation, aspirate the supernatant, taking care not to disturb the cell pellet, then resuspend the cells in 10 mL *stem cell medium* (without LIF) (SCM). Plate the cells at a concentration of 1–2 × 10^5 cells/mL in a vol 10 mL SCM in a 100-mm bacterial dish. This suspension culture will allow the cells to form cell aggregates called *embryoid bodies* (EBs).
4. Change the media every 2–3 d by transferring the EBs into a 15-mL conical tube and letting them settle out of suspension into the bottom of the tube. Aspirate the supernatant, add 10 mL fresh SCM, then transfer the EB suspension back into the bacterial dish.
5. After 7–9 d of culture, transfer the EB suspension into a 15-mL conical tube and again allow to them to settle out. Remove the supernatant, add 10 mL PBS, and allow the EBs to settle out. After the EBs have settled to the bottom, again remove the supernatant, add 3 mL of trypsin, and incubate for 3 min at 37°C. Following incubation, add 7 mL SCM to the trypsin solution and pipet up and down vigorously to dissociate the EBs. Pellet the cell suspension by centrifugation at 110*g* for 5 min. Remove the supernatant and resuspend the cells in 10 mL SCM. Transfer into two 100-mm tissue culture dishes and increase the vol to 12 mL SCM in each dish.
6. Examine for differentiated morphology daily and feed SCM every second day. Many different cell populations should become apparent, including blood islands and contracting myocytes. Additional details of IVD methods can be found in other chapters of this text.

Gene targeting ability and germline transmission

1. To test for the ability of your ES cells to undergo homologous recombination, a vector of known targeting frequency should be used. Electroporations are carried out as described below.

2. Ultimately, the novel ES cells must be capable of colonizing the germline of chimeric mice. The ES cells can be microinjected into blastocysts or aggregated with morula, according to standard protocols. Producing chimeras with host blastocysts or morula from strains different from the ES cells allows one to use coat color genetics to identify germline transmission of the ES cell genome.

Maintenance of ES Cells

Thawing ES cells

1. To prepare a fresh 100-mm PEF feeder plate, remove the old media, wash with 10 mL PBS, then add 15 mL SCML. Check the date on the feeder dish and examine to determine that feeder cells are healthy. Primary embryonic fibroblast feeders usually last 7–10 d. Put prepared feeders back into the incubator to equilibrate cells with higher serum concentration. (If you are thawing clones from an electroporation to expand, prepare a well in a 6-well dish.) These clones are 1/2 well of a 24-well dish when frozen.
2. Remove a vial of cells from the –150°C freezer and plunge into 37°C water bath, agitating the vial until the frozen suspension becomes a slurry. Sterilize the vial with 70% ethanol and transfer to a tissue culture hood.
3. Transfer the contents from the vial into the previously prepared PEF feeder plate. Most vials have enough cells to evenly plate a 100-mm dish with colonies (approx 4–6 × 10^6). Gently swirl the plate to distribute cells over the entire PEF feeder surface. Label the dish with the cell line, passage number, date, and then return the plate to the incubator.
4. Change the media the next morning, by removing the old media and replace with fresh SCML. Return the dish to the incubator and culture another day. If the cells recovered easily from the freeze–thaw, they should be ready to split approx 48 h after thawing.

Daily feeding of ES cells

1. Examine the dish for the condition of ES cell colonies and record observations. It is critical to monitor colony morphology, since this is the only gauge of culture conditions. Healthy ES cell colonies have smooth borders, the cells are tightly packed together so the individual cells are not detectable, and the entire colony has depth, giving a refractile ring around it.

2. Remove the media from the healthy cells and replace with SCML. Slowly aspirate the media down the side of the dish so that the cell layer is not disturbed.

Table 1.2. Media volumes and cell counts for ES cells in various different tissue culture and multiwell dishes

Dish size	*Media volume*	*Cell count*
4-well dish	0.5 mL	Embyros
24-well dish	1.0 mL	2.0×10^4
12-well dish	2.0 mL	3.0×10^5
6-well dish	5.0 mL	4.0×10^5
35-mm dish	3.0 mL	4.0×10^5
60-mm dish	5.0 mL	6.0×10^5
100-mm dish	15.0 mL	2.0×10^6
4-chamber slide	1.0 mL	1.4×10^5
8-chamber slide	0.5 mL	6.0×10^4

Subculture of ES cells

1. Change the media by replacing with 15 mL fresh SCML approx 1–2 h prior to passage and return the dish to the incubator.
2. Examine the dish for colony morphology, density, and size, and prepare feeder plates based on the determined split ratio. Decide the ratio to split the cultures based on the size and distribution of ES cell colonies. An even distribution of colonies averaging in size 200–400 μm in diameter and spaced around 400 μm apart in a 100-mm dish will have 1.0–1.5 $\times 10^7$ total cells. We typically split cultures at ratios from 1:6 to 1:8 resulting in approx 1.5–2.0 $\times 10^6$ cells to be plated in each new 100-mm tissue culture dish. Splitting ES cells will ensure healthy passage and no overcrowded or undercrowding.
3. Remove the media and wash with 10 mL PBS. Remove the PBS, add 2 mL trypsin (for a 100-mm dish; 0.5 mL/well of a 6- or 12-well dish; 4 drops/well of a 24-well dish), and incubate for 1–2 min, checking the dish every 30 s by tapping the dish against the palm of your hand to dislodge the colonies.
4. Once the cells are no longer attached, add 8 mL SCML to the trypsin cell suspension. Pipet up and down vigorously to dissociate cells. Then plate 2×10^6 cells to each prepared 100-mm dish and cryopreserve the remaining cell suspension.

Freezing ES cells

1. Transfer the remaining cell suspension into a 15-mL tube and pellet the cells by low-speed centrifugation at 110*g* for 5 min.
2. Remove the supernatant taking care not to disturb pellet. A 100-mm dish will yield enough cells to freeze 4–5 vials (approx 3–6 $\times$ 10^6 cells/vial).
3. Add 1 mL freezing medium (50% FBS, 40% SCML, and 10% DMSO) for each vial frozen based on cell number. Pipet up and down to dissociate the ES cells and transfer 1 mL cell suspension per cryovial.
4. Put cryovials into a Nalgene controlled-rate freezer box and then put the box into a –80°C freezer. The next day, transfer the vials of frozen ES cells into the –150°C freezer for long-term storage.

Electroporation of ES Cells for Gene Targeting

ES cell preparation

1. Thaw ES cells 4–5 d prior to electroporation.
2. Approximately 48 h after thawing, the cells should be ready to be split. Prepare two 100-mm feeder dishes with fresh SCML. Freeze the cell suspension or pellet for DNA as a control for wild-type..
3. Change the media on the ES cells with fresh SCML 1–2 h before electroporation. At the same time, dissociate the PEF cells from two 100-mm dishes and make 5 new dishes. This is done to minimize feeders rescuing ES cells during the selection process.
4. Prepare the ES cells from one of the two dishes made 2 d previously for subculture. While the cells are in trypsin, remove the old media from one PEF feeder dish, wash with PBS, and add 15 mL fresh SCML. Dissociate the cells. Transfer 1.0 mL trypsinized cell suspension (approx 2 $\times$ 10^6 cells) into the newly prepared feeder dish, which will be used as a control for selection, and transfer the remaining 8.5 mL ES cell suspension to a 15-mL centrifuge tube (approx 1–1.5 $\times$ 10^7 cells) for electroporation.
5. To the remaining dish, add 7 mL SCML and pipet up and down. Transfer the cell suspension to another 15-mL centrifuge tube for freezing. Pellet the contents of both tubes by centrifugation at 110*g* for 5 min.
6. Aspirate the supernatant and resuspend the cells to be electroporated in 10 mL SCML. Pellet again as in step 5. This is to ensure that all the trypsin has been removed.

Electroporation of ES cells

1. Ideal electroporation settings must be determined for each different type of instrument used. We use the BTX ECM-600 electroporator with the following conditions: set volts at 280 V, capacitance at 50 μf, and resistance timing at R8 (360 ohms). Turn BTX unit on and push reset button to clear. Place a 0.4-mm disposable cuvette into the hood taking care not to touch the rim of the lid or the metal sides of the cuvette.
2. Transfer 25 μg DNA into a microfuge tube. Care must be taken when removing the microfuge tube from the container so that sterility is maintained, therefore handle the tubes by the sides and avoid touching the inside of the cap or rim of the tube.
3. Remove the supernatant from the cell pellet in the 15-mL tube, then with a 1-mL pipet add 375 μL SCML to the DNA, and then pipet up and down to thoroughly mix the DNA and SCML. Transfer the SCML/DNA solution to the cell pellet and pipet up and down to ensure a single-cell suspension. Finally, transfer the cell suspension into a 0.4-mm cuvette. Replace the lid on the cuvette to maintain sterility.
4. Place the cuvette into the holding apparatus of the electroporator and make sure there is good contact to the electrodes. Push reset button to clear. To electroporate, press the "*automatic charge and pulse*" button. When electroporation is complete, record actual voltage and pulse length (time is in milliseconds.) Remove the cuvette from the holder and return to hood.
5. Following electroporation, set the cuvette off to the side to allow the ES cells to recover for approximately 10–15 min. Prepare the feeder dishes. Remove the media from the 4 feeder dishes that were previously prepared and add 15 mL fresh SCML to each dish. Also, transfer 12 mL SCML to a 15-mL tube and set aside.
6. Using the transfer pipet that came with the cuvette, aspirate a small volume of SCML from the 15-mL tube to wet the inside of the pipet so that the cells will not stick to the pipet. Now aspirate the electroporated cell suspension into the pipet slowly. Transfer the suspension to the 15-mL tube and repeat to ensure that most of the cells have been transferred to the tube. Using a 10-mL pipet, gently pipet up and down to disperse the cells, then transfer 3 mL cell suspension into each of the 4 new feeder dishes previously prepared. It is very important to pipet the newly

electroporated ES cells gently to ensure minimal cell damage. Incubate overnight in SCML.

7. The next morning, examine the dishes for colony morphology and cell survival. Record your observations. Remove the old media from the 4 dishes that contain the electroporated ES cells and the one selection control dish, and then replace with selection media. The selection medium used depends on the type of ES cell line and targeting vector used. HAT/SCML is used when the targeting vector restores the *hyposanthine phosphoribosyl transferase* (HPRT) function in HPRT-deficient ES cells, whereas 6-thioguanine/SCML is used when the targeting vector deletes the HPRT function in an ES cell line. G418-Gancyclovir/SCML is used for positive–negative selection when the targeting vector contains the neomycin resistance gene and the *thymidine kinase* (TK) gene. Positive selection selects for cells that are neomycin resistant, whereas negative selection selects for cells that have lost the TK gene during homologous recombination. Since prolonged use of gancyclovir is harmful, we only use it in our medium for the first 4 d of selection. Then, on d 5, we switch to G418/SCML and use this medium throughout the remainder of selection.
8. Examine all 5 dishes and record observations daily. Then remove the old media and replace with fresh selection media. Selection generally takes 7–9 d.

Picking ES cell colonies

1. Approximately 7–13 d following electroporation, the ES cell colonies are ready to be picked. Prepare 24-well feeder plates using one 100-mm PEF feeder dishes for each 24-well dish. Wash with 10 mL PBS, then add 2 mL 0.05% trypsin EDTA to each 100-mm dish. Incubate 1–2 min, then check for dissociation. Tap the dish against the palm of your hand to dislodge cells from the dish. If cells are not completely free-floating, incubate for another 30–60 s. When completely dissociated, add 22 mL sDMEM and pipet up and down, then transfer 1 mL to each of the 24 wells. Return the dishes to the incubator until ready to use.
2. When ready to pick colonies, remove the old media from each well of the 24-well feeder dish and replace with fresh selection medium. Prepare a 100-mm bacteriology dish with microdrops of PBS or SCML. These will be used to wash the pipet between picks. Make sure you have sterile drawn pipets to use for picking

and a filter on your mouth pipet tubing. This will help ensure the cultures remain free of contamination.

3. Place a dish with selected colonies on the microscope stage and examine it for colonies with the best morphology. Pick colonies that are approx 300 μm in diameter using a drawn mouth pipet.
4. Transfer the colony to one well in a 24-well dish and blow until bubbles appear in the well. Draw media from the well up and down in the pipet to transfer all ES cells into the well. Wash the pipet in a microdrop of PBS or SCML and pick next colony. We generally pick 48 colonies into two 24-well dishes over a 2- to 3-d period with DBA/1LacJ ES cells. However, with 129 ES cells, it is often better to pick all colonies the same day.

Expanding picked colonies into clonal ES cell lines

1. The days after you pick colonies, examine each well for the presence of ES cells. Observe each well to determine the average size of the surviving colonies. When the colonies are nearly 300 μm in diameter, dissociate them. If they are smaller and look fragile, change the media and leave the cells alone until the next day.
2. When the colonies are ready to dissociate (1–2 days after picking), remove the old media from each well. Wash by adding 0.5 mL PBS to each well, remove the PBS, then add 4 drops of trypsin solution per well, and incubate 1–2 min.
3. After incubation, vigorously tap the dish against the palm of your hand to dislodge the cells. Once the cells are completely dissociated, add 2 mL selection medium to each well. The next day, examine each well and record observation. Change the media in each well with 1.5 mL of fresh selection medium.
4. To keep ES cells undifferentiated, they must be dissociated every other day and the media changed daily. Dissociation and media changes may need to be done several times in the 24-well dish before there are enough ES cells to split 1:2 (half for freezing and half for DNA analysis). Not all clones grow at the same rate, therefore each clone must be handled as a separate cell line. When there are enough colonies (200–400 μm in diameter) to cover the dish, spaced 200–400 μm apart, they are ready to split.
5. Examine each well and mark the colonies that will be dissociated and left in the 24-well dish and the colonies that are ready to split 1:2 (half will be cryopreserved and half transferred into 12-

well dishes). Record the clone numbers in the data book. To prepare the 12-well dishes, follow the same protocol as for the 24-well dish. One 100-mm dish of PEFs (6–8 × 10^6 cells) will make two 12-well dishes. After trypsinizing the cells in the 100-mm dish, add 47 mL of sDMEM to the 2.0 mL of trypsin cell suspension. Pipet up and down and transfer 2 mL cell suspension/well into the two 12-well dishes. Let incubate about 1–2 h prior to use.

6. Before splitting the ES cells, change the media in each well of the previously prepared 12-well dishes and replace with 0.5 mL selection medium. Remove the media from the clones in the 24-well dish and wash with PBS. Add 4 drops of trypsin solution to each well and incubate 1–2 min at 37°C. After incubation, vigorously tap the dish against the palm of your hand to dissociate all the cells in the wells. For the wells that are just being dissociated and not split 1:2, add 2 mL selection media to each well. For each well to be split 1:2, aspirate 3 mL of selection medium into a 5-mL pipet. Transfer 1 mL of this medium into one well of the trypsinized 24-well dish and aspirate the entire contents of that well, then transfer to the 12-well dish. Pipet the entire volume up and down several times in the 12-well dish to ensure that all the ES cells are completely dissociated. Then transfer 1.5 mL of the cell suspension into the appropriate prelabeled cryovial, leaving the remaining cell suspension in the 12-well dish to continue growing for DNA analysis. When all the clones are transferred into the 12-well dish, fill each well with selection media to total 3 mL.
7. Pellet the cells in the cryovials by centrifugation at 110*g* for 5 min. Pour off the supernatant and add 0.5 mL freezing medium to each vial. Vigorously shake all the vials and place in a controlled-rate freezer at -80°C. The next day, transfer the vials into a liquid nitrogen freezer or –150°C freezer until the targeted clones are identified.
8. Prepare the ES cells for DNA analysis. Change the media on the ES cells for DNA analysis daily until they are overly confluent. At that point, remove the media from the cells and prepare for the extraction of genomic DNA for analysis by preferred method.
9. Once the targeted clones are identified, thaw those clones, except transfer the thawed cells into a prepared well of a 12-well dish (remember the frozen clone was half of a 24-well). When the cells are ready to be split, move half of the well into a 100-mm

dish on a new feeder in SCML. The remaining half in the 12-well dish can be grown for DNA. We do this routinely to confirm that the ES cells thawed are the targeted line. Once targeting is confirmed, all nontargeted ES cell isolates can be discarded.

10. The targeted cells in the 100-mm dish will most likely need to be split 11 the first time. With the next split, begin freezing vials. We typically freeze 8–10 vials of targeted ES cells from the first two splits.
11. Once targeting is confirmed, choose 2–3 targeted ES cell lines for karyotyping.

Preparation of ES Cells for Aggregations or Microinjection into Blastocyst Stage Embryos

Whole plate shake-off method

1. When ready to prepare ES cells for microinjection or aggregation, remove the media and wash with 10 mL PBS by tilting the dish and letting the PBS run down the dish to a pool in the bottom. Remove the PBS wash and add 2 mL trypsin solution. Immediately place the dish on the microscope stage and tap the dish gently while observing that some of the ES cell colonies will dislodge from the feeders. This process takes approximately 30 s if the trypsin is fresh.
2. When enough ES cell colonies are free from feeders, return the dish to the hood. Tilt the dish so the loose colony suspension pools at the lower edge. Using a 1-mL pipet, aspirate 0.5 mL of the colony suspension in trypsin and transfer to a 1.5-mL microfuge tube. If the cells are to be used for blastocyst microinjection, let the tube set for about 30–60 s to allow the cells to further dissociate from the colonies. Then add 1 mL M2 medium to the microfuge tube to inactivate the trypsin and pipet up and down to completely dissociate the ES cells into a single-cell suspension.

   If the cells are to be used for aggregations, after 15–30 s in trypsin, add 1 mL M2 medium to the microfuge tube to inactivate the trypsin. Pipet up and down several times so the cells are still in small clumps.
3. Allow the cells remaining in the original dish to finish dissociating in the trypsin solution (1–2 min total). When these cells are completely free-floating, aspirate 4 mL SCML into a 5-mL pipet, tilt the dish, and wash the cell population into a pool at the bottom of the tilted dish. Pipet up and down several times to

completely dissociate the cells, then transfer 1.5 mL of the cell suspension into a 1.5-mL microfuge tube to freeze for DNA. Transfer 1 mL cell suspension into the new PEF feeder (1:10 split). This dish will be used to carry the cells for additional microinjection or aggregation. Transfer the remaining 3 mL of cell suspension into a 15-mL conical tube for freezing if this is the first split. Only freeze the first split, then discard the surplus cell suspension thereafter.

4. Place the two microfuge tubes (ES cells to inject and cells to pellet for DNA) into the microfuge and spin for 5 min at 110*g*. To make sure trypsin is removed, aspirate the supernatant and add 1 mL M2 to the cells for injection and add 1 mL PBS to the cells for DNA. Repeat microfuge to pellet again. Aspirate the supernatant from the cells used for microinjection or aggregation and resuspend the cell pellet in 50 μL M2 by gently pipetting up and down to dissociate the cells. The cells are ready to be injected into blastocysts or aggregated with morula.

Notes

1. It is important to emphasize aseptic technique. Always scrub hands before handling dishes with 70% ethanol. Always douse bottles and vials with 70% ethanol before putting into hood. Always flame bottles before opening them. Never reenter a bottle with the same pipet more than once. A good motto for all tissue culture practices is "when in doubt throw it out". It takes several months to generate targeted clones, so aseptic technique cannot be overemphasized.
2. ES cells are very sensitive to pH and temperature changes, as well as overcrowding and undercrowding. Once the cells are thawed, you must be committed to caring for them every day and even over the weekends and holidays. There are no good short cuts for this routine care.
3. ES cells maintain their pH best if the CO_2 concentration is between 5–10%. Therefore, to reduce the risk of fluctuations above or below that range, we keep our incubators set at 6%. If an incubator is not very stable, consider replacing it.
4. The quality of the reagents used in tissue culture of ES cells is also critical. Where possible, purchase products that are qualified for ES cell culture. We found that improving reagent quality has increased our clone survival and targeting frequency.
5. ES cells from different mouse strains react differently to the FBS used in the medium. When working with ES cell lines from

different mouse strains, make sure to include all cell lines in your tests of different serum lots. We found this to be necessary even for ES cell qualified serum.

6. Keep in mind that ES cells grow more slowly following freeze–thaw and need time to recover. It is not unusual to dissociate the cells within the same dish or split 1:2. When this happens, an extra 2 d should be estimated into the time to thaw prior to electroporation, and use 2 μ gancyclovir for ± selection and only G418 for + selection.
7. Specific selection conditions need to be established for ES cell lines developed from different mouse strains. We use 300 μg G418/mL SCML ES cells from 129 mouse strains, and the 129 ES cell colonies are picked on d 9 and 10. For DBA1/LacJ ES cells, we use 175 μg G418/mL SCML, and ES cell colonies surviving selection are picked as early as d 7 and as late as d 13 following electroporation.
8. Typically, after electroporation and selection, you will notice differentiated cells and dead floating cells. Therefore, you must carefully choose the colonies to pick. Also, after you pick a colony, it sometimes dies. These we call "*tried but died*" and are probably being rescued from selective pressure by feeder cells in the 100-mm dish. Finally, we avoid large (>450 μm in diameter) "*perfect*" looking colonies, because of the observation that these colonies may have developed trisomy 8.
9. We store a pellet of ES cells, which were used to produce chimeras from each day of injections or aggregations, just in case there is a problem later and the resultant mice do not demonstrate the introduced genetic modification. This is an excellent control to have available, if the targeted mutation or germline transmission is not successful.

2

Serum-free Media

The removal of serum or other undefined components from cell culture medium involves a multifaceted approach. This is mainly due to the interactive nature of cell culture systems. Cell culture systems require various components: basal nutrients, a set of supplements that convey information, such as growth factors, mitogens, hormones, and cytokines, and for anchorage-dependent cells, a substrate. In designing any cell culture system attention must be paid to all these components and their interactions.

The role of the basal nutrient component needs to be addressed in a systematic manner. For normal eukaryotic cells, the optimization of the basal nutrients is an important aspect in the generation of serum-free media.

The removal of an essential nutrient, such as an amino acid, vitamin, or mineral, would effectively inhibit the activity of the cells in a serum-free culture environment. It has been demonstrated qualitatively that various cell types require the same nutrients, but that quantitatively there are significant differences. It is reasonable to assume that there is an optimized concentration for a set of nutrients for a particular function for a particular cell type.

Role of Serum and other Undefined Tissue Extracts in Cell Culture Systems

In media these undefined supplements (serum, pituitary brain, and other extracts) provide not only essential nutrients and attachment factors, but also act as detoxificants. The undefined nature of these supplements and their potential for carrying adventitious agents and determining the phenotype of cells, motivates their avoidance or minimization whenever possible. These supplements can contain inhibitors or promoters of the processes under study. In the bioprocessing industry the contribution of these supplements to the final product necessitates extensive downstream purification. In addition, the cell phenotype can be better controlled in the absence of serum.

Response Curves

Proliferating Cultures

The quantitative evaluation of nutrient requirements necessitates titration of the basal components in a systematic manner. The responses evaluated can be any cellular characteristic: proliferation, product secretion, or expression of a differentiated function. Two types of responses are possible: the essential nutrient response, and the non-essential nutrient response. The essential response is characterized by a zero response to zero concentration, followed by a positive increase to rising concentrations, a plateau region where no increase in response is observed with higher concentration, and finally an inhibitory region. The slope of the positive response, the length of the plateau, and the slope of the inhibitory regions are dependent on the nutrient and cell under study. The aim of this exercise is to determine the concentration of the nutrient that is represented by the middle of the plateau.

In optimizing the concentration of a nutrient, amino acids represent the largest group where interactions occur, because there are common transport systems for amino acids. These systems interact in a positive or negative manner. A consequence of this is that by changing the concentration of one amino acid it is possible to influence the response curve of another amino acid. By placing the concentration of a component in the middle of the plateau region, a shift in any direction will minimize the risk of entering a deficient or inhibitory area.

The non-essential condition is characterized by a significant response at zero concentration of the nutrient. There may also be a response to increasing concentrations of the nutrient. If the desired outcome is maximal proliferation, then the concentration of the non-essential nutrient also needs to be optimized. Examples of this are observed L-phenylalanine and L-tyrosine are titrated using normal human *aortic smooth muscle cells* (AOSMC). The L-tyrosine titration demonstrates that in the absence of L-tyrosine, proliferation close to maximum is observed. However, Lphenylalanine and L-tyrosine are titrated using normal human bronchial/tracheal epithelial cells (NHBE), L-tyrosine appears to be an essential nutrient. In whole animals L-phenylalanine can be converted to L-tyrosine by L-phenylalanine

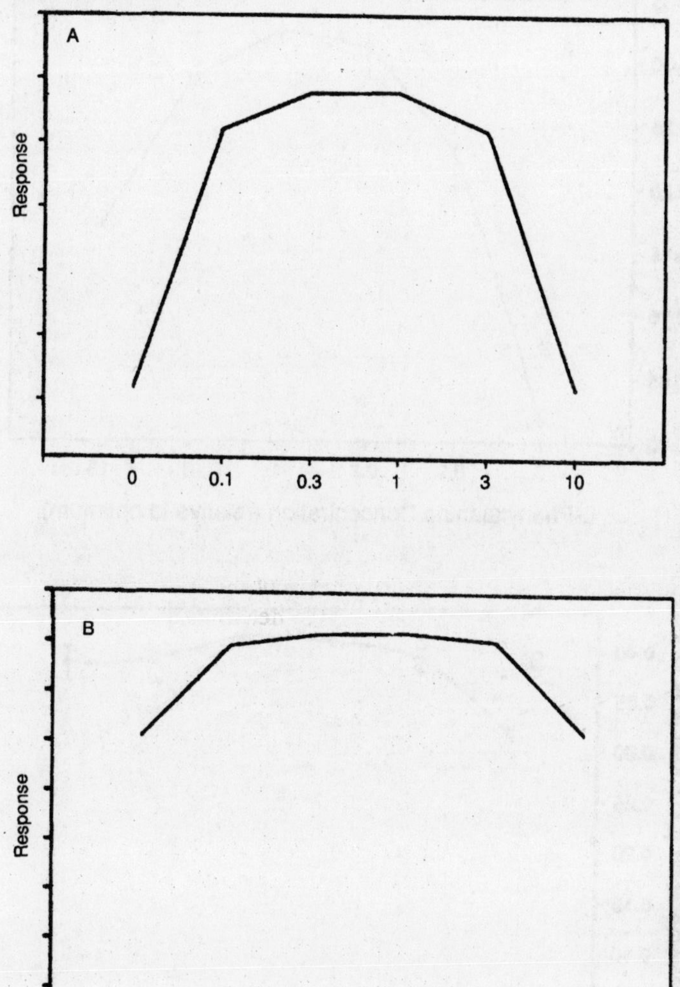

Fig. 2.1. Idealized response of an essential (A) and non-essential (B) nutrient.

hydroxylase. This metabolism has been observed in various other cell culture systems, but not with NHBE. In a similar fashion, titrating the amino acid pair L-methionine and L-cysteine, it is seen that L-cysteine is not an essential amino acid for AOSMC, whereas in NHBE, L-cysteine is essential. L-cysteine is synthesized from L-serine and L-methionine in the liver and brain and in fibroblasts expressing

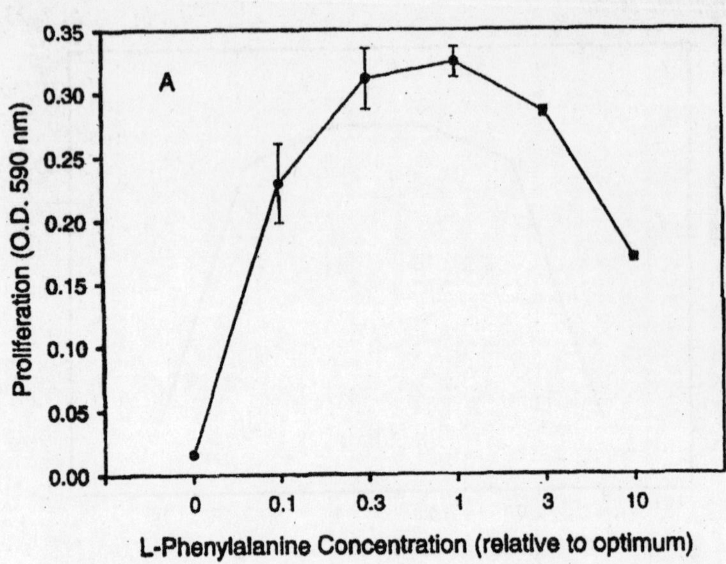

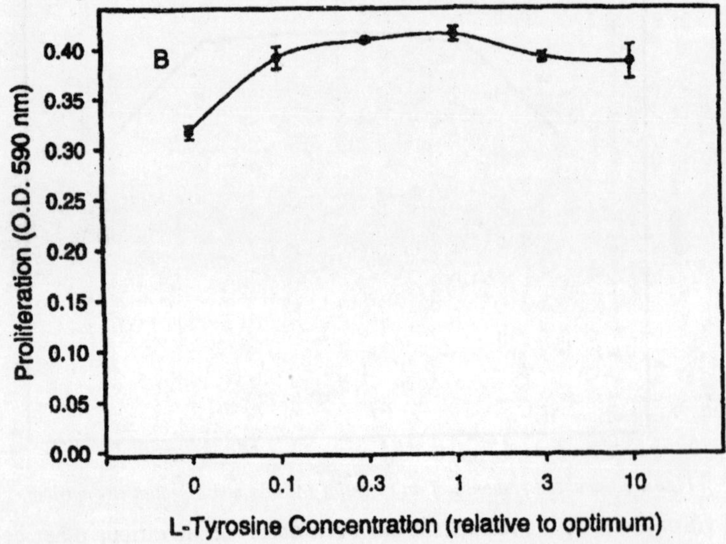

Fig. 2.2. Titration of L-phenylalanine (A) and L-tyrosine (B) using AOSMC.

cystathionine synthetase. Nutrient requirements for cells in culture are determined best in clonal growth assays. At high density, cells are able to condition the medium by excreting essential factors. In a

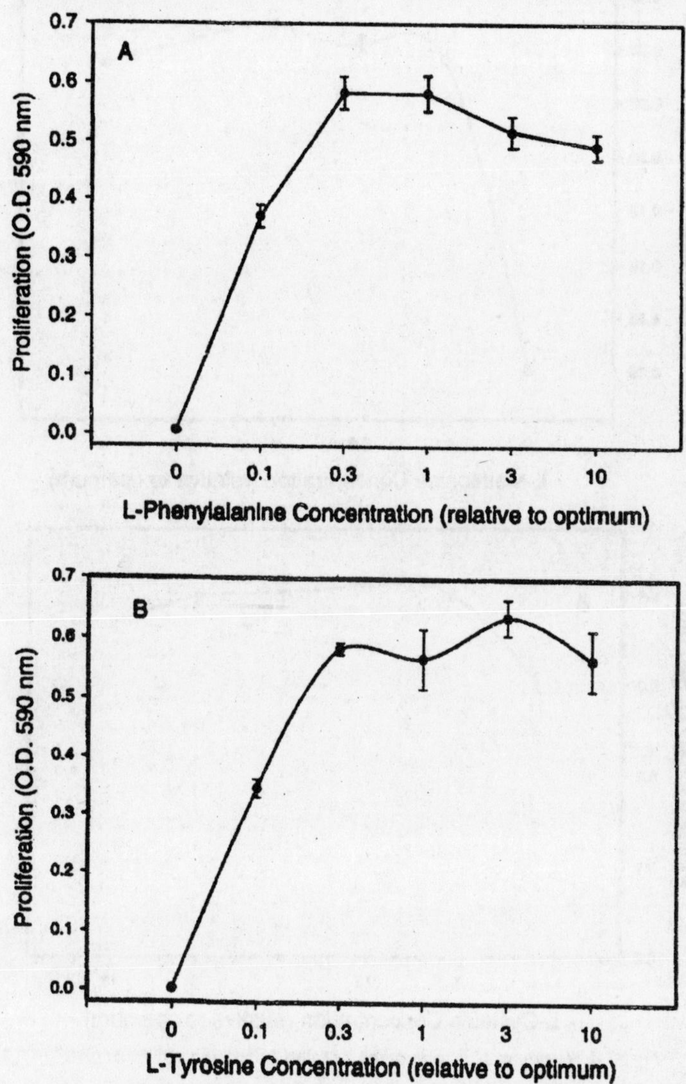

Fig. 2.3. Titration of L-phenylalanine (A) and L-tyrosine (B) in NHBE.

clonal growth assay the volume of medium is too large for the small number of cells to effectively condition the medium. Consequently, the cells are dependent on the nutrients present in the medium and a deficiency may be observed.

One basal component with profound effects on cell proliferation is sodium chloride (NaCl). The response is characterized by a relatively

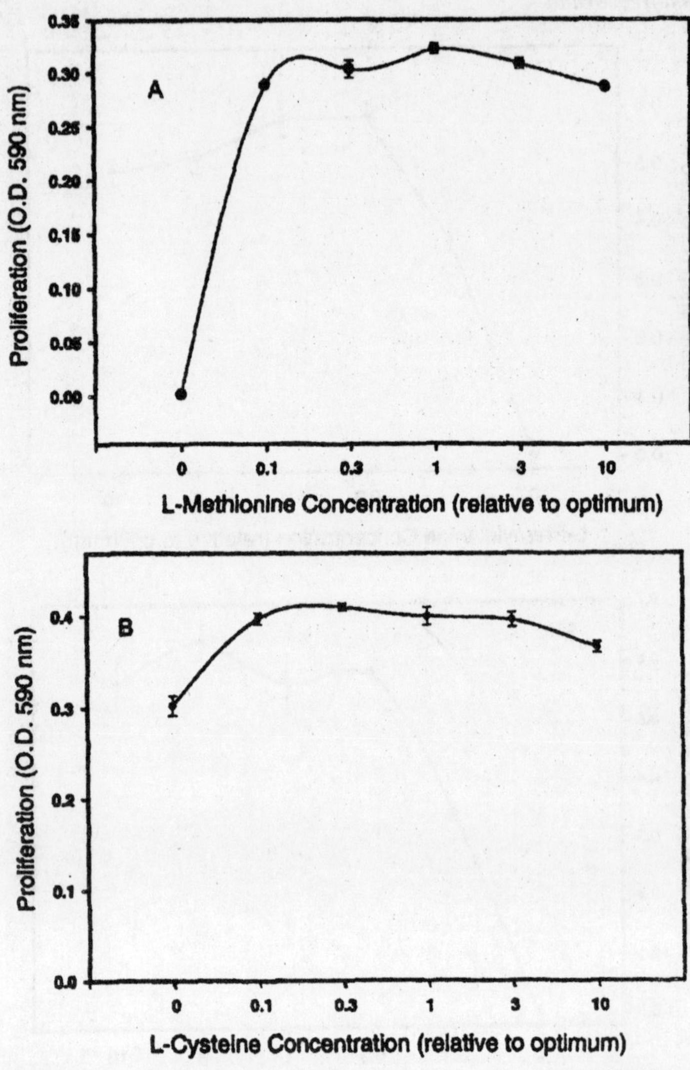

Fig. 2.4. Titration of L-methionine (A) and L-cysteine (B) in AOSMC.

sharp peak. The sensitivity to NaCl may be due to the energy expenditure associated with the maintenance of the Na^+-K^+-ATPase membrane transport.

Non-proliferating Cultures (Hepatocytes)

Normal adult rhesus monkey hepatocytes do not proliferate to any significant extent. A characteristic of hepatocytes is cytochrome P450

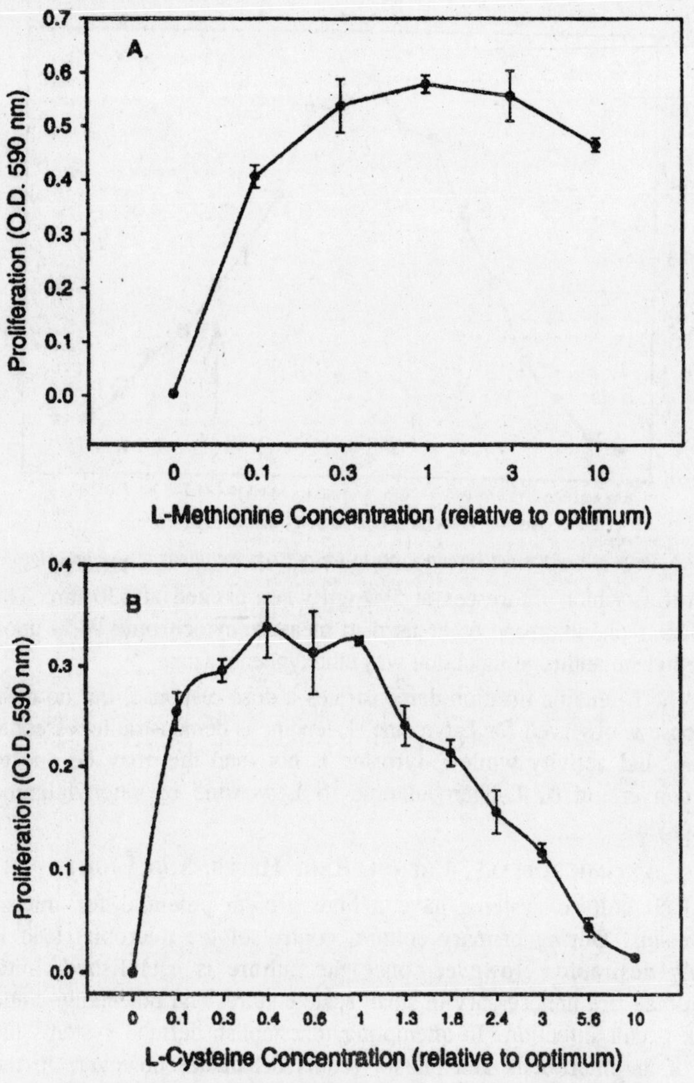

Fig. 2.5. Titration of L-methionine (A) and L-cysteine (B) in NHBE.

activity and the medium optimization uses this activity as an end-point. The titration of two amino acids is depicted using hepatocytes exposed for 48 hours to 3-methylcholanthrene. A homologous series of the n-alkyl ethers of phenoxazone (alkoxyresorufins) is available for tie study of various cytochrome P450s. The phenoxazone ethers are hydroxylated and 0-dealkylated to a common fluorescent metabolite,

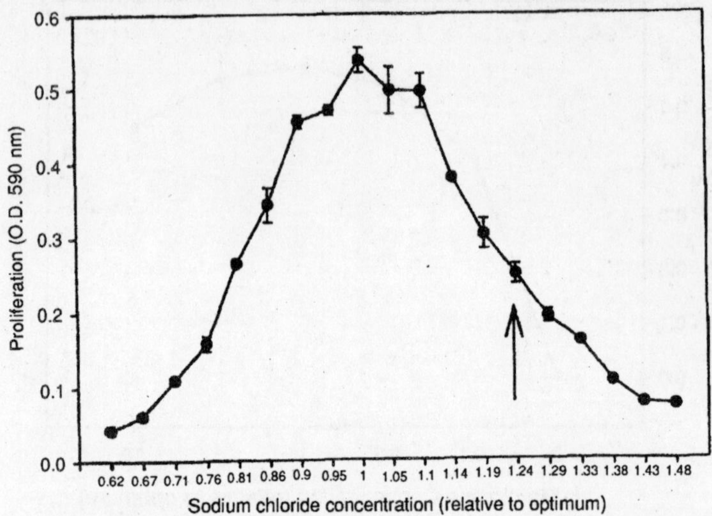

Fig. 2.6. Response of normal keratinocytes to various concentrations of sodium chloride.

resorufin, which fluoresces at 590 nm when excited at 530 nm. The particular phenoxazone ether used to measure cytochrome P450 upon 3-methylcholanthrene induction was ethoxyphenoxazone.

The L-leucine titration demonstrates a dose response, but no dose response is observed for L-tyrosine. L-leucine is demonstrating essential amino acid activity while L-tyrosine is not, and this may be due to the conversion of L-phenylalanine to L-tyrosine by phenylalanine hydroxylase.

Antimicrobials, Phenol Red, Hepes, and Light

Cell culture systems have a high growth potential for micro-organisms. During primary culture, control of the microbial load is highly desirable. However, once the culture is established, anti-microbials are unnecessary in small scale cultures and potentially could mask occult infection. In attempting to establish defined systems the use of antimicrobials and phenol red is debatable; however, in the bioprocessing industry where thousands of litres are at stake a microbial contamination would be a financial disaster.

Phenol Red

Phenol red is widely used as a pH indicator in cell culture media. Unfortunately, phenol red or contaminants can interact unfavourably with cultured cells, causing toxicity and oestrogenic activity. Sodium-potassium ion homeostasis is perturbed in serum-free medium containing

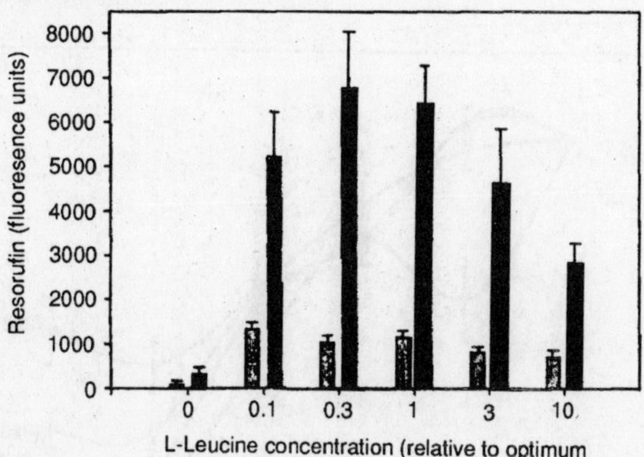

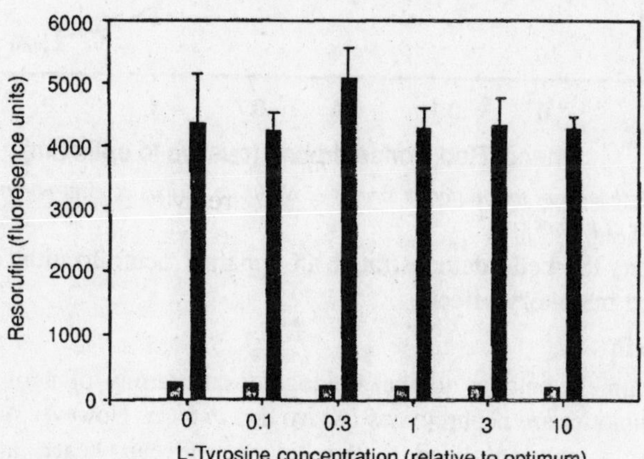

Fig. 2.7. Response of Rhesus monkey hepatocytes to varying concentrations of L-leucine and L-tyrosine.

phenol red and the effect is neutralized by serum or serum albumin. Grady *et al.* described a contaminant that became toxic when the pH rose above 7.4 but was virtually inert at lower pH values. HPLC analysis revealed many contaminating peaks and identified the cytotoxicity with a particular peak that differed from the oestrogenic activity. Anti-oestrogens frequently suppress growth below that of control cells not treated with oestrogen, even after careful steps have been taken to eliminate all oestrogen from the serum. The oestrogenic activity is due to a lipophilic impurity that binds to the oestrogen receptor with an affinity 50% that of oestradiol. These cells were cultured in the absence of *bovine pituitary extract* (BPE) and all experienced inhibition of proliferation. In the presence of BPE, at 3.3×10^{-6} M

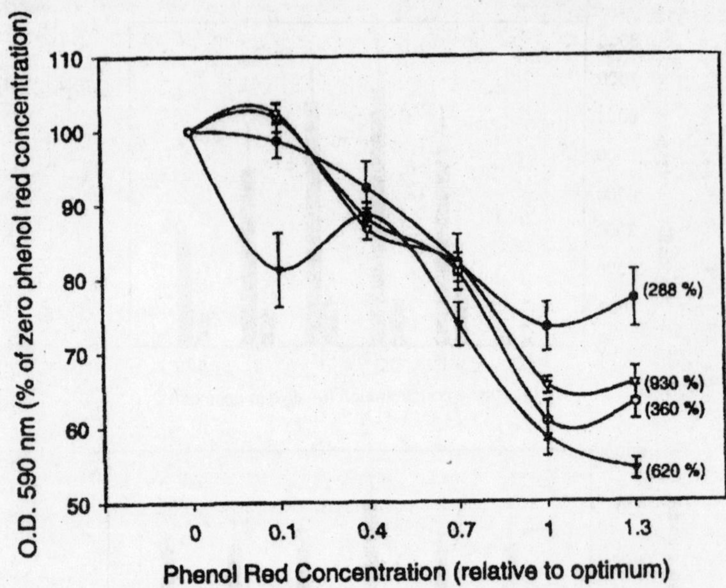

Fig. 2.8. Responses of four different strains of NHBE to various concentrations of one lot of phenol red.

phenol red, the cells demonstrated an apparent neutralization of the phenol red inhibitory effect.

Gentamicin

Gentamicin belongs to the aminoglycoside family of antibiotics. These antibiotics are nephrotoxins *in vivo* and *in vitro*. However, toxicity is not exclusively localized to the kidney: the ear, heart, and the cornea have all demonstrated susceptibility.

The mechanisms of toxicity range from free radical formation to calcium transport inhibition. Hepatocyte metabolic conversion of gentamicin to a compound toxic to sensory cells from the inner ear is another mechanism of toxicity. Gentamicin is toxic *in vitro* to corneal epithelial cells, fibroblasts, and hepatocytes.

Hepes

Hepes has been used extensively in cell culture due to its excellent buffering capacity in the physiological range and low binding of cations, but it can be toxic. For example, cytoplasmic vacuoles developed in chick embryo epiphyseal chondrocytes and membrane inclusion bodies in human dermal fibroblasts that resolved when Hepes was replaced by another buffer.

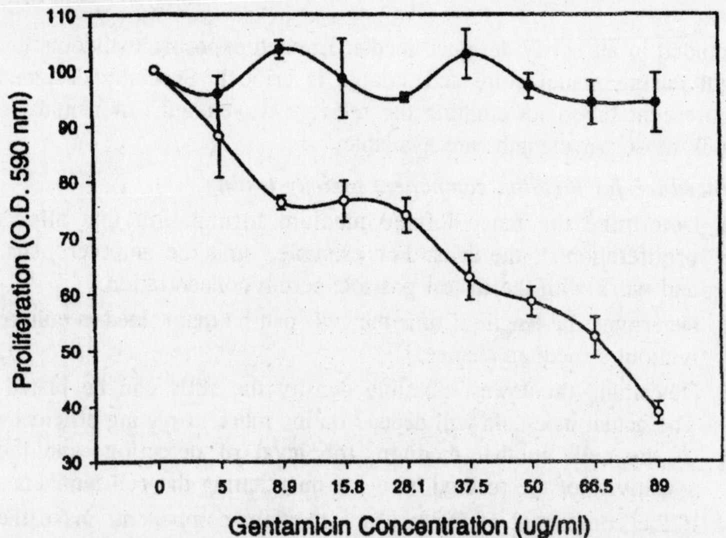

Fig. 2.9. The response of NHBE to various concentrations of gentamicin from two different lots.

The toxicity exhibited by Hepes appears to be mediated by the production of reactive oxygen species. The interaction of Hepes-buffered culture medium with fluorescent light at room temperature resulted in toxicity to V79 Chinese hamster cells. The toxicity was prevented by shielding or the inclusion of catalase in the medium, implicating extracellular hydrogen peroxide as the toxic agent. In endothelial cells, Hepes stimulates the production of toxic oxygen metabolites resulting in a decrease in growth. Quin2 is a transition metal ion chelator that potentiates iron-driven oxidant formation. Oxidants were detected in Hepes buffer. SIN-1 is an oxide-releasing compound and its toxicity to L929 cells is due to a cooperative action of hydrogen peroxide and reactive nitrogen species in the presence of Hepes.

Light

Fluorescent light can cause the deterioration of tissue culture medium, resulting in toxicity and mutagenicity. The deleterious effects are due to prolonged exposure, whereas short exposures can produce a mitogenic effect. The photo toxicity is due to wavelengths of light that can be absorbed by riboflavin. The pairing of riboflavin with tryptophan, Hepes, and guanine results in toxicity mediated by reactive oxygen species. Tryptophan and riboflavin are found in conventional and serum-free optimized media and because they are essential they will be included in all newly designed media. Limited exposure to fluorescent light during manufacture and storage is critical. Specially designed fluorescent tubes not emitting the relevant wavelengths or shields to block those wavelengths are available.

Procedure for medium component toxicity testing

1. Determine the most defined medium formulation that allows proliferation of the cells. For example, omit the antimicrobials, and work with the lowest possible serum concentration,
2. Determine the length of time the cells can be maintained in culture without a medium change.
3. Determine the lowest possible density the cells can be plated. The actual inoculum will depend on the inherent plating efficiency of the cells in that medium, the level of detection, and the sensitivity of the method used for quantitating the cell number.
4. If the component of interest is a medium component, make the medium from scratch minus that component. If the component is a supplement leave out the supplement.
5. Introduce the component at concentrations that flank the presumptive or existing concentration. Use logarithmic intervals, the size of which will be determined by the sensitivity of the response. The intervals need to be determined empirically.
6. Introduce the medium with the various concentrations into the plates and preincubate for at least 30 min before introducing the cells.
7. Incubate for the length of time determined in the above step.
8. Determine the relationship of component concentration to cell performance.

Purity of Components

In cell culture the largest component of the system is water. Consequently, a relatively small contaminant in the water can have a significant effect on the cells. On the other hand water may provide essential nutrients that become limiting as the purity of water is increased. One class of nutrients that could account for this phenomenon is the trace metals.

Ideally, in chemically defined media all the components are known and there are no contaminants present. Though the availability of recombinant proteins and peptides has made the pursuit of chemically defined media more attainable, most fall short of this ideal. Cells may need albumin or other serum fractions for the desired performance or stability. Also, manufacturers may not necessarily screen for all potential contaminants in their organic compounds, such as selenium contamination of a thyroxine preparation. The many surfaces with which the medium components come in contact during manufacturing may contribute contaminants.

The bioprocessing industry is redefining the notion of purity due to the potential infectious agents that may be present in medium components. It is desirable that all components are of non-animal origin. For example methionine and cysteine are generally isolated from animal hair, so new sources are required.

Fatty Acids

Serum provides lipids in various forms to cultured cells. These forms include cholesterol, phospholipids, triglycerides, fatty acids, fat soluble vitamins, and various esterified forms of these lipids.

The essential fatty acids that cannot be synthesized by animals are 18:2n-6 and 18:3n-3 ('18' refers to the number of carbon atoms and the number following the colon refers to the total number of double bonds in the molecule). Animals do not contain the enzymes necessary to place a double bond in these positions. Consequently, the acquisition of these molecules occurs only through an exogenous supply; however, animals (excluding true carnivores) contain desaturation and elongation/shortening enzymes that result in polyunsaturated fatty acids, most notably arachidonic acid, 20:4n-6, and eicosapentenoic acid, 20:5n-3, and fatty acids of higher degrees of unsaturation and elongation.

The removal of serum can result in an essential fatty acid deficiency since serum is the predominant lipid source for cultured cells. In the absence of a lipid source cells will synthesize, desaturate, and elongate fatty acids of the n-9 and n-7 families. Under these deficiency conditions there is a characteristic fatty acid profile represented by the polyunsaturated fatty acids of the n-9 and n-7 families. Cultured cells will utilize exogenous fatty acids preferentially to endogenously synthesized ones. The various fatty acid families are metabolized by the same enzymes but at different rates. The essential fatty acids are preferentially metabolized over the non-essential fatty acid families. This situation can be exploited to generate cells with a fatty acid profile favouring a particular fatty acid family.

Polyunsaturated fatty acids have an extensive metabolism; generating eicosanoids, epoxides with vascular function, anandamide (a ligand of the cannabinoid receptors), and non-enzymatic oxidation products of 20:4n-6, the isoprostanes, which are isomers of prostaglandins. Also, polyunsaturated fatty acids have functions that result from the fatty acid and not its metabolic product. For example, intracellular arachidonic acid opened a potassium-selective channel in neonatal rat atrial cells and under some conditions extracellular arachidonic and docosahexanoic (22:6n-3) acids blocked the major voltage-dependent potassium channel in cardiac cells. Fatty acids can also influence the physical characteristics of the membrane by influencing its fluidity. Polyunsaturated fatty acids can increase the membrane fluidity, as in J774A.1 cells, and decrease the uptake of acetylated low density lipoprotein.

The supplemented fatty acid is represented to a large extent in the *phosphatidylcholine* (PC) of fibroblasts. The cells are capable of metabolizing the precursor fatty acid by elongation and desaturation to other fatty acids. Cells with different fatty acid compositions may be functionally different. It may be tempting to consider the removal of serum the final step in the process of generating a scrum-free medium, but care must be taken to consider the fatty acid composition of the cells.

Complexing Fatty Acids to Albumin

1. Either bovine or human serum albumin can be used. The albumin needs to be close to fatty acid-free, and not just fatty acid-poor,
2. Run a toxicity curve to determine non-toxic concentrations.
3. Use the highest albumin concentration possible. For the chosen albumin concentration, calculate the fatty acid concentration that results in a mole ratio of fatty acid to albumin that is not greater than 2. This calculation determines the highest concentration of fatty acid possible in the system.
4. Dissolve the fatty acid in 100% ethanol of the highest purity. Calculate a fatty acid concentration so that once it is complexed to the albumin the final concentration of the ethanol the cells experience does not exceed 0.1%; however, each cell culture system needs to be tested.
5. Add the fatty acid dissolved in ethanol to the albumin solution in the smallest possible volume. Initially the solution will be cloudy but will clear upon gentle rotation at room temperature for approx. 15-20 min.
6. Maintain the smallest possible headspace and shield from fluorescent light. Ideally, flush the surface of the solution with nitrogen through a 25 mm 0.2 um sterile filter disk.
7. Aliquot the fatty acid-albumin complex and store at –70°C, Maintain low headspace.

3

Cell Cycle in Embryonic Stem Cell

Mouse Embryonic Stem Cells

The molecular mechanisms underlying self-renewal of pluripotent *embryonic stem* (ES) cells is still poorly understood. Deciphering these mechanisms is of prime importance for at least two reasons: (1) ES cells derive from, and are closely related to, the pluripotent stem cells of the blastocyst, the founder cells of the whole embryo proper. Hence, they constitute a unique model for studying embryonic development at the time of implantation, when embryos are inaccessible to experimental manipulation; and (2) Isolating and manipulating ES cells in species of economic or therapeutic interests is more difficult than in the mouse. It is likely that better defining their growth requirements will lead to major improvements in their culture conditions.

During the past 6 yr, intrinsic features of mouse ES cells regarding the regulation of their growth cycle have been pinpointed. These features may serve not only to understand how the cell cycle machinery of ES cells works, but also to better characterize ES cells isolated from embryos of other species. Hence, a striking feature of mouse ES cells is their unusual cell cycle distribution. The three phases of the cell cycle, G_1, S, and G_2/M, represent 15, 75, and 10%, respectively, of the total cell cycle, with a G_1 phase of approx 1 h. Hence, ES cells reenter the S-phase very shortly after exit from mitosis. These preliminary observations have paved the way to the analysis of cell cycle control in ES cells, focusing on the regulation of $G_1 \rightarrow S$ transition.

Retinoblastoma Pathway

The proliferation of mammalian cells is controlled largely during the G_1 phase of their growth cycles. The decision to initiate a new round of DNA synthesis is largely dependant upon the phosphorylation and functional inactivation of the *retinoblastoma* (RB) protein. This phosphorylation is driven by components of the cell cycle apparatus, specifically cyclins and *cyclin-dependent kinases* (CDKs). Of prime importance are complexes of D-type cyclins (cyclin D1, D2, and D3) and CDK4 or CDK6. Moreover, the cellular machinery that is organized to collect extracellular signals and transduce them via tyrosine kinase receptors and the SOS-RAS-MEKK-MAPK pathway seems to be dedicated largely to driving RB phosphorylation. This control circuitry appears to be operative in virtually all cell types. In contrast, the control of the ES cell mitotic cycle is likely to be markedly different. First, ES cells seem to lack the CDK4-associated kinase activity that characterizes all RB-dependent cells. They express very low levels of D-type cyclins, as a result of the very poor activity of the respective promoters. This is somewhat surprising as hypo-phosphorylated RB remains undetectable during the $M \rightarrow G_1 \rightarrow S$ transition, indicating that RB is rapidly rephosphorylated in G_1. Secondly, ES cells appear to be resistant to the growth inhibitory effect of the cyclin D:CDK4-specific inhibitor $p16^{ink4a}$, further suggesting that RB phosphorylation may not rely on proper CDK4-associated kinase activity in ES cells. Not surprisingly, induction of differentiation restores the expression of all three D-type cyclins, strong CDK4-associated kinase activity, and the sensitivity to the growth-inhibitory activity of $p16^{ink4a}$, suggesting that differentiating ES cells resume a normal cell cycle control.

Another important aspect of G_1 control lies in the regulation of cyclin D1 expression by the Ras→MAPK pathway. Phosphorylated ERKs activate cyclin D1 expression through fos and ets transcription factors. In ES cells, inhibition of ERK phosphorylation by wortmannin (an inhibitor of Ras activation) or by PD98059 (an inhibitor of *MEK*) neither inhibits background expression of cyclin D1 nor induces growth retardation. Induction of differentiation up-regulates the steady-state level of cyclin D1, whose expression then becomes sensitive to the inhibitors of the Ras→MEK→ERK cascade. Hence, cyclin D1 expression seems not to be regulated by Ras in ES cells. This regulation is likely to be restored upon differentiation.

Recently, it has been shown that Rb-E2F forms a transcriptional repression complex by recruiting histone deacetylase and SWI/SNF subunits. These large complexes are capable of blocking the transcription of cell cycle genes and remodeling chromatin. However, it is unclear if these large complexes have a specific role in chromatin organization of ES cells. Thus far, our preliminary immunoprecipitation experiments suggest that HDAC1 binds to the low amount of RB in ES cells. This could be a key aspect in the ES renewing cell cycle that should be investigated.

p53 Pathway

In somatic cells, cell cycle checkpoints limit DNA damage by preventing DNA replication under conditions that may produce chromosomal abberations. The tumor suppressor p53 is involved in such control as part of a signal transduction pathway that converts signals emanating from DNA damage, ribonucleotide depletion, and other stresses into responses ranging from cell cycle arrest to apopotsis. Stress-induced stabilization of nuclear p53 results in the transactivation of downstream target genes encoding, for example, the cyclin-dependent kinase inhibitor $p21^{cip1/waf1/sdi1}$ or Mdm2. $p21^{cip1/waf1/sdi1}$ inhibits RB phosphorylation, thereby preventing transition from G_1 to S. ES cells do not undergo cell cycle arrest in response to DNA damage (caused by γ-radiations, UV light, genotoxic agents) or nucleotide depletion. ES cells express abundant quantities of p53, but the p53-mediated response is inactive because of cytoplasmic sequestration of p53. Morevover, enforced expression of nuclear p53 still fails to induce cell cycle arrest, suggesting that, in addition to its cytoplasmic sequestration, p53 cannot activate the downstream targets required for growth arrest. One of these targets is $p21^{cip1/waf1/sdi1}$. ES cells do not express $p21^{cip1/waf1/sdi1}$, suggesting that the $p21^{cip1/waf1/sdi1}$ promoter is not responsive to p53 in ES cells. Therefore, it appears that ES cells have a very effective mechanism for rendering them refractory to p53-mediated growth arrest. Induction of differentiation restores the p53-mediated cell cycle arrest response.

Taken together, these results suggest fundamental differences in the regulation of cell proliferation in ES cells as compared to somatic cells. Firstly, they suggest that the complex apparatus that operates in most cells with extracellular mitogens, transducing signals through the SOS-RAS-RAF-MEKK-MAPK pathway and that ultimately leads to pRB phosphorylation is not engaged in ES cells. Induction of differentiation would reactivate this mechanism. Secondly, ES cells do not seem to have a p53-mediated checkpoint control. This control would also become operative when differentiation occurs.

MATERIALS

1. Feeder-independent ES cell line: CGR8.
2. Complete medium: *Glasgow's Modified Eagle's Medium* (GMEM) supplemented with 10% *fetal calf serum* (FCS), 2 m*M* L-glutamine, 1% nonessential amino acid solution, 1 m*M* sodium pyruvate, 0.1 m*M* 2-mercaptoethanol, 100 U/mL penicillin, 100 mg/mL streptomycin, and 1000 U/mL human *leukemia inhibitory factor* (LIF).
3. 0.25% (w/v) Trypsin in Phosphate-Buffered Saline (PBS).
4. 20 ng/mL Demecolcine.
5. 0.1% and 0.2% Gelatin dissolved in H_2O.
6. 5 m*M* BrdU (100× stock solution).
7. 1 mg/mL RNAse dissolved in PBS + 0.13 m*M* EGTA.
8. PBT: PBS + 0.5% *Bovine Serum Albumin* (BSA) + 0.5% Tween-20.
9. Anti-BrdU.
10. 100 μg/mL Propidium iodide (100X stock solution).
11. Sterile flasks and Petri dishes: sterile 5- and 10-mL pipets.
12. FACScan (fluorescence-activated cell sorter), equipped with a 15-mW 488-nm air-cooled argon-ion laser. Filters used: 530 nm *fluorescein isothiocyanate* (FITC), 585 nm (propidium iodide). Data aquisition and analysis are performed using CellQuest software.

METHODS

Synchronization of ES Cells by Mitotic Shake-Off

This protocol is intended to generate large numbers of synchronized ES cells exiting from mitosis, entering into G_1, and then into S phase, synchronously.

1. At d 1, trypsinize ES cells and seed at a density of 20–30 million cells in 25 mL complete medium in T160 flasks coated with 0.2% gelatin (gelatin is added to flasks at least 2 h before seeding the cells. Gelatin is thoroughly removed by aspiration just before seeding the cells). Incubate at 37°C in 7.5% CO_2.
2. At d 2, add 50 mL complete medium (removing the exhausted medium is not necessary) and incubate overnight.
3. At d 3, ES cells should form a confluent layer. Check that each flask is confluent. Discard those in which empty spaces are visible, as isolated clumps of cells are likely to detach from the flasks during the shake-off procedure. Then:

(a) Remove the loosely attached cells by preshaking the flasks 5 times by hitting the flasks against the palm of the hand.
(b) Quickly aspirate the medium and replace it with 25 mL complete medium containing 20 ng/mL demecolcine. Incubate for 3–4 h at 37°C in 7.5% CO_2.
(c) Shake the flasks 5 times by hitting them against the palm of the hand. Collect the medium in 50-mL disposal plastic tubes. From this step on, sterile manipulation is not required.
(d) Spin mitotic cells at 500*g* for 5 min. Aspirate the medium. Invert the tubes on absorbing paper for 5 min.
(e) Gently resuspend each pellet with 1 mL of prewarmed demecolcine-free medium using a P1000 Gilson pipet. Do not pipet the cells up and down more than required to get a single-cell suspension. Fill the tubes with complete medium.
(f) Spin at 500*g* for 5 min. Discard the medium. Invert the tubes onto absorbing paper to dry.
(g) Gently resuspend each pellet with 1 mL of prewarmed medium and pool into a single tube. Count the cells. This procedure yields approx 2 × 10^6 mitotic cells/T160 flask (i.e., approx 1% of the total number of cells).
(h) Prepare a cell suspension containing approx 10^6 mitotic cells/mL. Seed 6-cm dishes (coated with 0.1% gelatin) with 5 mL cell suspension. Incubate at 37°C in 7.5% CO_2.
(i) Collect the cells at various time points and analyze them for cell cycle distribution. Since mitotic cells usually take 4–5 h to attach to the dish, do not aspirate the medium. Any supplements should be added dropwise using 10X stock solutions.

Analysis of DNA Content in Synchronized ES Cells

As mitotic ES cells usually take 4–5 h to attach strongly to the Petri dish, the following protocol must be used to prepare a single-cell suspension suitable for flow cytometry:

1. Collect the cells by pipetting up and down approx 10 times with a P1000 Gilson to dissociate the loosely adherent cells. Trypsinization is not required. Transfer the suspension (>1 million cells) into a conical 15-mL tube.
2. Spin for 5 min at 500*g*. Discard the medium and wash in PBS. Repeat once.
3. After the last spin, resuspend cells in 100 μL of PBS. Pipet up and down with a P200 Gilson until clumps are no longer visible.

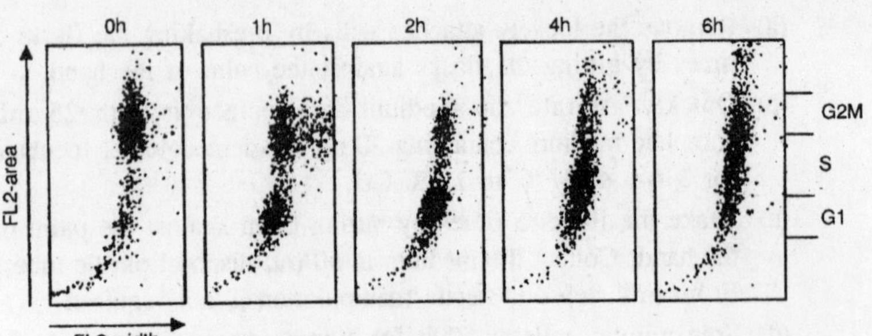

Fig. 3.1. Analysis of DNA content of ES cells synchronized by mitotic shake-off, determined according to the protocol.

Dropwise, add 1 mL of 70% ethanol at –20°C (1 drop/s to avoid formation of clumps of cells). Store the fixed cells at 4°C.

4. To analyze the DNA content, add 10 mL PBS directly to cells in ethanol. Incubate for 5 min at room temperature to allow cells to rehydrate.
5. Spin for 5 min at 500*g*. Resuspend the pellet in 100 μL of PBS. Add 10 mL PBS. Incubate for 5 min at room temperature.
6. Resuspend the pellet in 100 μL of 1 mg/mL RNase. Incubate for 20 min at room temperature. Store at 4°C if required (<24 h).
7. Add propidium iodide to a final concentration of 1 μg/mL. Incubate for 5 min at room temperature.
8. Analyze fluorescence using conventional setups.

Analysis of Cell Cycle Distribution in Exponentially Growing ES Cells

1. Refeed exponentially growing ES cells with complete medium. Incubate for 1 h at 37°C.
2. Add BrdU at a final concentration of 50 μ*M* and incubate for 30 min.
3. Trypsinize the cells and take 5 million cells for analysis of BrdU incorporation.
4. Spin for 5 min at 500*g*. Discard the medium and wash in PBS. Repeat once.
5. After the last spin, resuspend the pellet of cells in 100 μL of PBS. Dropwise, add 1 mL of 70% ethanol at –20°C (1 drop/s to avoid formation of clumps of cells). Store the fixed cells at 4°C for up to several weeks.
6. Add 10 mL PBS to ethanol-fixed cells. Incubate for 5 min at room temperature to allow cells to rehydrate.

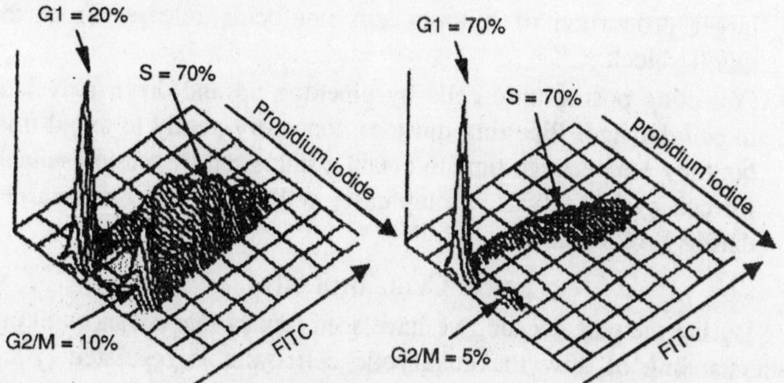

Fig. 3.2. Cell cycle distribution of ES cells and mouse embryonic fibroblasts.

7. Spin for 5 min at 500*g*. Resuspend the pellet in 200 μL of 2 N HCl and incubate for 20 min at room temperature.
8. Wash 3–4 times in 10 mL PBT.
9. Resuspend the pellet in 100 μL FITC-conjugated antibody raised to BrdU, diluted 110 in PBT, and incubate for 30 min at room temperature.
10. Wash 3–4 times in 10 mL PBT.
11. Resuspend the pellet in 100 μL of 1 mg/mL RNase. Incubate 20 min at room temperature. Store at 4°C if required (<24 h).
12. Add propidium iodide to a final concentration of 1 mg/mL. Incubate for 5 min.
13. Analyze fluorescence associated to FITC and to propidium iodide using conventional setups.

Notes

1. Check carefully that T-flasks are horizontal in the incubator, as uniformity is essential for recovery of pure populations of mitotic cells.
2. Do not leave the cells free of medium for more than 1 min, as this will lead to clumps of cells detaching during the shake-off procedure.
3. Following this protocol, one can obtain a population of ES cells containing >95% mitotic cells. Ninety percent of those cells will resume cell cycle progression within 1 h following incubation in demecolcine-free medium. ES cells will start entering the S-phase within 2 h, and the vast majority of them will be replicating their DNA at 4 h post-release from the mitotic block. Note that increasing the incubation time with demecolcine will result in a

larger proportion of mitotic cells not being released from the mitotic block.

4. Collecting post-mitotic cells by pipetting up and down may lead to cell damage. Pipetting must be done very gently to avoid this, but for a long enough time to obtain a single-cell suspension suitable for cell cycle analysis. In any case, cell debris will be discarded during FACS analysis.

Stem Cell Biology

During the past decade, we have seen remarkable advances in our understanding of how the eukaryotic cell cycle is regulated. As a result, we now have a detailed molecular description of the cell cycle machinery that controls both the G_1/S and G_2/M-phase transitions. Moreover, we have also seen the discovery of checkpoint controls that delay the cell cycle in response to intracellular perturbations including DNA damage, incomplete DNA replication, or a defective mitotic spindle. These allow time for repairs to be completed prior to chromosome segregation and cell division. Finally, the cell cycle can be regulated in response to extracellular signals generated by changes in nutrient status, pheromones, and mitogens.

Although we have made significant progress in our understanding of how the cell cycle is controlled in a variety of different organisms, little has been established on how the cell cycle is regulated in stem cells. Stem cells have an unlimited capacity for both self-renewal and production of differentiated progeny, and both processes must be tightly regulated to ensure the survival of the organism. Although it is still not clear what controls the decision to either self-renew or differentiate, or how the fate of a differentiating daughter cell is determined, the regulation of the cell cycle appears to play a key role in these processes. In this chapter, we provide a general overview of checkpoints and other cell cycle controls that operate in eukaryotic cells. This is followed by a discussion of how the cell cycle provides not only a means for controlling the numerical output of stem cells, but also a mechanism to regulate and implement developmental decisions.

Regulation of the Eukaryotic Cell Cycle

The eukaryotic cell cycle consists of alternating rounds of DNA replication (S phase) and cell division (M phase) separated by the gap phases G_1 and G_2. Progression through the cell cycle is controlled by the periodic activation of *cyclin-dependent kinases* (Cdks). Multiple Cdks have been identified in mammalian cells, whereas a single Cdk

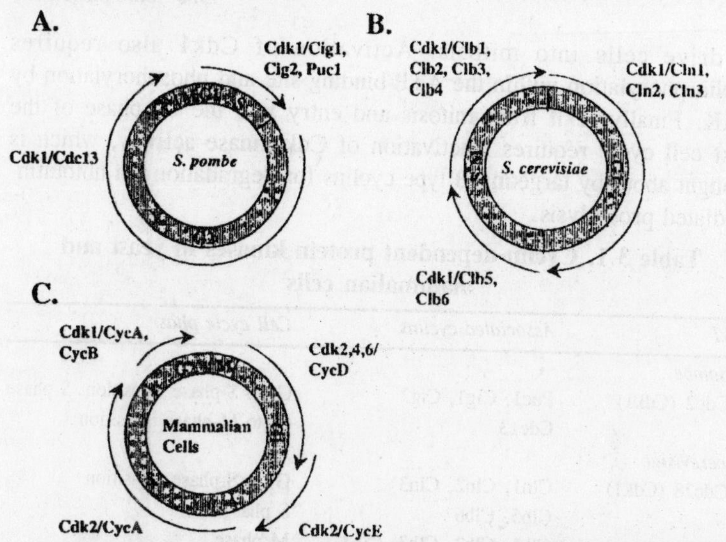

Fig. 3.3. Cell cycle regulation in yeast and mammalian cells: (A) S. prombe, (B) S. cerevisiae, (C) Mammalian cells.

is required for cell cycle progression in yeast. In general, Cdks can be regulated by at least three independent mechanisms. First, Cdk kinase activity is dependent on its association with distinct cyclin proteins, and formation of these complexes is driven by cycles of cyclin synthesis and degradation. Second, phosphorylation of Cdk kinase at a conserved threonine residue by *Cdk activating kinase* (CAK) is required for activity, whereas inhibitory phosphorylation within the ATP-binding site is regulated by the combined action of the Wee1 kinase and Cdc25 phosphatase. Finally, Cdk activity can be regulated by binding to a specific class of *Cdk inhibitors* (CKIs). As we discuss below, these multiple modes of Cdk regulation are involved in controlling cell cycle progression in response to cell cycle checkpoints and extracellular signals that regulate cell proliferation.

The first step in the cell division cycle is to replicate the genetic material. Initiation of DNA replication requires the activity of Cdk kinase (Cdk2 in mammalian cells), which is dependent on expression and accumulation of G_1-specific cyclins. Other posttranslational modifications of Cdk may be important during its activation at the G_1-to-S-phase transition, but the details of this regulation are still not clear. As cells proceed through S phase, B-type cyclins begin to accumulate in preparation for M phase. Accumulation of B-type cyclins and their association with Cdk1 kinase is essential, but not sufficient, to drive cells into mitosis. Activation of Cdk1 also requires dephosphorylation within the ATP-binding site and phosphorylation by CAK. Finally, exit from mitosis and entry into the G_1 phase of the next cell cycle requires inactivation of Cdk kinase activity, which is brought about by targeting B-type cyclins for degradation via ubiquitin-mediated proteolysis.

Table 3.1. Cyclin-dependent protein kinases in yeast and mammalian cells

CdkI	*Associated cyclins*	*Cell cycle phase*
S. pombe		
Cdc2 (Cdk1)	Puc1, Cig1, Cig2	G_1 to S-phase transition; S phase
	Cdc13	G_2 to M-phase transition
S. cerevisiae		
Cdc28 (Cdk1)	Cln1, Cln2, Cln3	G_1 to S-phase transition
	Clb5, Clb6	S phase
	Clb1, Clb2, Clb3, Clb4	M phase
Mammalian		
Cdc2	cyclins D1, D2, D3	G_1 phase
	cyclin E	
	cyclin A	
Cdk3	?	?
Cdk4	cyclins D1, D2, D3	G_1 phase
Cdk5	cyclins D1, D2, D3	G_1 phase
Cdk1	cyclins A and B	S phase; G2 to M transition

The cell cycle is also regulated by checkpoint controls that ensure the normal order of cell cycle events. Checkpoints were first identified in yeast as signaling pathways that delay cell cycle progression in response to DNA damage. Most of what we currently understand regarding the molecular basis of checkpoint controls is the result of genetic and biochemical analysis in yeast, where a number of checkpoint genes have been identified. These genes are highly conserved among all eukaryotic organisms, indicating that the basic checkpoint mechanisms have been preserved throughout evolution. In the following two sections, we describe the mechanisms of two checkpoint surveillance systems that monitor two fundamental cell cycle processes, DNA replication/repair and chromosome segregation.

DNA Damage-dependent Checkpoint in Yeast and Mammalian Cells

In the fission yeast, *Schizosaccharomyces pombe*, at least nine genes have been identified that are essential for the checkpoint control in response to DNA damage. Their gene products are thought to participate in a signal transduction pathway that inhibits activation of Cdk1 kinase at the G_2/M transition. Activation of the checkpoint pathway is thought to involve recognition and processing of a DNA lesion by "*sensor proteins*" generating a "*checkpoint signal.*" This signal is then transmitted via the Rad3 kinase to two additional kinases, Chk1 and Cds1. Both Chk1 and Cds1 phosphorylate and inhibit the Cdc25 phosphatase, which is required for activation of Cdk1 kinase and entry into mitosis. There is also evidence suggesting that Chk1 can stimulate the Wee1 kinase, a direct inhibitor of Cdk1. Although it initially appeared that Cds1 and Chk1 function redundantly, closer examination has revealed an interesting specificity regarding their activation during the cell cycle. Cds1 is only activated in response to DNA damage during S phase, whereas Chk1 is activated following DNA damage in G_2. Recently, it has been suggested that Cds1 inhibits Chk1 activity, which may explain why Chk1 remains inactive during S phase. What determines this specificity is not yet clear, but it has been suggested that Cds1 may be coupled to or activated by specific DNA structures that only form during DNA synthesis. Interestingly, Cds1 was originally identified as a multicopy suppressor of a temperature-sensitive mutant in DNA polymerase α, suggesting that these two proteins interact. Therefore, it is possible that Cds1 is directly associated with the replication machinery. Why Cds1 would inhibit Chk1 activation during S phase is not yet known. However, one possibility is that Chk1 might be required for a DNA repair process that is normally restricted to the G_2 phase.

In most eukaryotic cells, DNA damage delays cell cycle progression by inhibiting Cdk kinase activation at either the G_2-to-M or G_1-to-S-phase transition. However, depending on the organism or how the cell cycle is regulated, other mechanisms can be used to prevent segregation of damaged chromosomes. For example, in the budding yeast *Saccharomyces cerevisiae*, where S phase and certain mitotic events are initiated simultaneously early in the cell cycle, cells respond to DNA damage by inhibiting the metaphase-to-anaphase transition rather than by blocking entry into mitosis. Although this represents a very different physiological response to DNA damage, the checkpoint signaling pathway involved remains highly conserved. Activation of the checkpoint is still dependent on a DNA damage signal that is generated by sensor proteins, leading to the stimulation of the Rad3 homolog Mec1. Mec1 then activates two independent parallel pathways blocking cell division. One pathway involves activation of the Chk1

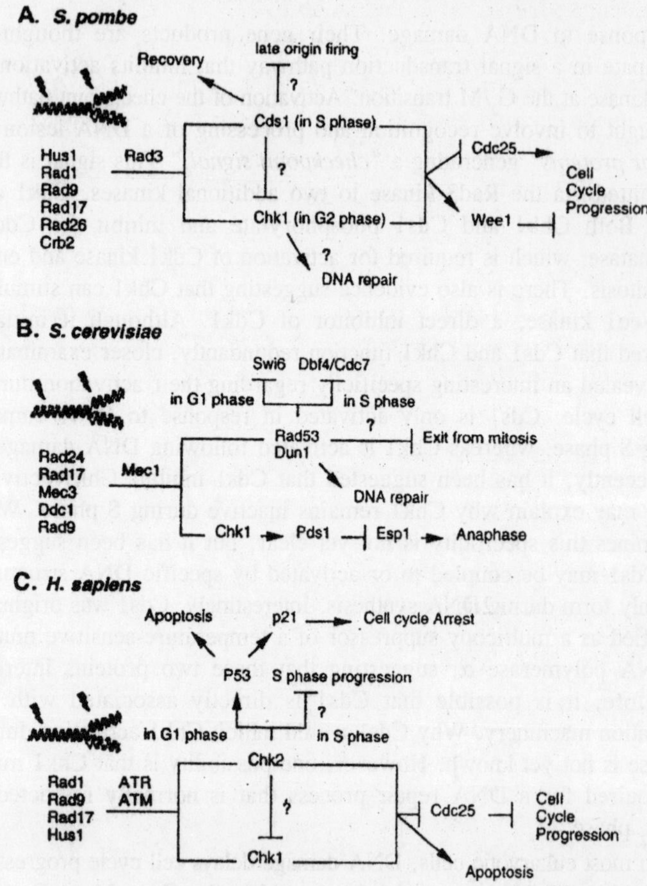

Fig. 3.4. Activation of checkpoint pathways in response to DNA damage. (A) In fission yeast, the sensor proteins Rad1, Rad9, Rad17, and Crb2 transmit the damage signal to protein kinase Rad3, which then phosphorylates the downstream kinases Chk1 and Cds1 in a cell-cycle-dependent manner. (B) In budding yeast, activation of the Rad3 homolog Mec1 is dependent on the sensor proteins Rad17/24, Rad9, Ddc1, and Mec3. (C) In human cells, homologous checkpoint pathways exist to block entry into S phase and mitosis.

kinase, which phosphorylates and stabilizes the metaphase-to-anaphase inhibitor, Pds1. The second pathway leads to activation of the Rad53 and Dun1 kinases, both of which contribute to the cell cycle delay by unknown mechanisms.

In addition to arresting cells in either G_2 or M phase, DNA damage can also delay entry into S phase. In *S. cerevisiae*, this checkpoint, sometimes referred to as the G_1 checkpoint, is also dependent on Mec1 and Rad53. Cell cycle arrest is thought to occur via Rad53 phosphorylation and inhibition of the transcription factor, Swi6, which is required for expression of G_1 cyclins.

In multicellular organisms, DNA damage activates a checkpoint control pathway that leads to either cell cycle arrest in G_1 or G_2, or programmed cell death, i.e., apoptosis. A key regulator in the decision to either arrest the cell cycle or undergo apoptosis is P53. P53 is a sequence-specific DNA-binding protein that activates the transcription of a variety of downstream effector genes. Stabilization of P53 following DNA damage is required for cell cycle arrest in G_1, or apoptosis from G_1 or G_2. Mammalian cells also arrest the cell cycle in G_2 in response to DNA damage, but this can occur independently of P53. Similar to *S. pombe*, arrest in G_2 involves inhibition of Cdc25 phosphatase by Chk1 and Chk2 kinases.

DNA damage-induced P53 stabilization requires a protein kinase cascade similar to one activated in response to DNA damage in both budding and fission yeast. In this case, the Rad3 homologs ATR and ATM phosphorylate and activate the downstream kinases Chk2 and possibly Chk1 (Chk2 is homologous to Cds1/Rad53). Both kinases have been shown to phosphorylate P53, leading to stabilization of the protein. P53 can then *trans*-activate a number of genes, including CIP1/WAF1/p21, a potent inhibitor of CDK activities required for the G_1-to-S-phase transition. The proposed role for p21 in cell cycle arrest is twofold; in addition to blocking CDK activity, p21 is also known to bind *proliferating cell nuclear antigen* (PCNA), a protein required for both DNA replication and DNA repair. Although p21 has been shown to inhibit SV40 replication in vitro, it is not yet known whether this is a primary function of the protein during DNA damage. Although P53-induced expression of p21 provides an attractive model for how cells arrest in G_1 following DNA damage, in at least some cell types, it is not the sole mediator of the checkpoint response. Other potential candidates that might participate in P53-dependent G1 arrest include the Gadd45, WIP1, Cyclin D1, and ABL genes.

It is still a mystery as to why, in multicellular organisms, some cells arrest the cell cycle in response to DNA damage whereas others undergo apoptosis. Although the basis for this difference is poorly understood, P53 has been shown to be essential for both G_1 arrest following DNA damage and apoptosis from G_1 or G_2. Apoptosis occurs through a pathway that involves activation of proteases of the ICE/ced3 class and its regulation by members of the Bcl2/ced9 family. P53 is known to *induce expression* of genes that promote apoptosis, including Bax, FAS, and insulin-like growth factor-1-binding protein-3 (IGF-BP3). However, there is also evidence that P53 can induce apoptosis in the complete absence of its transcriptional activity, suggesting that P53 is likely to have multiple functions. Interestingly, the Rad3 homolog ATM that is required for cell cycle arrest following DNA damage is not required for apoptosis. This suggests that the checkpoint pathways leading to cell cycle arrest or apoptosis, which are both dependent on P53 stabilization, are distinct processes. One model proposes that low-level DNA damage may signal cell cycle arrest via the ATM-dependent checkpoint, whereas more extensive damage may activate P53 by an alternative pathway, the latter leading to apoptosis.

Finally, the DNA damage checkpoint can also inhibit DNA replication initiation and S-phase progression. For example, in budding yeast, treatment of cells with DNA alkylating agents can slow DNA replication in a Mec1, Rad53-dependent manner. Whether this reflects inhibition of initiation or elongation is not yet known. Similarly, human cells defective for the ATM gene no longer delay S phase in response to DNA-damaging agents. This suggests that in addition to blocking mitosis, the checkpoint pathway can also target components directly involved in DNA synthesis. There is also evidence that the checkpoint control can delay initiation of DNA replication from late origins when cells are arrested in early S phase by treatment with hydroxyurea. This delay is dependent on expression of both Mec1 and Rad53, similar to what is observed following treatment of cells with DNA alkylating agents. In this case, the target of Rad53 is believed to be the Cdc7/Dbf4 kinase that has been directly implicated in replication origin firing.

Spindle Checkpoint

Checkpoints are not only involved with coupling DNA metabolic events to mitosis, but are also required to ensure the normal order of mitotic events. Segregation of chromosomes during cell division requires that all replicated chromosomes properly align themselves on the mitotic spindle during metaphase. In the presence of a disrupted or incomplete mitotic spindle, a checkpoint is activated that blocks the metaphase-to-anaphase transition and exit from mitosis. The so-called *spindle assembly checkpoint* was first characterized at the molecular level in budding yeast following the isolation of mutants that failed to arrest the cell cycle in response to microtubule-depolymerizing drugs. Seven genes were found to be essential for the checkpoint; these included

BUB1–3, *MAD1–3*, *MPS1*, and *PDS1*. Homologs for most of these genes have been found in several other organisms, including *S. pombe*, *Drosophila melanogaster*, *Caenorhabditis elegans*, and vertebrates. Similar to the DNA damage checkpoint, the spindle checkpoint is also thought to involve generation of a signal, in this case from the kinetochore that senses the existence of unattached chromosomes. Before discussing the mechanism of cell cycle arrest induced by the spindle checkpoint, we first briefly review how the later stages of cell division are regulated.

Separation of sister chromatids during anaphase relies on the activity of the *anaphase-promoting complex* (APC). This protein complex has a ubiquitin ligase activity that regulates the destruction of a specific set of proteins by ubiquitin-mediated proteolysis. Sister chromatids are held together by a protein complex known as *cohesin*, which is released from chromosomes upon anaphase entry. A key step in this process appears to be APC-dependent degradation of a protein called Pds1, which binds to and inactivates a protein called ESP1. Esp1 is essential for the proteolytic cleavage of a component of cohesin, a necessary step in the metaphase–anaphase transition. In addition to the degradation of Pds1, the B-type cyclin-dependent kinases are also targeted for degradation via the APC, and this is required for exit from mitosis.

Two pathways are activated in response to damaged spindles. One pathway involves the products of the *MAD1–3* and *BUB1* genes and is required for the inhibition of an APC accessory factor Cdc20, which

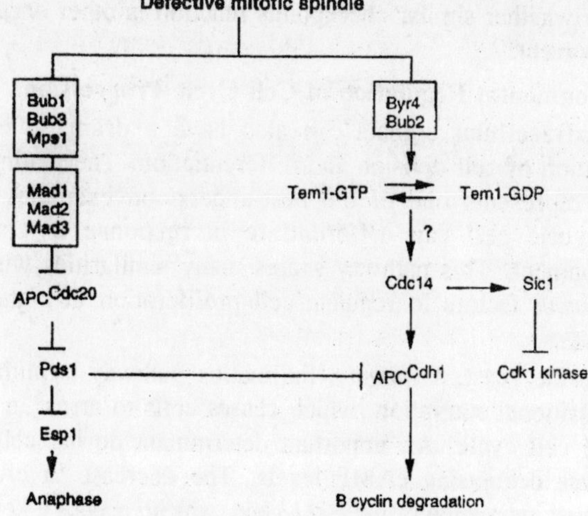

Fig. 3.5. The spindle assembly checkpoint.

normally targets the Pds1 protein for degradation. The second pathway requires the *BUB2* gene, which is thought to be a part of a two-component *GTPase activating protein* (GAP). In the presence of a damaged or an incomplete mitotic spindle, Bub2 is thought to stimulate the GTPase, Tem1. This leads to an accumulation of Tem1 in the inactive (GDP) form. In the absence of the active (GTP) form of Tem1, the Cdc14 phosphatase remains sequestered in the nucleolus, and cells are inhibited from exiting mitosis.

Cell Cycle Checkpoints in a Changing Cell Cycle

Analysis of stem cell divisions in the developing *Drosophila* embryo suggest at least three different checkpoint mechanisms are functioning during early fly development. During the first 13 cell divisions that occur in syncytium, DNA damage activates a checkpoint that blocks anaphase. During cycle 14, the *chk1* and *rad3* homologs *grapes* (*grp*) and *mei-41*, respectively, are required to coordinate completion of DNA replication to chromosome segregation, and following cycle 14, DNA damage results in the inhibition of entry into mitosis. Therefore, multicellular organisms like *Drosophila* appear to use a wide variety of checkpoint controls to maintain genome stability during development. These checkpoints share striking similarity to those used by both *S. pombe* and *S. cerevisiae*. However, whereas in yeast the checkpoint response differs depending on how the cell cycle is regulated, in *Drosophila*, the checkpoint response depends on the developmental stage at which DNA damage occurs. Additional studies will be required to clarify whether similar checkpoints function in other organisms during development.

Environmental Regulation of Cell Cycle Progression

Extracellular signals can also have a dramatic effect on the regulation of cell division and differentiation. The mating pathway in yeast represents one of the best-understood examples of a how a eukaryotic cell can differentiate in response to changes in its environment. This pathway shares many similarities with those used by growth factors to regulate cell proliferation in higher eukaryotic organisms.

In the yeast, *S. pombe*, the mating pathway is initially triggered by nutritional starvation, which causes cells to arrest in the G_1 phase of the cell cycle. An important determinant during cell cycle arrest involves decreasing cAMP levels. The decrease in cAMP leads to increased expression of the *ste11* gene, which encodes a key transcription factor required for expression of genes required for cell cycle arrest, conjugation, and meiosis. Mutational analyses of *S. pombe* genes encoding components of the cAMP cascade have shown that *S. pombe* cells stay in the mitotic cell cycle as long as the level of cAMP-dependent protein kinase activity remains high, but are committed to mating and meiosis if this activity is lowered. Mating pheromone, which is thought to activate a protein kinase cascade homologous to the MAP kinase cascade in mammalian cells, also contributes to the G_1 arrest.

In contrast to fission yeast, in *S. cerevisiae* the first step in the mating pathway does not require nutritional starvation, but only the presence of pheromone. The binding of pheromone to a cell-surface receptor is believed to induce a conformational change in a membrane-associated G (GTP-binding) protein complex that activates a mitogen-activated (MAP) kinase cascade required for gene activation, G_1 arrest, recovery from pheromone arrest, and nuclear and cellular morphological changes. Following formation of the diploid zygote, cells can resume vegetative growth. The second step in the mating pathway requires nutritional starvation, which triggers G_1 arrest, meiosis, and sporulation. As in *S. pombe*, cell cycle arrest in response to nutritional starvation is thought to involve regulation of cAMP-dependent kinase.

Responding to the Cellular Milieu

We have seen how yeast cells respond to external cues such as starvation or mating pheromone by arresting their cell cycle at G_1 preparatory to changing their cellular program. Although it is clear that *stem cells* (SCs) also respond to environmental factors by undergoing cell cycle arrest or activation, surprisingly little is known about the molecular mechanisms that regulate cell proliferation. The control of SC proliferation is especially critical given SCs' virtually unlimited capacity for the production of new cells and their role as the ultimate source of differentiated cells for the assembly and maintenance of multicellular organisms.

Organismal-level signals

Numerous organismal-level physiological stimuli have been identified that result in the induction or inhibition of cell cycle progression in SCs. However, since cell cycle status has frequently been monitored through production of genetically marked progeny or BrdU incorporation, the cell cycle phase from which proliferative entry and exit may take place is rarely known. In addition, changes in cellular output or BrdU incorporation could represent either changes in the length of a specific cell cycle phase or canonical cell cycle arrest/

activation. Yet, in a few systems we are beginning to see the outlines of pathways that begin with organismal-level cues and terminate in changes in the activity of the cell cycle machinery.

Activation or increase in SC division has been shown to take place under various conditions: transplantation into irradiated hosts; depletion of cycling cells by cytotoxins; in response to developmental programming; and even as a result of physical activity. The best-characterized system for transplantation and depletion studies is the hematopoietic SC or HSC. Both mammalian and human primitive HSCs give rise to long-term repopulation of all blood cell lineages and can be isolated from bone marrow as relatively quiescent G_0/G_1 cells. Upon injection into lethally irradiated mice, these HSCs provide multilineage long-term reconstitution of the peripheral blood.

The large percentage of donor-derived cells present in the blood 30 days after transplantation indicates that the rate of primitive HSC cell division must dramatically increase upon transplantation. This increase in HSC cell cycle activity is presumably required to replace the more mature lineage-restricted cycling precursors that die as a result of radiation damage. Whether the increase in HSC division occurs as a direct consequence of the loss of the precursors is not yet clear. The relative resistance of primitive HSCs to cycle-active cytotoxic agents suggests that most of these cells are dividing slowly, and indeed, long-term labeling with BrdU of up to 30 days is required before the vast majority of HSCs are observed to incorporate the S-phase label. This implies that primitive HSCs have an extremely elongated cell cycle resulting either from greatly extending at least one cell cycle

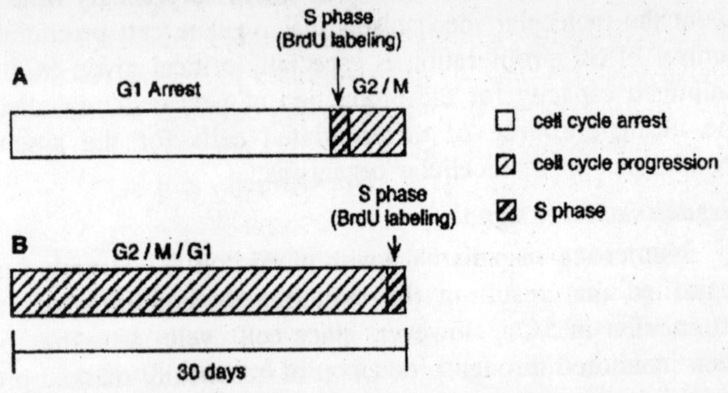

Fig. 3.6. Single stop-start versus slow cell cycle period. (A) Stop-start cell cycle shown with a long period of G_1 arrest followed by progression through S, G_2, and M. Stop-start cycles with G_0 or G_2 arrest are also possible. (B) Slow or extended cell cycle progression.

phase or from undergoing a stop–start cell cycle mode with rounds of transient cell cycle arrest followed by cell division.

Division of HSCs produces lineage-restricted precursors that in turn give rise to differentiated blood cells. Detection of donor-derived blood cells indicates unequivocally that quiescent HSCs have activated cell division upon transplantation. However, loss of donor-derived blood cell production could be due to either return of HSCs to mitotic quiescence or apoptosis of lineage-restricted precursors. Definitive studies to determine whether primitive HSCs return to their previously quiescent state after a period of rapid mitoses have yet to be performed.

A. Repopulation by transplanted HSC

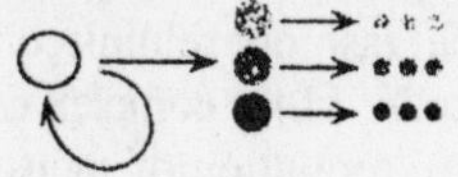

B. Mitotic arrest of HSC

C. Apoptosis of lineage-restricted precursors

Fig. 3.7. Loss of donor-derived peripheral blood cells posttransplantation.

Studies on the effect of depleting cycling precursors on SC division have also been carried out in the mammalian brain. The results suggest that forebrain neuroblasts transiently enter a more active mitotic state upon depletion of cycling progenitor cells; these conditions, however, do not address whether the transition is one between cell cycle arrest and activation, or between a very long and a very short cell cycle period.

The activation and arrest of SC division in response to developmental cues has been investigated primarily in mammalian and insect neuroblasts. In *Drosophila*, a number of lineage and BrdU studies have shown that a subset of neuroblasts undergo two bursts of mitotic activity, once in embryos and later during larval stages. The double burst of BrdU incorporation indicates initial embryonic division followed by relative quiescence, then reactivation for larval mitoses and a final arrest concomitant with differentiation. In this proliferation program, "*relative mitotic quiescence*" entails cell cycle arrest or passage through one very elongated cell cycle period.

Detailed genetic analyses in *Drosophila* have identified a number of genes required to control the temporal pattern of "on and off again" SC division. The first quiescence is achieved by inhibition of an initial mitogenic signal by the product of the *ana* gene, thus preventing neuroblasts from beginning S phase prematurely. The initiation of S phase at the appropriate time for larval division requires the product of the *trol* gene and the transcription repressor Even-skipped. Induction of S phase by *trol* occurs at a later stage of G_1 than the G_0/G_1 arrest mediated by *ana*, suggesting that *Drosophila* neuroblasts may activate cell division in a stepwise fashion reminiscent of the G_1 subphases described for mammalian cells. More complex control of G_1 progression is also observed during differentiation of mouse embryonic SCs. This involves up-regulation of cyclin D and E (two G_1-specific cyclins), Cdk 2 and 4, and the Cdk inhibitors p21 and p27, leading to an elongation of the G_1 period. For *Drosophila* neuroblasts, the transition to mitotic arrest is mediated by the developmental transcription factor Prospero. The molecular mechanisms by which some of these *Drosophila* genes regulate cell cycle progression are discussed below.

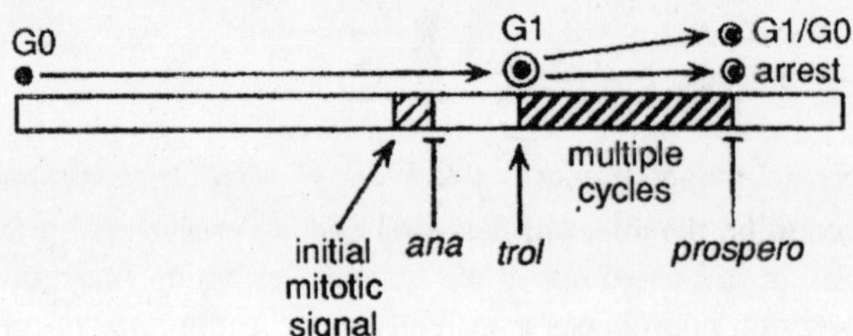

Fig. 3.8. Control of "on and off again" neuroblast division in Drosophila.

Finally, experiential cues have also been suggested to alter the mitotic activity of SCs in the adult mammalian brain. Enriched or complex environmental stimuli increased BrdU incorporation in the mouse hippocampus, as did increased physical activity. Interestingly, mice that score well in learning paradigms and normally have high levels of neurogenesis do not show increased BrdU incorporation in a more complex environment. These intriguing studies represent the beginning of a new line of investigation that will address whether experiential stimuli increase SC, precursor cell division, or both. Currently, we have no understanding of the molecular events connecting behavior to changes in the cell cycle.

Molecular signaling

For many of the organismal-level changes described above, the molecular basis underlying the phenomenology and how those changes are translated into molecular signals that can affect the cell cycle machinery in SCs are beginning to be addressed. Signal transduction pathways link many organismal-level changes to alterations in transcriptional or enzymatic activity. This robust area of research has been extensively reviewed; however, most of the studies occur in non-SC systems. Which signal transduction pathways operate specifically in stem cells and how they translate extrinsic signals into cell cycle arrest or activation remain to be determined. This is especially critical, since many extrinsic signals are known to produce a variety of different responses depending on cell type. Much of the SC-specific information at this level comes from in vitro analyses of cultured SCs and molecular genetic studies in *Drosophila*.

Studies in vitro have shown that the rate of cell division of freshly isolated G_0/G_1 primitive HSCs or the amplification of HSC-like cells increases upon addition of specific cytokines. HSC-like murine cells begin division in vitro in response to cytokines with the same kinetics as observed post implantation in vivo. These data suggest that cytokines may be directly linked to the increase in HSC mitotic activity in vivo. An interesting possibility is that cytokine signaling may couple decreasing lineage-restricted precursor pools to their regeneration by transplanted HSCs. Cytokines have been shown to activate a number of signal transduction pathways, prominent among them the Janus kinase/ signal transducer and activator of transcription (Jak/Stat) pathway. Although the Jak/Stat pathway is commonly considered to stimulate cell division, studies of SC proliferation suggest that it may function instead to maintain SC division or identity.

In vitro analyses of mammalian neuroblasts have also identified a number of factors such as *epidermal growth factor* (EGF) and basic fibroblast growth factor and their signaling pathways that are necessary to increase or decrease SC division. In vitro cultures of *Manduca* and *Drosophila* central nervous systems have identified the hormone ecdysone as a potential activator of neuroblast cell division through an as-yet-unidentified mechanism. However, as before, it should be noted that it is not clear whether these factors activate cell cycle progression in an arrested cell or dramatically increase the rate of cell division of cells progressing very slowly through the cell cycle. Given that it is not always clear whether SCs are cycling between cell cycle arrest and cell cycle activation or between short and very long cell cycle periods, a related issue is how the length of the SC cell cycle is regulated. Genetic studies of female *Drosophila* germ-line development have identified at least two signaling pathways that regulate the rate of SC division. The novel gene *piwi* is required both within an adjacent cell and in the SC itself to increase the division rate of germ-line SCs. In addition, inhibition of signaling by Dpp, a transforming growth factor-β-like molecule, results in slower cell cycles in germ-line SCs.

Linking molecular signals to cell cycle phase and the cell cycle machinery

We have surprisingly little information about the cell cycle phase in which signaling pathways function, let alone how they affect the cell cycle machinery to induce or inhibit cell cycle progression in SCs. Thus far, the best-understood system is the mitotic activation of *Drosophila* neuroblasts by *trol*. Preliminary evidence suggests that *trol* encodes the *Drosophila perlecan*, a coreceptor for FGF. In the absence of *trol* function, neuroblasts arrest in G_1 as indicated by decreased *cyclin E* mRNA levels and are able to enter S phase upon induced expression of either *cyclin E* or *E2F/DP* but not *cyclin B*. Genetic and molecular analyses suggest that the activity of Cyclin-Cdk complexes may also be stimulated by the Cdc25 protein phosphatase homolog string to promote the G_1-to-S-phase transition.

Cell cycle arrest of *Drosophila* neuroblast progeny in G_1 also occurs prior to differentiation. The dividing neuroblasts synthesize Prospero, which becomes asymmetrically localized into the ganglion mother cell upon cell division and is later translocated into the nucleus. Nuclear Prospero activates expression of *asense* and inhibits expression of *deadpan*, which encode two basic helix-loop-helix proteins. Altered levels of these two proteins result in the expression of *dacapo*, which encodes the *Drosophila* p21 homolog. An extrapolation from embryonic function would suggest that Dacapo then inhibits CyclinE-Cdk2 activity, thus arresting the immature neuron or glial cell in G_1. Studies of rat oligodendrocyte precursors also show cell cycle arrest in G_1 prior to differentiation by up-regulation of p21 and p27 through a cAMP-mediated pathway.

Although the molecular mechanisms through which organismal-level cues regulate SC proliferation are far from clear, the central portion of the framework, which includes the signal transduction pathways activated by molecular signals, has been well characterized. However, large gaps still exist in our understanding of how organismal changes, such as transplantation, translate into molecular signals that will activate the signal transduction pathways. Other gaps exist at the output end of the pathway: How do changes in the activity of multiple signal transduction pathways result in coordinated changes in cell cycle activity? Are there specific windows of opportunity during cell cycle progression in which signals must effect these changes? Are the input points where cell cycle progression is controlled the same as those used by canonical cell cycle checkpoints? These are just a few of the questions that still need to be addressed.

Changing the Program: Where Cell Cycle Regulates Development

We have seen ample evidence for changes in cell cycle activity as a result of environmental cues. However, it is also true that the cell cycle progression itself plays an important role directly and indirectly in determining the potency of a SC and the fate of its progeny.

Cell Cycle and Renewal of Multipotency

One of the most intriguing notions is that passage through the cell cycle can set the developmental fate of cells derived from SC division and renew SC multipotency. Transplantation studies suggest that shortly after S phase, the cell fate of the soon-to-be-born SC progeny is restricted by environmental factors, but passage through the next S phase restores SC multipotency, allowing new cues to dictate the identity of the resulting daughter. How might S phase renew SC multipotency? One hypothesis invokes changes in chromatin structure during DNA replication that can have dramatic effects on levels of gene expression and therefore cell fate. Recent studies in yeast reveal

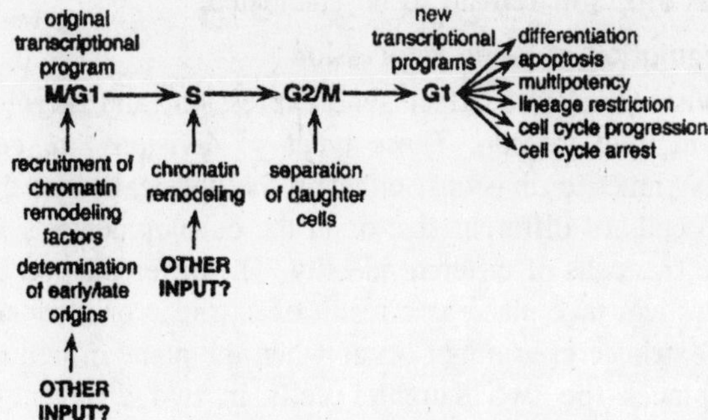

Fig. 3.9. Possible effects of changes in chromatin structure determined at earlier cell cycle phases on stem cell fates at later times.

that chromatin remodeling factors are recruited during M/G_1 and the chromatin is remodeled during S phase, resulting in changes in transcriptional programs in the subsequent G_1. This suggests that changes that occur at a specific cell cycle phase can affect decisions at a later time.

Direct Cell Cycle Regulation of Gene Expression

In the context of cell cycle control over development, it is reasonable to ask whether cell cycle progression is required for the normal developmental program of gene expression, or whether temporal patterning of gene expression will continue in cell-cycle-arrested cells. This question can be further subdivided to ask whether cytokinesis itself, the physical separation of two daughter cells, is sufficient or whether rounds of DNA synthesis are also essential. Much of our information here comes from analysis of *Drosophila* neuroblasts.

In vitro and in vivo studies show that some genes require neither cytokinesis nor cell cycle progression to achieve normal patterns of expression, whereas others require only cytokinesis, and a few require both. Surprisingly, the cell cycle dependence for expression of a certain gene in the same animal varies with the lineage examined. Analysis of neuroblasts in culture revealed that inhibition of M phase did not alter the expression of neurotransmitter synthetic pathway enzymes. Studies in vivo identified multiple cell cycle regulation patterns of neuroblast gene expression. Therefore, although cell cycle progression is required for the correct expression of some developmentally important genes, the inputs from cell cycle to gene expression are varied, and not all developmental genes require such input for their expression. Whether other cell cycle phases or events also trigger changes in the pattern of gene expression, and what the molecular mechanisms for such controls are, still remain to be elucidated.

Indirect Regulation of Gene Expression

Cell division can also affect gene expression, and therefore cell fate, in an indirect fashion. These types of divisions are generally defined as asymmetric divisions, either in the physical sense that they produce two cells of different size or in the developmental sense that they produce two cells of different identity. The differentiation between two cell types can take place as a result of extrinsic or intrinsic cues. Changes in extrinsic cues might occur when the plane of cell division physically places the two daughter cells in two different micro-environments, whereas changes in intrinsic cues may reflect unequal partitioning of internal factors between the two daughters.

The effect of microenvironment on cell fate is implied by in vitro studies where SC production is altered by different culture conditions and in vivo studies that correlate cell–cell interactions with cell fate. In vitro culture of HSCs demonstrates that the balance between the production of lineage-restricted progenitors versus HSCs depends on the relative concentration of cytokines in the media. Time-lapse microscopy studies of mammalian cerebral cortex show that changes in division plane, and therefore cell–cell or cell–extracellular matrix contacts, change the fate of the daughters. The change in division plane also correlates with symmetric or asymmetric distribution of the signaling molecule Notch1. In vivo studies of the *Drosophila* male and female germ line also correlate changes in SC niche with changes in the fate of SC progeny. Removal of surrounding somatic cells by ablation or by alteration of the expression of signaling molecules such as *piwi*, *fs(1)Yb*, and *Dpp* leads to the conversion of asymmetric to symmetric daughter cell fate. Similarly, stimulation or inhibition of the externally activated EGF receptor pathway can also lead to changes in cell fate.

Asymmetric distribution of internal factors upon SC division has been shown to affect the identities of resulting daughter cells. Perhaps the best-characterized examples of asymmetrically localized cell fate determinants in SCs are the analyses of Prospero and Numb in *Drosophila* neuroblasts. Both Prospero and Numb proteins are segregated into the lineage-restricted cell upon neuroblast division, where they are required for daughter cell fate as assayed by the identity of the progeny produced. Prospero and Numb asymmetric distribution is dependent on the Inscutable protein that is also responsible for spindle orientation during neuroblast division. Thus, the act of mitosis creates two cells with different levels of Prospero and/or Numb and ultimately leads to changes in daughter cell fate.

Canonical Cell Cycle Regulators and Development

Given that cell cycle progression affects SC gene expression both directly and indirectly, it seems reasonable that delay or arrest of the cell cycle by canonical cell cycle regulators such as starvation and DNA damage may also affect developmental events.

Serum starvation is a classic phenomenon that causes the arrest of mammalian tissue culture cells in early G_1. The organismal correlate is nutrient starvation in *Drosophila*, which leads to cell cycle arrest of larval neuroblasts and slowing of division in somatic SCs of the germ line. Nutrient-mediated arrest of larval neuroblasts occurs upstream and independent of *trol/ana* developmental regulation and late G_1 events, consistent with an early G_1 arrest. Cell culture experiments have shown that starvation-arrested neuroblasts can be activated by coculture with fat body, possibly by fat-body-derived growth factors that affect proliferation of imaginal disc cells. However, the signaling mechanism and cell cycle machinery targeted by nutritional arrest are not yet known.

As described above, in response to DNA damage, a checkpoint control can delay cell cycle progression in G_1 or G_2, followed by either DNA repair or apoptosis. In mice, overexpression of the *BCL2* gene can prevent apoptosis in HSC populations following irradiation or treatment with DNA-damaging agents. It is widely accepted that apoptosis offers a level of protection to the organism in providing a mechanism to remove unwanted damaged cells from circulation. It will be interesting to see how HSC development is affected by defects in the apoptosis or checkpoint pathways.

Stem Cells and Cell Cycle Control

With a few notable exceptions, our understanding of the mechanisms governing cell cycle regulation in SCs and the effect of mitotic control on SC development and biology currently centers on cues such as cytokine/growth factor-triggered signal transduction pathways and asymmetric distribution of intrinsic cell fate determinants. For the most part, the fascinating questions of how phenomena translate into distinct molecular signals and how those signals mesh with the intricacies of cell cycle phase and specific components of the cell cycle machinery have yet to be answered.

4

EMBRYONIC STEM CELL DIFFERENTIATION

With the attractive potential therapeutic uses of both endothelial and hematopoietic cells in medicine, much attention has been focused upon the differentiation of *embryonic stem* (ES) cells into mature cells of these lineages. In this chapter, we present a culture system that has successfully generated both complex endothelial structures and mature hematopoietic cells from undifferentiated ES cells in vitro. The simplicity of this system allows detailed analyses of factors and developmental steps essential in the generation of vascular structures and mature hematopoietic cell types.

Differentiation of Mesodermal, Endothelial, and Hematopoietic Cells in the Embryo

As the ES cell differentiation scheme presented in this chapter is based on developmental processes occurring within the embryo, an introduction to the generation of the mesoderm and its derivatives during embryogenesis is of use to the reader. The mesoderm is a sheet of cells located between the ectoderm and endoderm in the early developing embryo. In differentiating from an epithelial structure to a mesenchymal population, mesodermal cells lose expression of the cell adhesion molecule epithelialcadherin (E-cadherin). A subset of mesodermal cells up-regulate expression of vascular endothelial growth factor receptor 2, known in the mouse as *fetal liver kinase* 1 (Flk1). This Flk1$^+$ population includes endothelial cell precursors.

As the embryo develops, Flk1$^+$ cells in the yolk sac generate the blood islands that are composed of round clusters of hematopoietic cells surrounded by endothelial cells. Mice that lack Flk1 fail to generate blood islands and die early in development. Another mesodermal population, the paraxial mesoderm can be identified according to expression of platelet-derived growth factor receptor α (PDGFRα). While this population can generate endothelial cells, it does not generate hematopoietic cells. The Flk1$^+$ (lateral) mesoderm, on the other hand, is a source of both endothelial and hematopoietic cells. During vasculogenesis, endothelial cells derived from both paraxial and lateral mesoderm form a vascular plexus throughout the yolk sac and embryo body. This plexus is then modified into a highly complex branched vascular network during angiogenesis.

The first hematopoietic cells to appear, the primitive erythrocytes, are thought to arise directly from the lateral mesoderm. Definitive erythroid cells, myeloid, and lymphoid cells are generated from endothelial cells expressing vascular endothelial–cadherin (VE-cadherin). Mice that lack the putative transcription factor, AML1/Runx1/Cbfa2, cannot generate definitive hematopoietic cells from endothelial cells and die early in embryogenesis, clearly illustrating the importance of endothelial cells as a source of definitive hematopoietic cells. In early developmental stages, endothelial and hematopoietic cells co-express many markers, such as platelet endothelial cell adhesion molecule-1 (PECAM-1), CD34, AA4, and GSL I-B4 isolectin, demonstrating the close relationship between endothelial and hematopoietic cell development, and thus making the separation of either cell type by surface phenotype difficult.

Differentiation of Mesoderm and Mesodermal Derivatives from ES Cells

Induction of mesodermal cells in the absence of embryoid body formation

It has been previously demonstrated that ES cells, when induced to differentiate in culture, recapitulate early embryonic events in vitro. In these cultures, ES cells differentiate in complex three-dimensional cell aggregates termed *embryoid bodies* (EBs). To reduce the potential for interactions and, therefore, control more stringently the culture conditions, our group developed a culture system by which ES cells differentiated in the absence of EBs along a 2-dimensional plane. This system, therefore, lacked the structural complexity of the EB differentiation, and yet it effectively generated Flk1$^+$ putative mesodermal cells, indicating that the generation of mesodermal cells from ES cells did not require 3-dimensional cellular interactions or

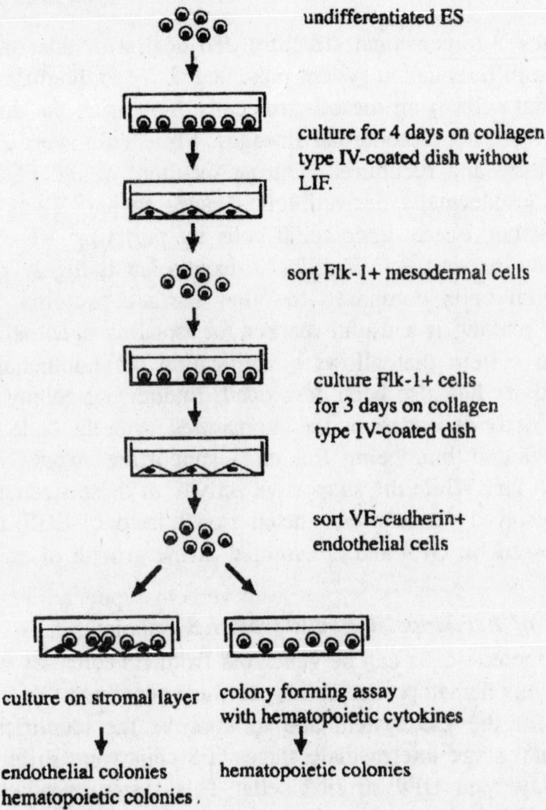

Fig. 4.1. In vitro differentiation of endothelial and hematopoietic cells from ES-derived mesodermal cells.

supportive feeder layers. Paraxial mesodermal cells (PDGFRα^+ E-cadherin$^-$) were also generated in this system, although the potential of these cells has not been investigated. As lateral mesodermal cells are a source of not just endothelial cells but also hematopoietic cells, we purified these cells using flow cytometry according to the expression of Flk1 in the absence of E-cadherin. The in vitro differentiation system used to generate mesodermal cells and, in turn, endothelial and hematopoietic cells. Such a system is far easier to monitor and manipulate than the EB system previously described.

Differentiation of endothelial cells from ES-derived Flk1$^+$ mesodermal cells

Endothelial cells can be generated from ES cells in EB. However, endothelial cell development cannot easily be observed or manipulated in a complex 3-dimensional structure. To deal with this, our group extended the differentiation system presented above to the differentiation of endothelial cells from mesodermal cells. To enrich the endothelial population from nonmesodermal lineages, $Flk1^+$ cells were sorted by flow cytometry and recultured. During reculture of the $Flk1^+$ cells, numerous mesodermally derived cell lineages appear. Therefore, we chose to further isolate endothelial cells by purifying VE-cadherin$^+$ cells by flow cytometry. VE-cadherin expression is highly restricted to endothelial cells compared to other surface proteins, such as PECAM-1, making it a useful marker for isolating endothelial cells. An in vitro system that allows quantification of endothelial colony forming activity has also been developed. Endothelial colony forming activity is strictly dependent on the stroma used, with the bone marrow-derived OP9 cell line being the most supportive stroma of those examined so far. While the supportive activity of these stroma has not been fully resolved, vascular endothelial growth factor (VEGF) is known to be expressed by OP9 and is essential in the growth of endothelial colonies.

Generation of hematopoietic cells from ES cells

Hematopoietic cells can be generated from ES cells. By culturing EBs, numerous hematopoietic cell types were generated. To avoid the complexity of the EB system and to improve the identification of developmental stage intermediate stages, ES cells were differentiated as a monolayer on OP9 stromal cells. This system resulted in the generation of all major hematopoietic cell types including lymphocytes. As mesodermal cells are the source of the hematopoietic lineage, we assessed our two-dimensional feeder-layer free differentiation system for hematopoietic ability. As ectodermal cells are known to inhibit *erythropoiesis*, $Flk1^+$ mesodermal cells were first isolated by cell sorting and then recultured. This resulted in the generation of hematopoietic cells in the absence of EB formation and feeder cells. As endothelial cells are know to be a source of definitive hematopoietic cells in the embryo, we also assessed the hematopoietic potential of VE-cadherin$^+$ cells from ES-derived $Flk1^+$ cells. VE-cadherin$^+$ cells, in contrast to their $Flk1^+$ precursors, required hematopoietic growth factors to generate hematopoietic cells.

MATERIALS

ES Cell Lines

With numerous ES cell lines currently available, each line should first be examined for its ability to generate $Flk1^+$ mesodermal cells.

The ES cell line most commonly used in our laboratory is CCE. The methods presented below are optimized for CCE.

Reagents

Cell culture media

1. ES cell differentiation medium:
   (a) Alpha Modified Eagle's Medium (αÁMEM) supplemented with 50 U/mL penicillin and 50 μg/mL streptomycin.
   (b) *Fetal calf serum* (FCS), 10% final concentration.
   (c) 2-Mercaptoethanol, 5×10^{-5} *M* final concentration.
2. OP9 culture medium:
   (a) αÁMEM supplemented with 50 U/mL penicillin and 50 μg/mL streptomycin.
   (b) FCS, 20% final concentration.

Induction and Purification of Mesodermal Cells

1. Collagen type IV-coated 6-well plates, Biocoat.
2. Cell dissociation buffer.
3. Phosphate-buffered saline (PBS) lacking Ca^{2+} and Mg^{2+}: 2.89 g Na_2HPO_4 (12 H_2O), 0.2 g KH_2PO_4, 8.0 g NaCl, 0.2 g KCl, and distilled water up to 1000 mL.
4. Normal mouse serum (NMS). NMS can be prepared from mice in-house or can be purchased.
5. Anti-Flk1 (AVAS12) monoclonal antibody (mAb), fluorescently labeled.
6. Anti-E-cadherin (ECCD2) MAb, conjugated to a second fluorochrome.
7. Hanks' buffered saline solution (HBSS) with 1% bovine serum albumin (BSA) (HBSS/BSA).
8. HBSS/BSA with 5 μg/mL propidium iodide (HBSS/BSA/PI).

Differentiation and Isolation of Endothelial Cells from ES-Derived Flk1$^+$ Cells

1. Fluorescently labeled anti-VE-cadherin mAb (VECD1).
2. HBSS/BSA and HBSS/BSA/PI.
3. Falcon 6-well polystyrene culture dishes.

Generation of Hematopoietic Cells from ES-Derived Endothelial Cells

1. Mouse cytokines: Erythropoietin (Epo), *stem cell factor* (SCF) interleukin-3 (IL-3), interleukin-7 (IL-7), Flt3 ligand.

METHODS

Induction and Purification of Mesodermal Cells

1. Into each well of a 6-well collagen type IV-coated plate, add 1×10^4 undifferentiated ES cells.
2. Add 3 mL of differentiation medium to each well. Leave undisturbed for 4 d in a 37°C incubator with 5% CO_2 environment.
3. After 4 d, harvest the cells by first removing the medium, washing with PBS, and incubating at 37°C with 2 mL cell dissociation buffer for 20 min. Pipet cells from the surface of the plate.
4. Incubate single-cell suspensions with NMS for 20 min on ice. Normally, we incubate 1×10^7 cells in 100 μL NMS.
5. Add an appropriate concentration of fluorescently labeled anti-Flk1 and anti-E-cadherin mAbs to cell suspension in NMS and incubate for 20 min on ice. Wash the cells twice with HBSS/BSA. Resuspend the cells in HBSS/BSA/PI for dead cell exclusion.
6. Sort the Flk1$^+$ E-cadherin$^-$ cells. Mesodermal cells express Flk1 but not E-cadherin.

Differentiation and Isolation of Endothelial Cells from ES-Derived Flk1$^+$ Cells

1. Add 3×10^5 Flk1$^+$ E-cadherin$^-$ cells to each well of a collagen type IV- or gelatin-coated 6-well plate.
2. Add 3 mL ES cell differentiation medium.
3. Incubate at 37°C for 3 d under 5% CO_2.
4. Harvest cells using cell dissociation buffer.
5. Block with NMS for 20 min on ice.
6. Stain with an appropriate concentration of labeled anti-VE-cadherin mAb.
7. Analyze or sort by flow cytometry.

ES-Derived Endothelial Colony Assay

1. Three days prior to commencing Flk1$^+$ cell culture, split one confluent 25-cm^2 flask of OP9 stromal cells into 3 × 6-well dishes.
2. Add 2 mL of OP9 medium to each well.
3. Three days later, when OP9 is confluent, add up to 5000 ES-derived Flk1$^+$ cells per well.
4. After 3 d, sheets of cells can be seen growing on the OP9 stroma. These cultures can be maintained for several more days, but should not be allowed to overgrow.

Generation of Hematopoietic Cells from ES-Derived Endothelial Cells

Generating primitive erythroid cells from ES cells

1. Sort Flk1$^+$ cells as described.
2. Add 1 × 10^4 cells to each well of a 6-well plate containing confluent OP9 stroma.
3. Add 3 mL ES cell differentiation medium/well supplemented with 2 U/mL Epo.
4. Analyze hematopoietic cells that have appeared after 4 d.

Generating definitive hematopoietic lineages from ES-derived VE-cadherin$^+$ cells

1. Sort VE-cadherin$^+$ cells as described.
2. Add 1 × 10^4 cells to a 25-cm^2 confluent flask of OP9 stromal cells.
3. Add 6 mL ES cell differentiation medium supplemented with cytokines. For erythroid-myeloid cultures, add 2 U/mL Epo, 100 U/mL SCF, and 200 U/mL IL-3. For B lymphoid cultures, we add 100 U/mL, 60 U/mL IL-7, and 50 U/mL Flt3 ligand.
4. Change the medium every 3–4 d. In general, definitive Ter-119+ erythroid cells first appear in culture after 2–3 d. After 5 to 7 d, mature myeloid cells appear expressing the markers Gr-1 and Mac-1. CD19$^+$ B lymphoid cells appear after 10–14 d of culture.

Notes

Mesoderm induction

1. CCE ES cells, when differentiated, generate Flk1$^+$ cells at a frequency of 30–40% of live gated cells.
2. We recommend using medium less than 4 wk old to obtain high yields of Flk1$^+$ cells.
3. FCS has a critical influence in obtaining high yields of Flk1$^+$ cells. Batch checks are therefore highly recommended for finding appropriate serum batches. Before commencing large-scale studies of ES cell differentiation into Flk1$^+$ cells, we checked 33 batches from 18 different companies. The frequency of Flk1$^+$ cells generated after 4 d in culture using different FCS batches ranged from 13 to 43%, highlighting the need for FCS batch checks.
4. 2-Mercaptoethanol is essential in obtaining both high yields of Flk1$^+$ cells and also later in hematopoietic cell differentiation, particularly during B lymphopoiesis.

5. It is highly recommended that FCS batch checks are conducted before large-scale use of OP9. The condition of OP9 is highly dependent upon FCS batch, and this stroma should be maintained carefully, as it easily loses its supportive activity. FCS can be checked by culturing OP9 for 10 passages in medium containing different FCS batches.

   The morphology of OP9 should be maintained over this period (i.e., large flat cells, not fibroblastic or transformed). Furthermore, the cell number should remain at 1–1.5 $\times$ 10^6 cells/25-cm^2 flask when confluent. Supportive activity is best analyzed by co-culture with $Flk1^+$ cells derived from ES cells, resulting in the generation of endothelial and hematopoietic cells as described in this chapter.

6. Several matrices have been analyzed for their ability to enhance mesoderm induction by ES cells including: gelatin, fibronectin, and types I and IV collagen. While all matrices generated mesodermal cells to some degree, type IV collagen resulted in the highest yield of $Flk1^+$ cells and was therefore routinely used.
7. NMS is used to prevent binding of the primary antibodies to the Fc receptors on the target cell surface.
8. $Flk1^+$ cells first appear after 3 d of differentiation. One day later, the frequency of $Flk1^+$ cells will have increased by 10-fold. By d 5 of culture, the frequency of $Flk1^+$ cells decreases, and endothelial markers begin to appear, indicating the differentiation of the mesodermal cells into endothelial cells. Therefore, d 4 was chosen as the optimal timepoint for isolating mesodermal cells. Within 4 d of differentiation, the cell number will increase 100-fold, from 1 $\times$ 10^4 cells/well to 1–3 $\times$ 10^6 cells/well.
9. Once sorted, $Flk1^+$ cells can differentiate into endothelial cells on either collagen type IV- or gelatin-coated dishes with similar efficiency. Gelatin has the advantage of being considerably cheaper than collagen type IV, as well as being easier for coating dishes. Gelatin solution (0.1%) can be prepared by dissolving 0.2 *g* of gelatin in 200 mL distilled water, followed by autoclaving. For coating dishes, apply enough solution to cover the bottom of the dish, leave for at least 2 h, then aspirate the gelatin, and use.
10. In general, 3 $\times$ 10^5 $Flk1^+$ cells cultured on collagen type IV dishes for 3 d yield approximately 3–5 $\times$ 10^5 cells, of which 15–20% are usually VE-cadherin$^+$, although this can vary between ES cell lines examined.

11. Flk1$^+$ cells growing on stromal cells can generate numerous cell lineages. Immunocytochemistry is the easiest method for demonstrating the presence of endothelial cells in the cultures. Several markers such as VE-cadherin, PECAM-1, and Flk1 are suitable for immunostaining of endothelial colonies. Immunostaining of colonies in the culture dish is an approach commonly used by our group.
12. Primitive erythropoiesis can be confirmed by retrieving the hematopoietic cells from the culture, preparing cytospots and immunostaining for embryonic hemoglobin. This system is particularly useful in analyzing the potential of genetically modified (such as null mutant or transgenic) ES cells to generate primitive erythroid cells.
13. Flt3 ligand has recently been shown to dramatically enhance the production of B lymphoid cells from ES cells cultured on OP9 stromal cells.
14. OP9 is a useful stroma, as it is deficient in macrophage-colony stimulating factor (M-CSF) production, resulting in few macrophages in the culture. If macrophages still grow excessively during ES cell differentiation into hematopoietic cells, hampering the development of other hematopoietic lineages, anti-M-CSF receptor (c-fms) monoclonal antibody, AFS98 can be added to inhibit macrophage development.

Gene Function During the Development of Adipose Cells

The white adipose tissue stores energy in the form of triglycerides in time of nutritional excess and releases free fatty acids during food deprivation. The adipose tissue mass is determined by the balance between energy intake and expenditure. Alterations of this steady state can lead to overweight and obesity, which is often accompanied by metabolic disorders associated with cardiovascular diseases such as hypertension and type II diabetes. The ongoing explosion in the incidence of obesity has focused attention on the development of adipose cells. Mainly, the in vitro system used to study adipogenesis is immortal preadipocyte cell lines. However, these systems are limited for studies of early differentiation because they represent already determined cells. The commitment of *embryonic stem* (ES) cells into the adipocyte lineage offers the possibility to study the first steps of adipose cell development. In addition, the combination of genetic manipulations of undifferentiated ES cells and in vitro adipocyte differentiation facilitates elucidation of

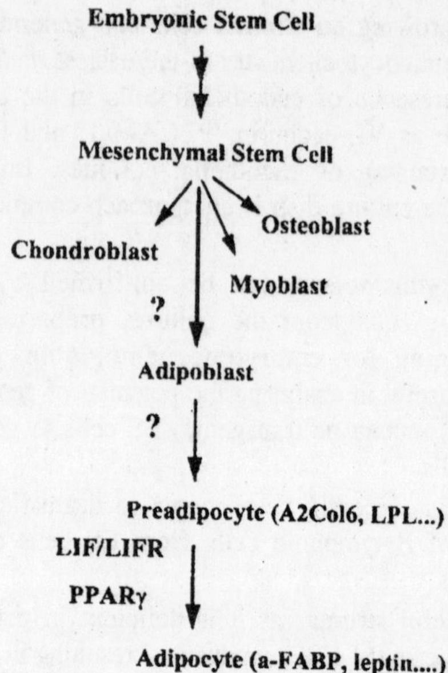

Fig. 4.2. Stages in the adipocyte development programm.

the role of genes expressed during adipose cell conversion. Terminal differentiation of preadipocytes into adipocytes is a multistep process. Several marker genes have been identified, and the hormonal regulation of these different genes has been studied in detail in recent years. However, the requirement for adipogenesis of genes known to be expressed during the different stages of differentiation remains to be investigated.

Peroxisome Proliferator-Activated Receptor γ (PPARγ) is member of the nuclear hormone receptor superfamily of ligand-activated transcriptional factors. PPARγ is expressed at high levels in adipose tissue, and several lines of evidence suggest that PPARγ plays a key role in the program of adipocyte differentiation. Recently, we observed that activation of the membrane receptor for the *Leukemia Inhibitory Factor* (LIFR) promotes adipogenesis. However, there has been no demonstration of the requirement of LIFR and PPARγ for differentiation. Both LIFR and PPARγ knock-out mice are not viable precluding the study of their role in the formation of fat. The generation of $LIFR^{-/-}$ ES cells or $PPAR\gamma^{-/-}$ ES cells combined with conditions of culture to commit stem cells into the adipogenic pathway allowed us and Rosen and colleagues to circumvent the lethality problem and to provide evidence that LIFR and PPARγ are important for the development of adipose cells. The capacity of ES cells to undergo differentiation into several cell types in vitro enables the investigator to evaluate the specificity of the effect of the mutation. As the adipoblast and the skeletal myoblast come from the same mesenchymal stem cell precursor, it is interesting to compare the adipogenic and the myogenic capacity of mutant ES cells. Studies of the levels of expression of *adipocyte-Fatty Acid Binding Protein* (a-FABP), an adipocyte-specific gene, and of Myogenin, a skeletal myocyte gene, in outgrowths derived from LIFR-mutant ES cells and in those derived from wild-type ES cells, allowed us to evaluate in vitro the effect of LIFR mutation on the commitment of stem cells into the adipogenic and the skeletal myogenic lineages.

The use of in vitro differentiation of ES cells to analyze the effects of mutations on adipogenesis is just beginning. However, the capacity of ES cells to undergo adipocyte differentiation in vitro provides a promising model for studying early differentiative events in adipogenesis and for identifying regulatory genes involved in the commitment of multipotent mesenchymal stem cell to the adipoblast lineage. In this chapter, an improved protocol to commit mouse ES cells into the adipogenic lineage at a high rate is detailed.

MATERIALS

Maintenance of Mouse ES Cells

1. Complete growth medium (1X, stored at 4°C): to 300 mL autoclaved Milli-Q-plus water add:
   (a) 40 mL of 10X Glasgow minimal essential medium (GMEM)/ BHK21, stored at 4°C.
   (b) 13.2 mL of 7.5% Sodium bicarbonate, stored at 4°C.
   (c) 4 mL of 100X Nonessential amino acids, stored at 4°C.
   (d) 8 mL of 200 m*M* glutamine, 100 m*M* sodium pyruvate, stored at –20°C.
   (e) 0.4 mL of 0.1 *M* 2-Mercaptoethanol (stored at 4°C).
   (f) 40 mL of Selected *fetal calf serum* (FCS), stored at –20°C.
2. 0.1 *M* 2-Mercaptoethanol: add 100 μL 2-mercaptoethanol to 14 mL of sterile H_2O. Store up to 3 wk at 4°C.
3. LIF.
4. *Phosphate-buffered saline* (PBS) calcium and magnesium free: 0.17 *M* NaCl, 3.4 m*M* KCl, 4 m*M* $Na_2H PO_4$, and 2.4 m*M* KH_2PO_4, pH 7.4. Filter-sterilized.

5. Trypsin solution: Trypsin 1X is prepared by adding 1 mL of 2.5% trypsin, plus 1 mL of 100 m*M* EDTA and 1 mL of chicken serum to 100 mL PBS. Aliquot (10 mL) and store at –20°C. Thawed aliquots are stored at +4°C.
6. Gelatin 0.1%: purchase gelatin 2% and dilute to 0.1% with PBS. Store at 4°C.
7. Tissue culture 25-cm^2 flasks.

Differentiation into Adipocytes

1. Retinoic acid: All *trans retinoic acid* (RA) is diluted in the dark into *dimethyl sulfoxide* (DMSO) to prepare a 10 m*M* stock solution. Aliquot and store at –20°C. Subsequent dilutions of RA are performed in ethanol and used for one experiment only. After dilution into the culture media, the concentration of ethanol never exceeds 0.1%.
2. Differentiation medium: This medium consists of growth medium with selected serum and is supplemented with antibiotics, 0.5 μg/mL insulin and 2 n*M* triiodothyronine.
3. Insulin is prepared at 1 mg/mL in cold 0.01 N HCl. Mix gently and sterilize by filtration. Aliquot (1 mL) and store at –20°C. Thawed aliquots are stored at 4°C.
4. Triiodothyronine is prepared at 2 m*M* in ethanol (stock solution). Store at –20°C.
5. Antibiotics: 1000X penicillin–streptomycin (5000 IU/mL–5000 UG/mL). Aliquot (1 mL) and store at –20°C.
6. Bacteriological grade 100- and 60-mm Petri dishes.
7. Tissue culture 100-mm dishes.

X-Gal and Oil-Red O Staining

1. Fix buffer: 0.25% glutaraldehyde in PBS supplemented with 2 m*M* $MgCl_2$ and 5 m*M* EGTA (pH 8.0). Store at 4°C.
2. Wash buffer: PBS supplemented with 2 m*M* $MgCl_2$. Store at 4°C.
3. 5-Bromo-4-chloro-3-indolyl-β-D-galactopyranoside (X gel) staining solution:
   (a) 1 mL of Potassium ferrocyanide (0.105 mg/mL).
   (b) 1 mL of Potassium ferricyanide (0.082 mg/mL).
   (c) 1 mL of X–gal (50 mg/mL).
   (d) 50 mL of Wash buffer. Filter to remove crystals and store in the dark at 4°C.

4. Oil-Red O solution. Stock saturated solution: 0.5 % Oil-Red O in isopropanol. Working solution: mix 6 volumes of stock solution with 4 volumes H_2O. Mix and filter. Store at room temperature.
5. Store solution: 70% glycerol in H_2O (v/v).

RNA Preparation from Differentiating Embryoid Body Outgrowths

1. Guanidinium lysis buffer: 4 *M* guanidinium thiocyanate dissolved in 25 m*M* sodium citrate, pH 7.0, 0.5% sarcosyl. Just before to use, add 2-mercaptoethanol to a final concentration of 0.1 *M*.
2. Sterile 2.2-mL Eppendorf tubes.
3. Phenol saturated solution, pH 4.0 containing 0.1% 8-hydroxy-quinoline.
4. Chloroform.
5. Isopropanol
6. 100% Ethanol.
7. 5 *M* NaCl. Sterilize.
8. TES buffer: 10 m*M* Tris-HCl, pH 7.4, 0.1 m*M* EDTA, 0.1% *sodium dodecyl sulfate* (SOS). Sterilize. Store at room temperature

Reverse Transcription-Polymerase Chain Reaction

1. Reverse transcription polymerase chain reaction (RT-PCR) kit.
2. Pairs of primers used for detecting:
   (a) a-FABP, an adipocyte-specific gene; 5'-GATGCCTTTGTG GGAACCTGG-3' and 5'-TTCATCGAATTCCACGCCCAG-3'.
   (b) Myogenin, a skeletal myocyte-specific gene: 5'-AGCTCCCTCA ACCAGGAGGA-3' and 5'-GGGCTCTCTGGACTCCATCT-3'.
   (c) Chain α2 of collagen type VI (A2COL6), a gene preferentially expressed in mesenchymal cells: 5'-AACTTCGCCGTGGTCAT CACTGACG-3' and 5'-AGGAATCTCCAGGCAGCTCACC TTG-3'.
   (d) *Hypoxanthine phosphoribosyltransferase* (HPRT), as a standard to balance the amount of RNA and cDNA used, except for RNA prepared from E14TG2a ES cells, which are HPRT-deficient cells: 5'-GCTGGTGAAAAGGACCTCT-3' and 5'-CACAGGACTAGAACACCTGC-3'.

METHODS

Maintenance of ES Cells

Several previously published protocols describe procedures to maintain ES cells on fibroblast feeder layers. The conditions outlined below are applicable for the maintenance of feeder layer-independent ES cell lines, such as CGR8, E14TG2a, or Zin40. These cells can be grown on gelatin-coated tissue culture flasks and maintained in a multipotent undifferentiated state providing they are exposed to LIF.

1. For a 25-cm^2 flask, aspirate the medium off and wash twice with 5 mL of PBS. Aspirate off the PBS and add 1 mL of trypsin solution. Ensure the trypsin covers the cell monolayer and incubate at 37°C, 5% CO_2, for 2 to 3 min. Check under an inverted microscope that cells are correctly dissociated.
2. Add 5 mL of complete growth medium to stop trypsinization and suspend the cells by vigorous pipetting. Transfer the cells to a sterile tube and centrifuge at 250*g* for 5 min at room temperature.
3. Aspirate the medium off and resuspend the cell pellet with 5 mL of complete growth medium by pipetting up and down 2 to 3 times. Count the cells.
4. Add 10^6 ES cells into 10 mL of prewarmed complete growth medium containing LIF, then transfer to a freshly gelatinized 25-cm^2 flask.
5. Change the medium every day.
6. Trypsinize the cultures 2 d later as in step 1. Cultures should be subcultured before cells have reached confluence.

Differentiation of Embryoid Bodies into Adipocytes

Multilineage differentiation of ES cells is initiated by aggregation. The aggregates form structures known as *embryoid bodies* (EBs). To induce adipocyte lineage, the hanging drop method for the formation of EBs is routinely used.

1. Change media on ES cells with complete growth medium supplemented with LIF 2 h before subculture.
2. Aspirate medium off and wash twice with 5 mL of PBS. Aspirate off the PBS and add 1 mL of trypsin solution. Incubate at 37°C, 5% CO_2, for 2 to 3 min.
3. Add 5 mL of complete growth medium to stop trypsinization and suspend the cells by vigorous pipetting. Transfer the cells to a sterile tube and centrifuge at 250*g* for 5 min at room temperature.
4. Aspirate the medium off and resuspend the cell pellet into 10 mL of complete growth medium without LIF, but supplemented with antibiotics. After cell counting, adjust the suspension to a concentration of 5×10^4 cells/mL.

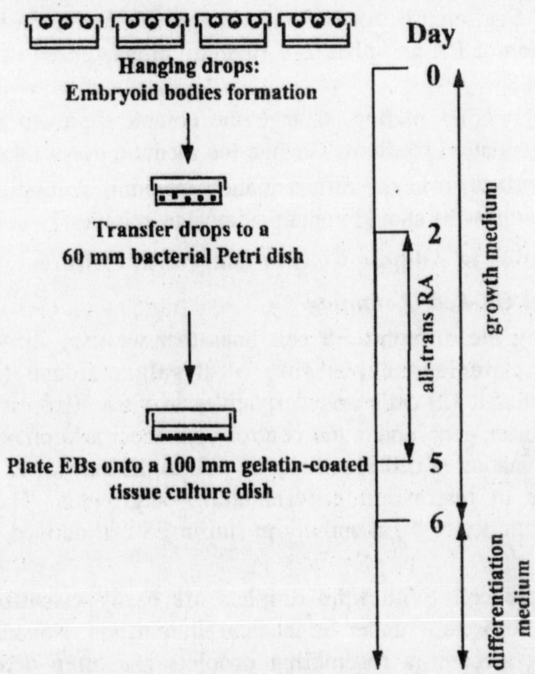

Fig. 4.3. Experimental protocol used for the commitment of ES cells into the adipocyte lineage.

5. Place aliquots of 20 μL of this suspension onto the lid of bacteriological grade dishes. This is defined as day 0 of EB formation.
6. Invert the lid and place it over the bottom of a bacteriological Petri dish filled with 8 mL PBS containing few drops of FCS. It is essential to cover the bottom of the dish with the liquid to prevent the evaporation of the hanging drops. When the lid is inverted, each drop hangs, and the cells fall to the bottom of the drop where they aggregate into a single clump (EB).
7. Two days later, remove the lid, invert it, and collect drops containing EBs in a conical sterile tube. Let it stand for 5 min at room temperature to allow the aggregates to sediment. Aspirate the supernatant and resuspend the pellet in 4 mL of complete growth medium supplemented with 10^{-7} *M* RA. Transfer the suspension into 60-mm bacteriological grade Petri dishes.
8. Incubate for 3 d in the presence of RA, changing the medium every day.

9. At d 5 after EB formation, change the medium without the addition of RA and plate 2–4 EBs/cm^2 in gelatinized-tissue culture dishes.
10. The day after plating, change the complete growth medium to differentiation medium. Change the medium every other day.
11. After 10–20 d in the differentiation medium, at least 50–70% of EB outgrowths should contain adipocyte colonies.

Identification of Adipose Cells in Outgrowth Cultures

X-Gal and Oil-Red O staining

Among the different ES cell lines that we use, Zin40 displays ubiquitous nuclear expression of β-galactosidase (β-gal) in undifferentiated ES cells and derivatives in vitro. Expression of the LacZ reporter gene under the control of a specific promoter enables the determination of cell specificity as well as the kinetics of expression of a gene of interest in differentiating outgrowths. Therefore, a procedure to detect β-gal activity in situ in ES cell-derived cultures is described.

Adipose cells with lipid droplets are easily visualized microscopically, especially under bright field illumination. Nonadipose cells containing structures resembling droplets are often detectable in untreated and RA-treated cultures. Therefore, it is essential to identify droplet-like structures as triglyceride droplets. Staining of cultures with Oil-Red O, a specific stain for triglycerides, gives a good indication of adipocyte differentiation.

A simple procedure to detect both β-gal activity and adipocytes containing triglycerides in situ in outgrowth cultures is described.

1. Aspirate medium and wash cells once with PBS.
2. Fix cells for 15 min at room temperature.
3. Wash twice for 10 min.
4. Stain with X-gal solution overnight at 37°C.
5. Aspirate X-gal solution and wash with H_2O.
6. Stain with Oil-Red O solution for 15 min.
7. Wash twice with H_2O. Cover cells with a film of storage solution (70% glycerol in H_2O). β-gal and Oil-Red O staining can be seen either with phase contrast or bright field illumination.

RNA preparation from EB outgrowths

The procedure described is an adaptation of the single-step method previously published by Chomczynski and Sacchi.

1. Wash cells with PBS.
2. Add 1.6 mL of guanidinium lysis buffer per 100-mm tissue culture dish containing EB outgrowths. Split cell lysate into 2 sterile 2.2-mL Eppendorf tubes. Vortex mix vigorously.
3. Add 0.8 mL of phenol-saturated solution, pH 4.0, per tube and mix by inversion. Then add 0.2 mL of chloroform. Shake vigorously for 10 sec and keep on ice for 15 min.
4. Centrifuge at 10,000*g* for 20 min at 4°C.
5. Carefully transfer the aqueous phase containing RNA (usually the upper phase) to a fresh 2.2-mL tube. Take care to avoid any traces of interface material.
6. Repeat from step 3.
7. Transfer the aqueous phase to a fresh 2.2-mL tube. Add 0.8 mL of isopropanol, mix, and place at –20°C for at least 12 h.
8. Centrifuge at 10,000*g* for 20 min at 4°C.
9. Aspirate the supernatant off and collect the RNA pellets from both tubes in 0.3 mL of guanidinium lysis buffer.
10. Precipitate with 0.6 mL of ethanol at –20°C for 12 h.
11. Centrifuge at 10,000*g* for 20 min.
12. Carefully pour off the supernatant and dissolve the RNA pellet in 0.5 mL of TES (10 m*M* Tris-HCl, pH 7.4, 0.1 m*M* EDTA, 0.1% SDS).
13. Determine the RNA concentration. Store RNA at –20°C.

Analysis of adipocyte gene expression

Expression of a-FABP and Myogenin in 20-d-old EB outgrowths can be detected by Northern blotting using 20 μg of total RNA. However, detection of the expression of these genes in early differentiating outgrowths requires a more sensitive method such as RT-PCR. Table 4.1 gives the temperatures of annealing for the PCR and the size of expected cDNAs, as well as the size of genomic DNA-derived contaminated bands. Gene expression can be detected either by a diagnostic ethidium bromide band or after blotting and hybridization with appropriate cDNA probes.

Notes

1. ES cell cultures may contain a proportion of "*differentiated*" cells that have lost their pluripotency. It is crucial to minimize this proportion of differentiated cells. This is achieved by the addition of LIF in a high quality culture medium, i.e., an adequate batch

Table 4.1. Detection of adipocyte- and skeletal-myocyte specific genes by PCR

	Base Pairs		
Gene	*cDNA*	*Genomic*	*Annealing temperature (°C)*
a-FABP	213	2400	56
Myogenin	500	1500	60
A2COL6	300	500	55
HPRT	249	1100	60

of serum. Identification of pluripotent stem cells is difficult unless one is familiar with the appropriate cellular morphology. Pluripotent stem cells: (1) are small, (2) have a large nucleus containing prominent nucleoli structures, and (3) have minimal cytoplasm. Pluripotent stem cells, in contrast to differentiated cells, grow rapidly. Our experience has been to select a batch of serum, which is able to support the growth of stem cells, plate 10^6 cells/25-cm^2 flask in 10% of each set of FCS, supplemented with LIF, and subculture the cells every 2 d for 4 passages. For a high quality serum, a flask should yield 5–10 × 10^6 cells at each passage. Furthermore, no toxicity of the selected serum should be observed at a 30% concentration.

2. LIF is required to maintain pluripotent ES cells and is omitted to induce the commitment of ES cells towards the adipogenic lineage. To produce LIF, Cos7 cells are transiently transfected with a LIF-expressing construct (using standard techniques), and after 4 d of growth, the medium is collected. A titration of LIF activity in the medium conditioned by transfected Cos7 cells is then performed by testing several dilutions of the medium on ES cells plated at clonal density in 24-well plates. LIF is also commercially available.
3. Serum added into the differentiation medium is preselected to support the terminal differentiation of preadipocyte clonal cell lines. The addition of a PPARγ activator, such as 0.5 μM thiazolidinedione BRL49653 in the differentiation medium dramatically stimulates the terminal differentiation of RA-treated EBs into adipocytes. This compound is not commercially available. The regimen used to promote differentiation of 3T3 to L1 preadipose cells has recently been tested on RA-treated EBs. With this regimen, 17-d-old EB outgrowths are treated with *dexamethasone* (DEX) (400 ng/mL) and *methylisobutylxantine* (MIX) (500 n*M*) for 2 d. Then, DEX and MIX are removed from the culture media,

and outgrowths are maintained in differentiation medium. As expected, hormonal treatments, which have been previously proved to promote terminal differentiation of preadipocytes into adipocytes, are also efficient to induce terminal differentiation of RA-treated EBs. It is important to note that the treatment of EBs with RA is a prerequisite.

4. 8-hydroxyquinoline stains phenol yellow, which allows the unambigous identification of the phenol phase and the aqueous phase.
5. Attachment of ES cells to the substratum is susceptible to change according to the tissue culture material. The author uses Corning or Greiner tissue culture flasks and dishes. Coating is performed by covering the surface of a 25-cm^2 flask with 5 mL of 0.1% gelatin for 15 min at room temperature, followed by careful aspiration of the gelatin solution.
6. It is critical to produce a single cell suspension for subcultures. This is achieved by knocking the flask several times to ensure complete dissociation during the trypsin treatment.
7. The formation of EBs in mass culture (by maintaining pluripotent ES cells in suspension at a high density, i.e., 5×10^5 cells/mL in bacteriological grade Petri dish) leads subsequently to a low number of outgrowths containing adipocyte colonies.
8. We use a multipipettor with a sterile combitip dispensor (Eppendorf). Approximately 80 drops can be fitted on the lid of a 100-mm Petri dish.
9. Owing to the high instability of RA, the concentration of RA able to commit ES cells into the adipogenic lineage at a high rate should be determined for each new preparation of RA (try 10^{-8} to 10^{-6} *M*). RA is light-sensitive.
10. Bacterial grade Petri dishes are used to prevent cell attachment to the substrate. EBs have a tendency to attach to the bottom of the plastic dish. This phenomenon is reduced by changing the media daily. EBs that are firmly attached to the dish should be eliminated, as these kind of EBs seem to have no adipogenic capacity.
11. A higher density of EBs can lead to a decrease in the number of EB-containing adipocyte colonies. Two-day-old EBs from 4 lids of 100-mm Petri dishes are pooled into one 60-mm bacteriological grade Petri dish, then, after RA treatment, are plated into one 100-mm tissue culture dish.
12. A wide variety of differentiated derivates, such as neurone-like cells, fibroblast-like cells, and unidentified cell types, appear over

this period. Spontaneously beating cardiomyocytes should appear 1–5 d after plating the control culture (untreated with RA). At least 40% of EBs should contain beating cardiomyocytes. In contrast, few EBs should contain beating cells from RA-treated cultures. Large adipocyte colonies appear late in the RA-treated culture.

13. We routinely get 100–200 μg RNA from one 100-mm tissue culture plate containing 20-d-old EB outgrowths.

Embryonic Stem Cell Differentiation and the Vascular Lineage

The ability of mouse *embryonic stem* (ES) cells to undergo differentiation *in vitro* complements their ability to contribute to numerous tissues *in vivo* and provides a unique model system for aspects of early mammalian development. ES cells are differentiated in two major ways: (i) unmanipulated differentiation involves the removal of differentiation inhibitory factors, allowing the ES cells to undergo a programmed differentiation to form multiple cell types that provide developmental cues to each other; and (ii) manipulated differentiation begins with the removal of differentiation inhibitory factors, but at some point the cells are usually disaggregated and cultured with specific added factors to purify or to increase the proportion of cells that acquire a particular developmental fate. Examples of both kinds of differentiation are found in this volume. The protocols provided here are for unmanipulated differentiation, which reproducibly results in the development of a primitive vasculature. Endothelial cells typically comprise 15–20% of the differentiated ES cells.

Mouse ES cells were first differentiated in vitro using small clumps of ES cells in suspension culture in media containing human cord serum. The ES cell clumps formed *embryoid bodies* (EBs) containing an outer and inner layer, and some EBs expanded to form a large lumen and were called *cystic embryoid bodies* (CEBs). These CEBs had hemoglobinized areas indicative of hematopoietic development. It was suggested that vascular development occured, analogous with the development of blood islands in the yolk sac. Subsequently the presence of mouse endothelial cells and a primitive vasculature in CEBs was shown, and the requirement for human cord serum was eliminated by using lot-tested *fetal calf serum* (FCS). A further technical modification involved reattaching the EBs to tissue culture plastic prior to overt differentiation of vascular tissue. This allowed for more sophisticated

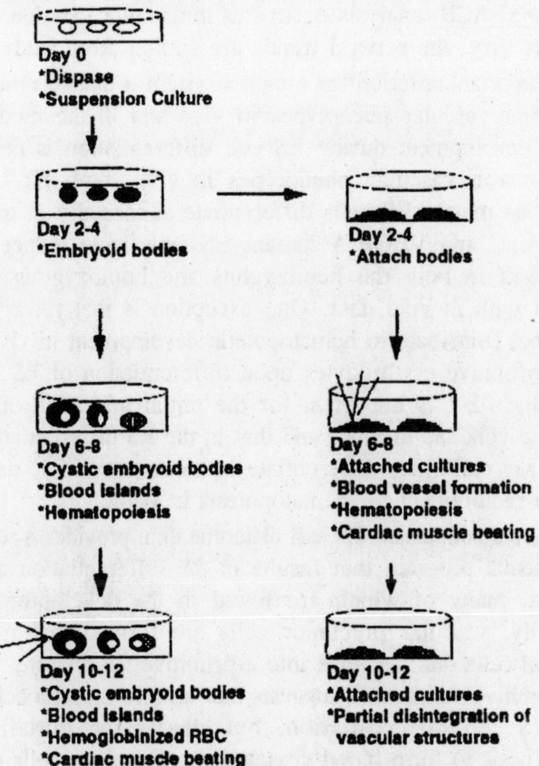

Fig. 4.4. In vitro differentiation of ES cells.

experimental manipulations of the early vasculature and for the introduction of quantitative assays of vascular development.

A number of markers for the early mouse vasculature were identified and used to further characterize mouse vascular development in vivo and in ES cell cultures and CEBs. *Platelet endothelial cell adhesion molecule*-1 (PECAM-1) (also called CD31) has proven particularly useful, since it is a transmembrane protein expressed abundantly early in vascular development. Other early markers are the vascular endothelial cell growth factor-A (VEGF) receptors flk-1 (VEGFR-2) and flt-1 (VEGFR-1), SCL/TAL, and vascular endothelial cadherin (VE-cadherin). Vascular markers that are expressed later, as angioblasts organize into blood vessels containing endothelial cells, include *intercellular adhesion molecule*-2 (ICAM-2), CD34, and Tie-2. Several groups have analyzed vascular marker expression as a function of time during ES cell differentiation using either *fluorescence-activated cell sorter* (FACS) analysis or in situ immunolocalization, and while the details vary, the general trends are similar from study to study.

An important criterion for a model system is that specific mutations have similar vascular phenotypes in vivo and in the model system. Vascular development during ES cell differentiation is perturbed by mutations with vascular phenotypes in vivo: both flk-1 and flt-1 homozygous mutant ES cells differentiate abnormally in the vascular compartment, and VEGF-A mutant ES cells have reduced vascular development in both the hemizygous and homozygous mutations, consistent with in vivo data. One exception is that flk-1 mutant ES cells do not contribute to hematopoietic development in vivo, but can produce primitive erythrocytes upon differentiation of ES cells. It is thought that flk-1 is important for the initial migration of precursor cells to the yolk sac in vivo, and that in the ES differentiation model, the yolk sac cell types differentiate in close proximity, negating the migration requirement for hematopoiesis in vitro.

Thus, unmanipulated ES cell differentiation provides a reproducible developmental program that results in the differentiation of multiple cell types, many of which are found in the developing yolk sac. Specifically, vascular precursor cells are formed that mature into endothelial cells that organize into a primitive vasculature. The shape of the primitive vasculature suggests that little or no remodeling occurs during ES cell differentiation, but clearly the initial stages of vasculogenesis to form blood vessels from precursor cells occur, and expansion of the primitive vasculature via angiogenic sprouting is also most likely a process that occurs during ES cell differentiation. The disadvantage of unmanipulated differentiation is that multiple cell types impose a requirement for careful in situ expression analysis, so vascular immunofluorescence staining and quantitation protocols are included in this section. However, we feel that by approximately recapitulating the in vivo environment in which vascular development occurs, with its attendant cell types producing important developmental signals, it is possible to analyze the roles of specific molecules in vascular development in an important but accessible context.

Materials

General Tissue Culture Materials (All Sterile)

1. 24-Well tissue culture dishes.
2. 6- and 10-cm Tissue culture dishes.
3. Disposable pipets of sizes: 1 mL, 5 mL, 10 mL, and 25 mL.

4. Bacteriological Petri dishes—these dishes are the best we found to prevent sticking of EBs during the initial culture period.
5. Ca^{+2} and Mg^{+2}-Free *phosphate-buffered saline* (PBS).
6. 0.1% Gelatin Type A, porcine, Bloom factor 200, in PBS.
7. Monothioglycerol cell culture tested. 100X stock is 32.5 μL to 50 mL PBS. Store at 4°C.
8. 1X Trypsin-EDTA: Store at –20°C long term. Thawed aliquots can be kept at 4°C for several weeks.
9. Gentamicin (Gibco-BRL) (50 mg/mL gentamicin sulfate). 1000X stock, store at 4°C.

ES Cell Culture

All reagents and materials are sterile.

1. ES cell culture medium: 67% 5637 cell conditioned medium (5637 human bladder carcinoma cell line [ATCC], source of *leukemia inhibitory factor* [LIF], for collection protocol), *Dulbecco's modified Eagle medium* (DMEM)-H (Gibco-BRL), 17% FCS (lot tested from different manufacturers), 82 μM monothioglycerol, 1X gentamicin (50 μg/mL). Store at 4°C and use within 1 mo.
2. Trypsin-EDTA solution: 0.25X trypsin-EDTA (diluted in PBS). Store stock at –20°C and working solution at 4°C for several wk.
3. Trypsin stop medium: 5637 cell conditioned medium: FCS (1:1).

Enzymatic Disruption and In Vitro Differentiation

All reagents and materials are sterile.

1. ES cell differentiation medium: DMEM-H, 20% lot-selected FCS, 150 μM monothioglycerol (2 mL 100X stock/100 mL medium = 2X concentration), 1X gentamicin (50 μg/mL). Store at 4°C and use within 1 mo.
2. Dispase: (Dispase Grade II, 2.4 U/mL). Dilute 1:1 in PBS just prior to use, so final concentration is 1.2 U/mL.
3. Medi-droppers: autoclaved.

Materials for Fixation, Antibody Staining, and Imaging

Fixation

1. Methanolacetone (1:1). Methanol and acetone are stored separately at –20°C and mixed just prior to use.
2. Fresh 4% *paraformaldehyde* (PFA): 2g PFA powder in 50 mL PBS (use a 50-mL conical tube). Heat solution to 60°C, shaking occasionally. When most of the solid is in solution, cool to room temperature, then filter through 0.45-μm filters fitted to a syringe.

Extra 4% PFA can be stored at -20°C in aliquots and thawed on ice just prior to use.

Antibody staining

1. Dilution buffer for all antibodies: 3% FCS, 0.1% NaN_3 in PBS.
2. Primary antibodies (partial list): rat anti-mouse PECAM-1 at 11000; rat anti-mouse ICAM-2 (3C4) at 1500; rat anti-mouse flk-1 (Ly-73) at 1100.
3. Secondary antibodies (partial list): donkey anti-rat B-phycoerythrin cross-absorbed at 1300 (for PECAM-1 and ICAM-2 detection); and donkey anti-rabbit IgG (H+L) tetramethyl rhodamine isothiocyanate (TRITC) cross-absorbed at 1100.

Imaging

1. Olympus IX-50 inverted microscope with epifluorescence and camera hook-up.
2. Photographic film (black and white).
3. Computer, Adobe Photoshop and plug-ins.

Methods

ES Cell Culture

1. Collection of 5637 cell condition medium (CM): 5637 cells are grown to confluence in 15-cm tissue culture dishes in DMEM-H, 10% FCS, 1X monothioglycerol, and 1X gentamycin. At confluence, the medium is changed to the same formulation but with 5% FCS instead of 10% FCS. After 48–72 h, remove dishes from incubator to hood, where 50-mL conical tubes with loose caps are set up. Tilt each dish and remove the medium with a 25-mL pipet, and put into the 50-mL tubes. Add 30 mL fresh medium (DMEM-H, 5% FCS, 1X monothioglycerol, 1X gentamicin) to each dish and return to incubator. Balance the volumes in the 50-mL tubes and spin at 2000*g* for 10 min at 4°C. In the hood, remove the supernatant from each tube and filter using a 0.45-μm filter. When all the medium is filtered, place 3 mL in a 6-cm dish in the incubator and check after 24 h for any contamination. Do the collection every 2 to 3 d until you have 5 or 6 collections, then discard cells, pool collections, aliquot, and freeze at -20°C.
2. Prepare 0.1% gelatinized dishes by adding 2 mL/6-cm dish. Tilt the dish to cover the bottom. Incubate at 37°C for 1 h up to 1 wk. Check each dish under the microscope for bacterial or fungal contamination prior to use. If uncontaminated, aspirate the excess

gelatin solution and add the appropriate amount of ES cell culture medium to each dish. Return to the incubator.

3. The ES cells should be passed every 3 to 4 d, when they are seen in shiny tight clumps. Remove the dishes from the incubator and aspirate off the medium. Wash 2 times with PBS and aspirate. Add 0.25X Trypsin-EDTA (diluted from 1X stock in PBS) to just cover the bottom of the dish (i.e., 1 mL for a 6-cm dish).
4. Place at 37°C until the majority of the ES cell clumps come up with gentle agitation (1–3 min). Stop the reaction by adding Trypsin stop solution (4 vol stop solution/vol Trypsin-EDTA). Gently draw the solution up and down a 10-mL pipet a few times to break up the clumps of cells.
5. Put 2 to 3 drops of the cell suspension into the 6-cm dish with medium in it. Observe the size of the clumps under a microscope. Try to have the cell clumps average about six cells each. If they are significantly larger, pipet further to break them up a bit. Move the dishes gently to disperse the ES cell clumps evenly throughout the dish and return to the 37°C incubator.

Enzymatic Disruption and In Vitro Differentiation

Enzymatic disruption

All volumes are given assuming that one 6-cm dish of ES cell colonies is being processed. If using a larger dish or more dishes, adjust the volumes accordingly.

1. Choose the dish that has the best ES cell clumps to process for in vitro differentiation—the cell clumps should be flattened and differentiated on the very edge but round and shiny (aspects of undifferented ES cells) in the middle. Plates are incubated for 5 to 6 d without feeding after normal passage of ES cells.
2. Aspirate the medium from the dish. Rinse twice with PBS and aspirate. Add 1 mL cold Dispase. Let sit at room temperature for 1 to 2 min. Check periodically by shaking the dish to see if the cell clumps have detached from the bottom.
3. When the majority of cell clumps have detached, use a 10-mL pipet to gently transfer the cell clumps to a 50-mL conical tube containing 35 mL PBS at room temperature. Invert the tube once gently to mix. Rinse the dish with 5 mL PBS and add to the contents of the 50-mL tube.
4. Let the tube sit until the cell clumps have settled to the bottom, then aspirate the liquid carefully to avoid disturbing the cell clumps.

Add 30–40 mL PBS down the side of the tube, swirl gently to redistribute the cell clumps, and let settle again. Aspirate, repeat the PBS wash and aspiration, then add 5 mL differentiation medium down the side of the tube. With the residual PBS, the volume in the tube will be 6 to 7 mL.

5. Pipet 10 mL of differentiation medium into each of two bacteriological dishes. Pipet the cell clumps and medium from the 50-mL tube into a 25-mL pipet. Transfer them as equally as possible to the two bacteriological dishes, pulling in an air bubble to keep the clumps distributed in the medium.
6. Check the number of cell clumps in the dishes. The goal is to get very approximately 100 clumps per dish, since fewer is not an efficient use of medium, and more may promote aggregation of cell clumps. Incubate at 37°C in a humidified incubator with 5% CO_2.

In vitro differentiation

1. Change medium as required, at least every other day after Dispase treatment. To feed EBs, gently swirl the dish in a circular manner so that the EBs go to the center of the dish. Carefully aspirate most of the old medium from the dish. Gently add 10 mL fresh differentiation medium to each dish, being careful not to disrupt the EBs.
2. To form attached cultures, use EBs that have been in suspension culture for 3 d. Swirl the bacteriological dish so the EBs are fairly close together.
3. Dispense 1.5 mL of differentiation medium into each well of a 24-well tissue culture dish that is to be seeded. Use the sterile medidropper to move EBs from the dish to the wells of the 24-well dish. Generally, try to get 10–20 EBs into each well. The bodies can be estimated by covering the dish and holding it up to look through the bottom.
4. Spread the EBs evenly in the well. If they do not spread evenly by gentle rocking of the dish, it can be done by gently pipetting medium up and down in a well (use a Pipetman with a sterile tip). Keep the plate level as you return it to the incubator, as the EBs usually attach where they are left. Attachment generally occurs within a few hours. Incubate at 37°C in a humidified incubator with 5% CO_2 and feed with 2 mL fresh differentiation medium at least every other day until fixation.

Immunolocalization

1. Take dishes with attached cultures and aspirate the medium. Rinse twice with PBS and aspirate, then add 1 mL of the cold fixative/well of the 24-well dish. Incubate 5 min at room temperature.
2. Aspirate the fixative and rinse twice more in PBS. Either store at 4°C in PBS until ready to stain, or proceed with the staining protocol.
3. Add 1 mL of dilution buffer to wells in which the PBS has just been aspirated and incubate at 37°C for 30 min to 1 h.
4. Aspirate dilution buffer and add fresh dilution buffer, in which the primary antibody has been diluted. Incubate at 37°C for 1 to 2 h.
5. Rinse in three changes of dilution buffer, aspirating carefully each time.
6. Add fresh dilution buffer, in which the secondary antibody has been diluted, and incubate at 37°C for 1 h.
7. Rinse in three changes of PBS and store in PBS at 4°C.
8. To analyze the vasculature for quantitation, set up an inverted microscope outfitted with epifluorescence and a camera. Set up a protocol so that, using a 4× objective, you can take 6 or more frames of each well that are nonoverlapping.
9. To determine the percentage of vascular area, convert the data into digital images in Adobe Photoshop.

Notes

1. ES cells maintained off feeder layers often look somewhat differentiated, especially around the edges of colonies. In our hands, different ES cell lines look more or less differentiated when kept under these conditions. We find that the passage of ES colonies in this state does not usually compromise the experiments—the differentiated cells presumably do not expand while the true ES cells remain pluripotent. While this protocol is not recommended for maintaining ES cells that will be reintroduced into mice, the in vitro differentiation process is not affected unless the majority of the cells are differentiated prior to enzymatic treatment.
2. It is imperative to use bacteriological dishes to prevent sticking of the cell clumps, so that they will form EBs in suspension culture. If this source is not available, check the dishes from 2 or 3 sources for sticking, since there are large variations in this parameter. Having said this, some of the cell clumps invariably

stick to the bottom, and they are discarded after the EBs are moved to dishes for attachment.

3. It is important to use a 25-mL pipet here to prevent further mechanical disruption of the cell clumps.
4. We have found that timely feeding with fresh medium is the most important parameter for good differentiation and minimal cell death. We monitor the pH of the medium with phenol red, and if the medium on the cultures is light orange to yellow after 24 h, it is changed every day. We find 24-h feedings necessary for densely seeded wells, and sometimes for late days of a time course when there are a large number of cells in each well.
5. For CEB production, leave the EBs in the bacteriological dish and feed every other day (or every day) by swirling EBs/CEBs to the middle and aspirating off the old medium. If a dish has many cell clumps that have attached to the bottom, we sometimes transfer the EBs to a new bacteriological dish after 3 to 4 d of culture using a medidropper. After 3 to 4 d, there should be very few additional cells that stick to the bottom of the dish.
6. EBs can be plated into tissue culture dishes at any point, from right after the Dispase treatment (d 0) to d 4. By d 5, the EBs have usually started to become cystic, and, in any case, do not readily stick to the tissue culture dish and spread. In our hands, the best differentiation and development occurs when EBs are plated at d 3, but earlier times can be assayed as long as all cultures to be analyzed and compared are plated into tissue culure dishes on the same day.
7. Attached cultures can be set up in any size tissue culture dish from a 10-cm to a 48-well dish. The number of EBs is adjusted to fit the surface area. In general, we find that the area of a 24-well dish is suitable for antibody staining or *in situ* hybridization. The smaller wells sometimes have more lifting of the cell layer from the edges of the well after fixation and during the staining process, while larger wells waste expensive antibodies. However, for RNA analysis, we routinely plate in larger dishes, and for-high-throughput screening, the smaller wells suffice.
8. Using our protocols, we routinely visualize vascular development in cultures that have been differentiated for 8 d (using the day of Dispase treatment as d 0). We first see angioblasts at d 4–6, vessels forming d 6–8, and some expansion after d 8. A typical time course to analyze vascular development would cover d 5–8.

9. We use fresh cold methanol:acetone whenever possible, since this fixative produces good cell permeation, and it results in minimal lifting of the cell layers off the bottom. The PECAM, ICAM-2, and flk-1 antigens are all stable in this fixative. The fresh 4% PFA is less efficient at producing a permeable cell, and the cell layers are more prone to lift off the bottom of the dish (a problem that usually decreases with increasing age of the culture, as the attached areas get larger). However, certain antigens, such as CD34 and Mac-1 (to visualize macrophages), require this fixative.
10. As alluded to in Note 9, one common problem is lifting of the cell layers off the dish. This can occur at any time from the fixation step through the final wash of the antibody staining. This problem is most prevalent at early time points and when using the PFA fixative. We have found that the careful aspiration and addition of reagents, often using a pipet with a bulb rather than a vacuum trap, can minimize lifting. At early time points, we try to fix numerous wells so as to have staining choices. Finally, if the layers lift up, but can be kept relatively intact, the antibody staining will still work. When most of the PBS is carefully removed from the well, the layers will sit on the bottom and can be visualized microscopically.
11. You can either use black and white film (400 ASA) or a digital camera.
12. Make sure that each field has 100% cell coverage, since the total area will be considered the denominator to determine the percentage of vascular area. By using the protocol, a given area should be photographed even if there are no vessels or if they are not centered—this is for quantitative data, not aesthetics.
13. Using the appropriate tools in Adobe Photoshop, remove any light areas that are not vascular. It is important that all the light area be bona fide vascular staining. A good antibody stain with low background is essential, but even with that, we have noticed that the domes of cells that stay relatively thick, where the EBs first attach, often pick up nonspecific reactivity. This is hazy compared to the crisp vascular staining, so we have no trouble discerning the difference and removing the nonspecific areas. Of course, any antibody staining protocol should have control wells that are incubated with secondary antibody only for comparison.
14. Using the plug-ins provided by Reindeer Games, change the image to a binary mode. The computer can then determine the percent

of the area that is white (corresponding to stained area) vs black (nonstained area).

15. The percentage stained area is averaged for all the fields of a given well, with no attempt to use error bars. It is a given that there are variations in vascular coverage from field to field in a well, and the average is to prevent that bias. Once the averages for three or more wells of a given condition are calculated, the overall average and standard deviation are calculated using standard formulas.
16. The percentage area that is stained usually corresponds fairly well to independent assays of endothelial cell number such as FACS analysis. This is somewhat surprising given that most "areas" have two layers of endothelial cells, because most vascular structures have a lumen. However, most areas are also likely to have multiple layers of nonendothelial cells as well. We never use the percentage area stained as an absolute indicator of the amount of vasculature, however, except in comparison to other wells analyzed the same way. We find significant changes among different mutant ES cell cultures compared to wild-type, and while the absolute numbers sometimes vary a bit from experiment to experiment, the trends do not change.

5

Two-Hybrid System in Embryonic Stem Cells

The *two-hybrid system* (THS) is a molecular genetic screen that detects protein– protein interactions. The protein specified by the yeast *GAL4* gene activates the transcription of genes involved in galactose metabolism. It has two functional domains, a DNA binding domain, $Gal4_{BD}$, and a transcriptional activating domain, $Gal4_{AD}$, which interact with DNA sequences in the promoter regions of *GAL1*, *GAL2*, and *GAL7* to stimulate transcription. The screen involves two plasmids; one carries the $GAL4_{BD}$ sequence fused, in-frame, to a sequence coding for a "*bait*" protein, and the other carries $GAL4_{AD}$ sequences, fused to "*prey*" sequences from a cDNA library. The two plasmids are introduced, typically by transformation, into a yeast strain carrying a reporter gene coupled to a *GAL1*, *GAL2*, or *GAL7* promoter. If the proteins encoded by the bait and prey sequences interact to allow correct positioning of the $Gal4_{AD}$ and $GaL4_{BD}$ moieties, the reporter gene is activated. Transformants are plated on medium that allows the detection of reporter gene activation. The plasmid carrying the $GAL4_{AD}$:*cDNA* plasmid can be recovered, and the positive cDNA isolated and characterized.

Numerous modifications of the THS system have been developed, many of them employing positive selection of reporter gene activation. In our laboratory, we use yeast strains with several reporter genes. Primary selection is for one or more nutritional markers fused to a galactose gene promoter, e.g., *GAL1-HIS3* or *GAL2-ADE2* or both, and positive transformants are then tested for the activation of *GAL1*-

lacZ or *GAL7-lacZ*. Specific plasmid–yeast stain combinations allow the detection of protein-protein interactions that require phosphorylation.

Large numbers of transformants are required to screen a mammalian cDNA library by the yeast THS, and this necessitates efficient transformation protocols. We have developed the LiAc/SS-DNA/PEG transformation protocol to generate the numbers of transformants required for such screens. A recent study reports the use of the yeast mating system to combine the bait and prey plasmids in a single diploid cell. Transformations were carried out in microtiter plates—a protocol for this procedure can be found in Gietz and Woods.

A THS screen involves the following steps:

1. Preparation of the *GAL4BD* bait plasmid carrying the gene of interest.
2. Transformation of the bait plasmid into a reporter yeast strain and checking for autoactivation of the reporter gene(s).
3. Preparation and amplification of the cDNA library in the $GAL4_{AD}$ prey plasmid.
4. Transformation of the reporter yeast strain carrying the bait plasmid with the prey plasmid and screening for positive transformants.
5. Identification of true positive interactions and isolation of the appropriate prey plasmids.

Materials

1. β-Mercaptoethanol (β-ME).
2. Glass beads (425–600 μm).
3. Lithium acetate dihydrate.
4. Micotiter plates (96 well).
5. Microtiter plate replicator (96 well).
6. N,N-dimethyl formamide.
7. ONPG (o-nitrophenyl-β-D-galactopyranoside).
8. Polyethylene glycol (PEG) 3350.
9. Salmon sperm DNA.
10. Tris-EDTA (TE) buffer: 10 m*M* Tris-HCl, 1 m*M* Na_2 EDTA, pH 8.0.
11. X-GAL (5-bromo-4-chloro-3-indolyl-β-D-galactopyranoside).
12. Yeast cracking buffer: 10 m*M* Tris-HCl, pH 8.0, 100 m*M* NaCl, 1 m*M* EDTA, 2% (v/v) Triton X-100, 1% (w/v) *sodium dodecyl sulfate* (SDS).

Yeast Strains

The genotypes, reporter genes, and nutritional markers used for plasmid selection of five yeast strains commonly used for THS.

Two-Hybrid Plasmid Vectors

All of the $GAL4_{BD}$ plasmids are selected using the *TRP1* marker and all the $GAL4_{AD}$ plasmids using *LEU2*. Several plasmids of both types contain the *hemaglutinnin* (HA) tag, which allows for the immunological detection of the fusion protein with appropriate antibodies.

Bacterial Strains

Plasmids are routinely amplified in and purified from the *Escherichia coli* strain DH5α *(F-/endA1 hsdR17 glnV44 thi-1 recA1 gyrA relA1 Δ[lacIZYA-argF]U169 deoR[ϕ80dlacΔ(lacZ)M15])*. Positive prey plasmids are recovered into *E. coli* strain KC8 *(hsdR leuB600 trpC9830 pyrF*::Tn5 hisB463 lacΔX74 strA galU galK) and then electroporated into DH5α for plasmid DNA preparation. Procedures, media, and solutions for bacteriological techniques and plasmid manipulation can be found in Ausubel et al. or Sambrooke et al..

Yeast Growth Media

Yeast extract-peptone-adenine-dextrose medium

Yeast strains are routinely grown on or in yeast extract-peptone-adenine-dextrose (YPAD) medium; the adenine is added to decrease the selective advantage of *ade2* to *ADE2* reversions. Double-strength YPAD, 2X YPAD broth, reduces the doubling time of yeast strains and increases transformation efficiency.

Component	*YPAD agar*	*2X YPAD broth*
Difco Bacto Yeast Extract	6 g	12 g
Difco Bacto Peptone	12 g	24 g
Glucose	12 g	24 g
Adenine hemisulphate	60 mg	60 mg
Difco Bacto Agar	10 g	—
Distilled–deionized water	600 mL	600 mL

Place a 1.0-L medium bottle or Erlenmeyer flask containing 600 mL water and a magnetic stir bar on a stir plate, add the ingredients, except agar, and mix until dissolved. Add the agar for YPAD agar medium and autoclave for 20 min. After autoclaving, swirl the bottles of YPAD agar to ensure even distribution of the agar. Equilibrate the bottles of YPAD agar to 55°C in a water bath and use each bottle to pour twenty 100 × 15 mm standard Petri plates. Allow the poured plates to dry overnight, and then store them in plastic sleeves in a refrigerator or cold room. The 2X YPAD broth should also be stored in the cold. Yeast extract-peptone-dextrose (YEPD) agar and broth media can be purchased from Becton Dickinson Microbiology Systems, Cockeysville, MD 21030, USA (BBL YEPD agar and YEPD broth). Adenine hemisulfate should be added to these media to make YPAD agar and 2X YPAD broth.

Synthetic complete medium

The plasmids used in THS carry one or more of the following selectable markers: *URA3* (uracil requirement), *TRP1* (tryptophan requirement), *HIS3* (histidine requirement), *LEU2* (leucine requirement), and *ADE2* (adenine requirement). *Synthetic complete* (SC) selection medium is based on Difco Yeast Nitrogen Base (without amino acids) with the addition of a mixture of amino acids, purines, pyrimidines, and vitamins. Selection for particular genetic markers is achieved by the omission of these specific components from the mixture.

Ingredient	*SC selection medium*
Difco Yeast Nitrogen Base without amino acids	4.0 g
Amino acid mixture	1.2 g
Glucose	12.0 g
Difco Bacto agar (agar is omitted to make liquid SC selection medium)	10.0 g
Distilled–deionized H_2O	600.0 mL

Place a 1.0-L medium bottle or Erlenmeyer flask containing 600 mL water and a magnetic stir bar on a stir plate, add the ingredients and mix until dissolved. Adjust the pH to 5.6 with 1.0 N NaOH. Add the agar for SC agar medium and autoclave for 15 min. After autoclaving, swirl the bottles of agar medium to ensure even distribution of the agar. Equilibrate the bottles of agar medium to 55°C in a water bath and use each one to pour twenty 100 × 15 mm standard Petri plates. One bottle of SC selection medium will pour 7–9 of the 150 × 15 mm standard Petri plates used for the screening transformation. Since SC selection medium is light sensitive, the plates should be dried in the dark at room temperature overnight and then stored in sealed bags in the dark at 4°C.

Amino acid mixture

Add the following ingredients to a polypropylene 250-mL bottle and mix by thorough shaking with several glass marbles.

Adenine SO_4	0.5 g	Methionine	2.0 g
Arginine	2.0 g	Phenylalanine	2.0 g
Aspartic acid	2.0 g	Serine	2.0 g
Glutamic acid	2.0 g	Threonine	2.0 g
Histidine HCl	2.0 g	Tryptophan	2.0 g
Inositol	2.0 g	Tyrosine	2.0 g
Isoleucine	2.0 g	Uracil	2.0 g
Leucine	4.0 g	Valine	2.0 g
Lysine HCl	2.0 g	p-aminobenzoic acid	0.2 g

Omit the ingredients in bold type to select for specific plasmid markers (*ADE2*, adenine SO_4; *HIS3, histidine HCl; LEU2*, leucine; *TRP1*, tryptophan; *URA3*, uracil).

Solutions

Lithium acetate (1.0 M)

Add 5.1 g of lithium acetate dihydrate to 50 mL of water in a 100-mL medium bottle and stir on a magnetic stir plate until dissolved. Sterilize by autoclaving for 15 min and store at room temperature.

PEG MW 3350 (50% w/v)

Add 50 g of PEG 3350 to 30 mL of distilled–deionized water in a 150-mL beaker and mix on a stirring hot plate with medium heat until dissolved. Allow the solution to cool to room temperature and make the volume up to 100 mL in a 100-mL measuring cylinder. Seal the cylinder with Parafilm and mix thoroughly by inversion. Transfer the solution to a glass medium bottle and autoclave for 15 min. Store securely capped at room temperature. Evaporation of water from the solution will increase the concentration of PEG in the transformation reaction and severely reduce the yield.

Single-stranded carrier DNA (2.0 mg/mL)

Add 200 mg of salmon sperm DNA to 100 mL of TE buffer in a beaker and stir at 4°C for 1 to 2 h. Store at –20°C in 1.0-mL samples. The carrier DNA must be denatured in a boiling water bath for 5 min before use and immediately chilled in ice-water. It can be boiled 3 or 4 times without the loss of activity.

Z Buffer

Dissolve 13.79 g of $NaH_2PO_4{\bullet}H_2O$, 750 mg of KCl, and 246.0 mg of $MgSO_4{\bullet}7\ H_2O$ in 1000 mL of ddH_2O, and titrate to a pH of 7.0 with 10 N NaOH. Before use, add 270 mL of β-ME to 100 mL of Z buffer.

X-Gal (20 mg/mL)

Dissolve 1.0 g of X-Gal in 50 mL of N,N-dimethyl formamide and store at –20°C.

Z Buffer/β-Me/X-Gal

Add 270 μL of β-ME and 1.67 mL of X-gal solution to 100 mL of Z buffer immediately before use.

ONPG (4 mg/mL)

Dissolve 200 mg of ONPG in 50 mL sterile double-distilled water. Store at –20°C in aliquots.

METHODS

Preparation of the GAL4$_{BD}$:GOI Fusion Plasmid

Your "*Gene of Interest*" (*GOI*) encoding the bait protein must be cloned into a suitable THS vector in-frame with the binding domain fragment of the Gal4 protein. The DNA sequences and fusion reading frames of the *multicloning sites* (MCS) in these vectors are given in Figure 5.1.

A number of cloning strategies can be used to clone YFG into a *GAL4$_{BD}$* vector:

1. *Polymerase chain reaction* (PCR) amplification of the *GOI open reading frame* (ORF) with the addition of unique *GAL4BD* vector restriction sites to the ends of the primers.
2. Using existing *GAL4$_{BD}$* vector MCS restriction sites within *GOI* ORF.
3. Blunt-end ligation of a restriction fragment containing the *GOI* ORF into the appropriate frame in the plasmid.

If the *GOI* has identifiable protein domains or motifs, these can be fused to the *GAL4$_{BD}$*, especially if the *GOI* is relatively large. The most important aspect of this cloning is to ensure that the *GOI* is in-frame with the *GAL4$_{BD}$*, so that a fusion protein can be produced. Finally, the fusion junction of any plasmid constructed using the blunt-end ligation strategy should be sequenced prior to performing any screen to confirm the ORF fusion. The *GAL4$_{BD}$* sequencing primer, 5'-TCA TCG GAA GAG AGT AG-3', can be used for *GAL4$_{BD}$* vectors such as pGBT9, pAS1, and pAS2. More details concerning cloning strategies can be found in Gietz et al. and Parchaliuk et al.

Transformation of the GAL4$_{BD}$:GOI Fusion Plasmid into Yeast

The first step in a THS screen is to transform the constructed *GAL4$_{BD}$*:GOI fusion plasmid into the appropriate yeast strain. The rapid

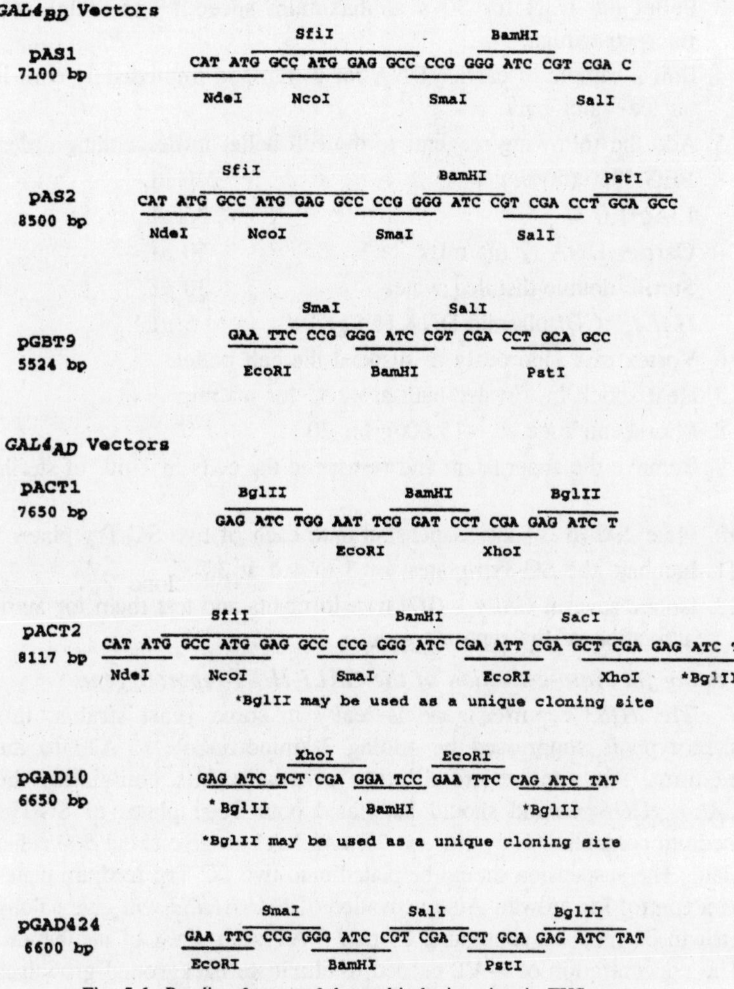

Fig. 5.1. Reading frames of the multi-cloning sites in THS vectors.

LiAc/SS carrier DNA/PEG transformation protocol yields sufficient transformants for the isolation of single plasmid transformants. Although actively growing yeast cells give the highest yields, cultures that are several days or even weeks old can be used. You should set up separate transformations for each *GAL4*$_{BD}$:*GOI* fusion construct you have prepared.

1. Inoculate the yeast strain onto a YPAD plate and incubate overnight at 30°C.
2. Scrape a 20-μL sample of yeast from the plate and resuspend the cells in 1.0 mL of sterile water.

3. Pellet the cells for 30 s at maximum speed (~15,000*g*) in a microcentrifuge.
4. Boil a sample of carrier DNA for 5 min and immediately chill in an ice-water bath.
5. Add the following reagents to the cell pellet in descending order.

PEG 3350 (50% w/v)	240 μL
LiAc 1.0 *M*	36 μL
Carrier DNA (2 mg/mL)	50 μL
Sterile double-distilled water	30 μL
$GAL4_{BD}$:*GOI* plasmid DNA (1 μg)	5 μL

6. Vortex mix vigorously to suspend the cell pellet.
7. Heat shock in a water bath at 42°C for 40 min.
8. Microcentrifuge at ~15,000*g* for 30 s.
9. Remove the supernatant and resuspend the cells in 1 mL of sterile water.
10. Plate 200 μL of the suspension onto each of two SC-Trp plates.
11. Incubate the SC-Trp plates for 3 to 4 d at 30°C.
12. Isolate several $GAL4_{BD}$:*GOI* transformants and test them for auto-activation of the reporter genes.

Testing for Auto-activation of the GAL1-HIS3 reporter gene

The *HIS3* reporter gene is leaky in some yeast strains; this phenotype is suppressed by adding 3-aminotriazole (3-AT) to the medium. An appropriate dilution of yeast cells containing the $GAL4_{BD}$:*GOI* plasmid should be plated onto Petri plates of SC-His medium containing 1, 5, 10, and 25 m*M* 3-AT to give about 500 cells/plate. The suspension should be plated onto two SC-Trp medium plates as a control for growth. Auto-activation of *GAL1-HIS3* will give colony growth on medium containing 25 m*M* 3-AT after 1 wk of incubation. The concentration of 3-AT needed to eliminate background growth is plasmid- and strain-dependent. Higher levels of 3-AT may be required to suppress background *GAL1-HIS3* expression when screening plasmid constructs based on pAS1 or pAS2, or when using the strain Y190.

Testing for auto-activation of the lacZ reporter gene

You can use the two control plates of SC-Trp from the previous test.

1. Place a sterile 75-mm circle of Whatman no. 1 filter paper on top of the colonies on SC-Trp medium and ensure that the paper makes good contact with the colonies. Punch through the filter in

an asymmetric pattern with an 18-gauge needle to mark the orientation of the paper.

2. Remove the paper and immerse it in liquid nitrogen for 10–15 s.
3. Remove the filter and allow it to thaw on a sheet of plastic wrap, colony side up. Freeze and thaw the paper twice more.
4. Soak another circle of filter paper in 1.25 mL of Z buffer/β-ME/X-gal in a Petri dish.
5. Place the first filter, colony side up, onto the filter soaked with Z buffer/β-ME/X-gal.
6. Put the lid on the Petri dish, seal it in a plastic bag and incubate it at 37°C for 120 min.
7. Place the filters on filter paper soaked in 1.0 *M* Na_2CO_3 and incubate for 5 min. Strong auto-activation of *lacZ* will give a blue color within 1 to 2 h.

Expression of the Fusion Protein

The steady state expression of the $GAL4_{BD}$:*GOI* fusion protein should be assayed by Western blotting. This can be done with a specific antibody for the product of the *GOI*. The vectors pAS1, pAS2, pACT1, and pACT2 contain the HA tag, which is recognized by the 12CA5 mAb available from Pharmacia or equivalent. A $Gal4_{BD}$ antibody from Santa Cruz Biotechnology can also be used.

Library Transformation Efficiency Test

Before embarking on a large-scale THS screen, you should perform a $GAL4_{AD}$:*cDNA* plasmid library transformation efficiency test. This will ensure efficient use of the library plasmid DNA and allow you to plan on a specific number of transformants for THS screening. Too much library DNA will result in multiple $GAL4_{AD}$:*cDNA* library plasmids in a single yeast cell; this will complicate the analysis of two-hybrid positives.

The high efficiency LiAc/SS carrier DNA/PEG transformation protocol can be used to transform 0.1, 0.5, 1, and 2 μg, of the $GAL4_{AD}$:*cDNA* library plasmid DNA into the THS yeast strain containing the $GAL4_{BD}$:*GOI* plasmid.

Day 1

1. Inoculate the yeast strain carrying the $GAL4_{BD}$:*GOI* plasmid into 25 mL of SC selection medium and incubate at 30°C on a rotary shaker at 200 rpm overnight. The culture should reach a titer of 1 to 2 × 10^7 cells/mL.
2. Incubate a bottle of 2X YPAD broth and a 250-mL flask at 30°C.

Day 2

1. Determine the titer of the overnight culture. This can be done by measuring the optical density at 545 or 600 nm of a 10^{-2} dilution of the culture in sterile water. For most yeast strains, a suspension containing 10^6 cells/mL has an OD_{545} of 0.1. Alternatively, you can count the number of cells with a hemocytometer. For accurate determination of the cell titer, you should determine the relationship between OD or hemocytometer count and colony counts for your yeast strain.
2. Dispense 50 mL of prewarmed 2X YPAD into the prewarmed 250-mL flask and return it to the 30°C incubator.
3. Calculate the volume of suspension that contains 2.5×10^8 cells and transfer to a 50-mL centrifuge tube. Pellet the cells at 3000*g* for 5 min. Resuspend the cells in 10 mL of the prewarmed 2X YPAD and transfer into the flask. The starting titer will be 5×10^6 cells/mL.
4. Incubate the culture at 30°C on a rotary shaker at 200 rpm for 4 to 5 h and determine the cell titer. When the cells have divided at least twice (cell titer $\geq 2 \times 10^7$/mL) harvest the cells by centrifugation at 3000*g* for 5 min in a 50-mL centrifuge tube.
5. Boil a sample of carrier DNA for 5 min and chill in an ice-water bath.
6. Wash the cells in 25 mL of sterile water and resuspend them in 1 mL of sterile water.
7. Transfer the suspension to a 1.5-mL microcentrifuge tube, centrifuge at ~15,000*g* again, and discard the supernatant.
8. Add water to a final volume of 1 mL and vortex mix vigorously to resuspend the cells.
9. Pipet 100-μL samples (approx 10^8 cells) into four individual 1.5-mL microfuge tubes, one for each plasmid concentration, centrifuge at ~15,000*g* for 20 s, and remove the supernatant.
10. Make up sufficient transformation mixture, lacking plasmid DNA and water, for the planned number of transformations plus one extra. For four transformations, make sufficient for five: 1200 μL of PEG, 180 μL of 1 *M* LiAc, and 250 μL of boiled carrier DNA. Keep the transformation mixture minus DNA in ice-water.
11. Add 326 μL of transformation mixture minus plasmid DNA to each transformation tube.

12. Add 0.1, 0.5, 1.0, and 2.0 μg of the *GAL4*$_{AD}$:cDNAcf:L, to the transformation tubes, and resuspend the cells by vortex mixing vigorously.
13. Incubate the transformation tubes at 42°C for 40 min.
14. Microcentrifuge at ~15,000*g* for 20 s and remove the transformation mixture.
15. Pipet 1 mL of sterile water into each tube. Loosen the pellet by stirring with a micropipet tip, and then vortex mix vigorously.
16. Dilute the suspensions 10^{-2} (10 μL into 1 mL), vortex mix thoroughly, and plate duplicate 10- and 100-μL samples onto plates of appropriate SC selection medium. The 10-μL samples should be pipetted into 100-μL puddles of sterile water.
17. Incubate the plates at 30°C for 3 to 4 d, and count the number of transformants on the plates.

Calculate the transformation efficiency (transformants/1 μg plasmid/10^8 cells) and transformant yield (total number of transformants/transformation). Scaling up the volumes of transformation mixture 60-fold and using 1 μg plasmid DNA per "*unit transformation*" should result in an overall yield of 120 × 10^6 transformants (2 × 10^6 × 60). This should be sufficient for the most demanding two-hybrid screen.

THS Library Screen

The High Efficiency LiAc/SS Carrier DNA/PEG Transformation Protocol can be scaled up to ensure that an adequate number of transformants is screened.

Day 1

1. Inoculate the yeast strain carrying the bait plasmid *(GAL4*$_{BD}$*:GOI)* into SC selection medium. The volumes of medium and flask sizes for 30-, 60-, or 120-fold scale-up are listed below:

	Scale-up		
	30X	*60X*	*120X*
SC selection medium	50 mL in 250 mL	100 mL in 500 mL	200 mL in 1 L

2. Incubate at 30°C on a rotary shaker at 200 rpm overnight.
3. Warm an appropriate vol of 2× YPAD and a culture flask(s) at 30°C overnight.

Day 2

1. Determine the titer of the overnight culture and calculate the volume required for regrowth from a starting titer of 5 × 10^6

cells/mL. The numbers of cells, volumes of overnight culture, and the vol of 2X YPAD and flask sizes for regrowth are as follows:

		Scale-up	
30X	*60X*	*120X*	
Cells required	7.5×10^8	1.5×10^9	3.0×10^9
Volume of SC culture (approx)	40 mL	80 mL	160 mL
2X YPAD for regrowth	150 mL	300 mL	600 mL
Flask size for regrowth	1000 mL	2 × 1000 mL	3 × 1000 mL

(A larger number of flasks can be used. The volume of medium should be no more than one fifth of the flask volume.)

2. Use a sterile pipet or measuring cylinder to measure the appropriate volume of overnight culture, and transfer it to an appropriate number of sterile centrifuge tubes.
3. Pellet the cells at 3000*g* for 5 min and discard the supernatant.
4. Resuspend the pellet in warm 2X YPAD and dispense into the flask(s) for regrowth. Make up to the volume(s) indicated in step 1.
5. Incubate the flasks at 30°C on a rotary shaker at 200 rpm until the cell titer reaches 2×10^7/mL. This may take 4 to 5 h.
6. Boil sufficient carrier DNA for 5 min and chill in ice-water until required.
7. Harvest the cells by centrifugation at 3000*g* and discard the supernatant.
8. Wash the cells twice in half the regrowth culture volume of sterile water and transfer the suspension to a single 50-mL centrifuge tube (30X and 60X scale-up) or divide it between two tubes (120X scale-up).
9. Centrifuge and discard the supernatant.
10. Prepare the transformation mixture for the appropriate scale-up:

		Scale-up	
	30X	60X	120X
PEG 50%	27.20 mL	14.40 mL	28.40 mL
LiAc 1 M	1.08 mL	2.16 mL	4.32 mL
Carrier DNA 2 mg/mL	1.50 mL	3.00 mL	6.00 mL
Plasmid DNA plus sterile water	1.02 mL	2.04 mL	4.08 mL
Total volume	10.80 mL	21.60 mL	42.80 mL

11. Add the transformation mixture to the cell pellet(s) and vortex mix vigorously until the pellet is completely resuspended.

12. Incubate the tubes at 42°C for 45 min. Mix by inversion at 5-min intervals to ensure temperature equilibration.
13. Centrifuge at 3000*g* for 5 min and remove the supernatant.
14. Add sterile water to the cell pellet(s) (30X, 20 mL; 60X, 40 mL; 120X, 40 mL) and resuspend the cells by pipetting up and down and then vortex mixing vigorously.
15. Spread 400-μL samples of the cell suspension onto 150-mm plates of appropriate SC selection medium. You will need 50 plates for a 30X scale-up and 100 plates for 60X and 120X.
16. Incubate the plates at 30°C for 3–10 d until colonies have grown.
17. Score the plates for colonies that show an interaction between the proteins specified by the $GAL4_{BD}$:*GOI* and $GAL4_{AD}$:cDNA plasmids.

Check the plates and subculture positives to fresh SC –Trp –Leu –His + 3-AT after 4 d. Continue checking and subculturing for a maximum of 2 wk.

Assay for lacZ Reporter Gene Activity

Once positives have been subcultured and duplicates set aside, *lacZ* gene activation can be assayed. The positives should be maintained on the appropriate SC selection medium and tested.

Storage of the THS Positives

Positives that activate the *lacZ* reporter should be stored frozen. Streak them onto fresh SC-Trp-Leu-His + 3-AT plates and incubate at 30°C for 24–48 h. Suspend a 20-μL sample of cells in 1 mL of sterile 20% glycerol in a 1.5-mL microcentrifuge tube or cryotube and store at –70°C. Large numbers of positives can be inoculated in a grid pattern onto 150-mm SC-Trp-Leu-His + 3-AT plates and stored frozen in microtiter plates. Sterilize a 96-well microtiter plate replicator by flaming in ethanol and then allow the replicator to cool. Gently place the sterile replicator onto a plate of SC selection medium and press gently to mark the surface of the agar. Patch the positive colonies onto the imprints and incubate the plate overnight. The next day, flame and cool the replicator, and lower onto the plate of patched colonies so that the prongs make contact with each inoculum. Move the replicator gently to transfer cells to the prongs. Remove the replicator, and lower into a sterile microtiter plate containing 150 μL of sterile 20% glycerol in each well. Jiggle the replicator using a gentle rotating action to suspend the cells in the glycerol. Remove the replicator, replace the lid on the microtiter plate, and store it, sealed in a plastic bag, at –70°C.

Characterizing THS Positives

The $GAL4_{AD}$:cDNA plasmids in the THS positives that activate the reporter genes can be isolated by the method of Hoffman and Winston.

1. Inoculate the THS positives onto SC-Trp-Leu-His + 3-AT plates in 2-cm^2 patches and incubate at 30°C overnight.
2. Scrape 50-μL samples of cells from the plate and resuspend them in 500 μL of sterile water in 1.5-μL microcentrifuge tubes, pellet at top speed in a microcentrifuge for 1 min, and discard the supernatant.
3. Add 200 μL of Yeast cracking buffer, and gently resuspend the cell pellet using a micropipet tip.
4. Add 200 μL of 425 to 600-μm glass beads and 200 μL of buffer saturated phenol : chloroform.
5. Vortex mix each sample vigorously for 30s and then place on ice. Repeat twice, leaving the samples on ice for 30s between treatments.
6. Microcentrifuge at ~15,000*g* for 1 min.
7. Remove the aqueous phase to a fresh tube and precipitate the nucleic acids with 20 μL of 3.0 *M* sodium acetate (pH 6.0) and 500 μL of 95% ethanol. Incubate at -20°C for 30 min, and then microcentrifuge at ~15,000*g* for 5 min at 4°C. Wash the pellet of plasmid DNA with 100 μL of 70% ethanol and dry it for 5 min at room temperature. Dissolve the pellet of plasmid DNA in 25 μL of TE (10 m*M* Tris-HCl, pH 8.0, 1 m*M* EDTA).

The plasmids should be electroporated into an *E. coli* strain containing a *leuB* mutation, such as KC8 or HB101. This will allow for the direct selection of the yeast *LEU2* gene on the $GAL4_{AD}$:cDNA plasmid.

Reconstruction of THS Positives

It is essential to reconstruct each THS positive before proceeding with further analysis. $GAL4_{AD}$:cDNA plasmid DNA isolated from each THS positive strain should be transformed back into the THS yeast strain containing the $GAL4_{BD}$:*GOI* plasmid by the protocol described above. Transformants can then tested for the activation of the *HIS3* reporter gene by plating onto SC -Trp-Leu-His + 3-AT medium, and for *lacZ* activation by the procedure described above.

Failure of THS positives to reconstruct

A single yeast cell can be simultaneously transformed by several $GAL4_{AD}$:cDNA plasmids. At high plasmid concentrations, the frequency of such co-transformation can be as high as 30% (our unpublished observations). If the original THS positive isolate contained several $GAL4_{AD}$:cDNA plasmids, only one of which was responsible for reporter gene activation, it is likely that a failure to reconstruct the positive phenotype is because the plasmid used is not the one responsible for reporter gene activation. Isolate an additional 20 plasmids from original THS positives and characterize them by digestion with *Eco*RI. Repeat the reconstruction with each distinct plasmid type that you are able to identify. In some cases, a THS positive can result from deletions between direct repeats within the bait gene, giving rise to an auto-activating $GAL4_{BD}$:*GOI* bait plasmid. For this reason, the $GAL4_{BD}$:*GOI* bait plasmid isolated from the original THS positive should be tested for auto-activation.

Characterizing the Strength of the THS Positive Interaction

The activation of the *lacZ* reporter can be quantified using the liquid ONPG assay modified for application to yeast.

1. Inoculate the THS positives onto SC-Trp-Leu-His + 3-AT plates in 2-cm² patches and incubate at 30°C overnight. Scrape 50-μL samples of cells from the plate and resuspend them in 1 mL of sterile water in 1.5-mL microcentrifuge tubes. Dilute 5 μL of suspension into 1 mL of water and measure the OD_{600}. Pellet the remainder of the suspension at ~15,000*g* in a microcentrifuge for 1 min and discard the supernatant.
2. Resuspend the cell pellet in 100 μL of Z buffer and add glass beads to one half the volume of the resuspended cells.
3. Chill on ice and vortex mix for 30 s followed by a 30-s incubation on ice. Repeat 4 times.
4. Mix duplicate samples of 50 μL of extract with 450 μL with Z buffer/β-ME, and equilibrate in a 37°C water bath for 5 min.
5. Add 160 μL of ONPG (4 mg/mL) and incubate at 37°C. As soon as a yellow color is evident, add 0.4 mL of 1.0 *M* Na_2CO_3 to terminate the reaction, and note the elapsed time. Incubation times can range from 1–60 min.
6. Centrifuge at ~15,000*g* for 1 min and measure the absorbance at 420 nm. The activity of β-galactosidase in Miller units is calculated using the formula:

$$\text{Units} = (A_{420} \times 1000) / (t \times V \times OD_{600})$$

where t = elapsed time (min), V = volume of culture used (1 mL), and OD_{600} = optical density of cell suspension.

Deletion Mapping of Interacting Domains

The identification of protein motifs responsible for the interaction can be accomplished by deletions of the *GOI* and library cDNA genes using restriction sites or with a PCR strategy. Begin by deleting the 3' ends of both genes, to preserve the fusion junctions of the translated proteins. Altering the fusion junction can dramatically affect the steady state level of fusion protein expression. When deleting the 5' end of a gene in a bait or prey plasmid, it is recommended that the steady state level of expression of the fusion protein expression be determined by Western blot to ensure accurate interpretation of the *lacZ* assay.

False Positives

A successful THS screen may result in the isolation of several hundred apparent positives. Many of these will be true positives, in which activation of the reporter genes requires an interaction between the proteins encoded by the *GOI* and the *cDNA* sequence, but some will almost certainly be false positives. Three classes of false positives can occur in a THS screen. They can be tested for at the same time that you carry out the reconstruction of THS positives. Type 1 false positives arise when the $GAL4_{AD}$:*cDNA* library plasmid activates reporter genes without interacting with the $GAL4_{BD}$ plasmid. They can be identified by transforming the $GAL4_{AD}$:*cDNA* plasmid into the THS yeast strain lacking the $GAL4_{BD}$ plasmid. If activation occurs, it is the result of auto-activation by the $GAL4_{AD}$:*cDNA* plasmid. Type 2 false positives occur when the $GAL4_{AD}$:*cDNA* plasmid activates the reporter genes in the presence of any $GAL4_{BD}$ plasmid. They can be identified by co-transformation of the THS yeast strain with the $GAL4_{AD}$:*cDNA* plasmid and an unrelated $GAL4_{BD}$ plasmid. Growth on SC-Trp-Leu-His + 3-AT indicates a Type 2 false positive. Type 3 false positives result from the interaction of a $GAL4_{AD}$:*cDNA* plasmid with an "*empty*" $GAL4_{BD}$ plasmid. They can be identified by co-transformation of the THS yeast strain with the $GAL4_{AD}$:*cDNA* plasmid and an empty $GAL4_{BD}$ plasmid. Growth on SC-Trp-Leu-His + 3-AT indicates a Type 3 false positive.

Notes

1. Extending the duration of incubation at 42°C to 180 min increases the number of transformants to approx $1 \times 10^6/\mu g$ plasmid/10^8 cells with some strains.
2. If you use the wrong amino acid mixture for your yeast strain–plasmid combination, you will obtain:

(a) No transformants if your yeast strain requires a nutrient that is absent from the amino acid mixture.

(b) Confluent growth if your yeast strain does not require the component missing from the amino acid mixture.

3. Auto-activation by the $GAL4_{BD}$:*GOI* construct can usually be overcome by cloning your bait gene into a different vector, such as pGBT9, if pAS1 or pAS2 were used previously. Alternatively, the construct can be modified by deletion to remove the region responsible for the auto-activation. An alternative approach to dampen the auto-activation of specific $GAL4_{BD}$:*GOI* constructs uses the SSB24 repressor sequence fused to the $GAL4_{BD}$ in pGBT9.
4. Completion of two divisions is required for high transformation efficiency. Yields remain high for at least another cell doubling. High efficiencies are also obtained if the cells are grown overnight from a small inoculum, 2×10^4 cells/mL, and harvested at mid to late log phase, $>1.0 \times 10^8$ cells/mL.
5. For example, if the transformation reaction containing 1.0 μg of plasmid gave an average of 200 colonies on the two 10-μL sample plates, the calculation is:

$$200 \times 100\ (10^{-2}\ \text{dilution factor}) \times 100\ (10\ \mu\text{L plating factor}) \times 1\ (\text{plasmid factor})$$

Transformation Efficiency $= 2 \times 10^6/\mu g$ plasmid DNA/10^8 cells

6

STEM CELL IN VIVO

Adult multicellular organisms are massively larger than the eggs from which they arise. To a significant degree, the final shapes of tissues, organs, and, ultimately, bodies are dictated by highly regulated cell growth processes that spatially and temporally control the rates of cell division within tissue primordia. Biologists have made great advances in understanding this fundamental problem during the last ten years. The molecular regulation of the cell cycle has been largely deciphered, and many aspects of intercellular signaling, which regulate growth spatially, have also been clarified. Now, in favorable systems such as the *Drosophila* embryo and larval imaginal discs, as well as in vertebrate organ primordia, complex mechanisms that integrate patterning and growth-promoting signals are being learned. The major signaling systems mediated by hedgehogs, epidermal growth factors, bone morphogenetic proteins, fibroblast growth factors, and wingless/wnts frequently stimulate, modulate, or repress the growth rates of receptive cells. Eventually, proliferation ceases when a programmed size and morphology has been achieved, presumably due to changes in the signals and in cellular responsiveness.

On the basis of the studies described below, we have come to regard adult stem cells as a special case of developmental growth regulation, albeit a particularly interesting and potentially informative one. We define an adult stem cell as a cell residing within an adult tissue that divides, either autonomously or in response to regulated signals, to produce cells that contribute to organismal homeostasis. Stem cells make it possible for tissues to recover from injuries and to continue functioning well beyond the average lifetime of their component cells. They make possible organ systems such as epithelial and germinal tissues which regularly produce progeny that are shed from the body. Our studies of *Drosophila* stem cells suggest that as the adult form is realized and most cells cease division and complete their differentiation, stem cells continue to receive and respond to internal or external growth-promoting signals. Moreover, differentiation has contrived to stabilize the anatomical context and signaling environment experienced by these cells. Daughter cells move away, leaving the microenvironment within and around the stem cells unchanged, ensuring that their activity will continue.

This view of stem cells raises a variety of obvious questions. Do the cells that become adult stem cells require a particular developmental history, or can they be selected from among a wide variety of relatively undifferentiated embryonic cells? Do they express characteristic stem cell genes, or do they simply reside in a special anatomical context or niche? Are there special signals that act in the permanent context of an adult stem cell, or do they utilize the same growth-promoting signals as in developing tissues? Other than the requirement to disperse and differentiate to avoid tumor formation, do stem cell progeny play any role in regulating their activity?

Likewise, do processes internal to stem cells such as differential divisions play a necessary role in programming daughter cells to take on different fates? How long do individual stem cells live, and can they be replaced during adult life? How variable are the mechanisms that program and maintain stem cells? Are some common to stem cells in different tissues, or between related tissues of evolutionarily distant species? To address all these questions, it is necessary to have systems where stem cells and their neighbors can be identified and genetically perturbed. In this chapter, we describe studies on one such system, the stem cells of the *Drosophila* ovariole, which can be studied with a high level of anatomical precision and genetic sophistication.

Ovary: A Stem Cell-based Tissue

The ovary requires highly active stem cells because it is a large-scale egg production machine. *Drosophila* females are able to produce massive numbers of eggs continuously (more than half the adult's weight per day) during adult life. The prodigious reproductive capacity of ovaries results from a highly parallel design. Each of the paired ovaries contains about 16 ovarioles that each independently produce eggs. At its anterior end, every ovariole contains a special region known as the *germarium*, where permanent *germ-line stem cells* (GSCs) and *somatic stem cells* (SSCs) reside, followed by increasingly older and more mature germ cells, followed by a string of progressively older ovarian follicles. Young females of most strains contain two or three GSCs at the anterior tip of the germarium. *Drosophila* follicles contain not just an oocyte, but a cyst of 16 interconnected germ cells. Cysts form near the anterior of the germarium, and then move posteriorly, first in a double file and then a single file, until they finish acquiring a somatic follicular layer and bud off the posterior end of the germarium. Throughout life, follicles continue to mature, move posteriorly in the ovariole, and, when complete, exit the ovariole at its posterior end. The loss of completed oocytes is balanced by the continued production of new follicles in the germarium, and the entire process from stem cell to mature egg requires only about 8 days to complete.

The distinctive cell biology of early germ cell development greatly aids the study of ovarian stem cells and their progeny, which would otherwise be hard to distinguish in the germarium. After a GSC divides, one daughter remains as a stem cell at the tip while the other daughter differentiates into a cystoblast. Cystoblasts subsequently undergo four rounds of synchronized cell division without cytokinesis to form 16-cell cysts, whose individual germ cells are interconnected by ring canals and a fusome.

Fusomes are specialized vesicular organelles that are known to be present primarily in germ cells. They contain membrane skeletal proteins including α- and β-spectrins, and Hu-li Tai Shao (Hts), that may form a cytoskeletal meshwork to retain the vesicles. Completed 16-cell cysts become wrapped with somatic follicle cells as they pass through the middle portion of the germarium. These somatic cells are produced by exactly two SSCs located on opposite sides near the middle of the germarium that divide about every 9.6 hours. So far, neither fusomes nor any other distinctive cytoplasmic structures have been observed in somatic stem cells.

The ovariole provides at least two major advantages for stem cell studies. First, the linear organization and steady posterior movement of cells in the ovariole allows the past activity of the stem cells to be deduced for many days after it has occurred. Second, individual stem cells can be precisely recognized, and distinguished from subsequent developmental stages. Ovarian GSCs can be distinguished from neighboring cells by their size, location, and the precise structure of their fusome. GSCs are the largest and most anterior germ-line cells in the germarium. GSCs and cystoblasts contain a round fusome (also called a *spectrosome*), whereas germ cells in developing and completed cysts contain branched fusomes. The exact structure of the fusome in stem cells and cystoblasts can even indicate their cell cycle stage. A round fusome accompanied by a small "*plug*" of fusome material in the ring canal joining it with its daughter cystoblast indicates a G_1 or early S-phase stem cell. Later in G_1 and S phase, the stem cell fusome extends to join old and new fusome components, creating an elongated fusome. Finally, in late S or G_2 phase, the fusome breaks in two, reverts to a round shape, and the stem cell and daughter cystoblast lose cytoplasmic contact. During G_1 and early S phase, cystoblasts resemble stem cells in their fusome morphology, but can be recognized since they are located more posteriorly and do not contact the cap cells. Later in their cell cycle, cystoblasts never display an elongated fusome; the old fusome and newly synthesized plug simply fuse and the ring canal does not break down. Still older germ cells are easily distinguished from stem cells and each other by fusome morphology and location.

It is equally important that the somatic cells surrounding GSCs and SSCs have been well defined by their unique morphologies and locations, and by using molecular markers, which allow the relationships between GSCs and their surroundings to be precisely studied.

Experimental Methods for Studying Stem Cells

Before describing the molecular biology of ovarian stem cells in detail, it is worthwhile to briefly mention the techniques that allow their properties to be characterized. The major experimental approaches used for studying stem cells are the same ones that make *Drosophila* a premier system for studying many other aspects of cell and developmental biology. The *Drosophila* genome has been completely sequenced, and mutations in thousands of different genes have been identified. Large-scale genetic screens can be carried out that have the potential to identify any *Drosophila* gene by a loss-of-function or gain-of-function phenotype. Special techniques have been developed to misexpress genes in a regulated manner, to construct mosaic animals that lack gene function in special cell subgroups, and recently, for homologous gene knockouts. These resources can be applied more quickly and with less expense than in vertebrate models. Finally, a large cadre of *Drosophila* researchers have already generated extensive knowledge about fundamental processes likely to be relevant for understanding stem cell regulation.

Several techniques are used heavily in stem cell studies. One is the ability to mark cells genetically using site-specific recombination to activate a *lacZ* gene within random members of cell populations. GSCs can be unambiguously marked, and the number and development of their labeled progeny can be studied so that the self-renewal capacity and division rate of stem cells can be measured quantitatively. By creating marked stem cells that are also homozygous for a known mutation, it is possible to test the role of specific genes in stem cell function. Mutants that have defects in stem cell maintenance and differentiation can be identified among collections of candidate mutations isolated in genetic screens using this test. A second important technique is the ability to visualize the stem cells and their surrounding cells by confocal microscopy.

Somatic stem cells are more difficult to identify than GSCs. SSCs strongly resemble the flattened, *inner germarium sheath* (IGS) cells among which they reside. Currently, there are no known morphological or molecular markers that distinguish SSCs from their neighboring IGS cells. SSCs do not contain a detectable fusome or other cytological specialization. However, SSCs can be genetically marked using the same techniques described above for GSCs, and the division and differentiation of their marked progeny can be studied. Under these conditions, an SSC can be recognized as the most anterior *lacZ*-positive cell of a somatic stem cell clone, and its precise location relative to nearby cells can be tested by double labeling using appropriate markers. SSC behavior can then be tested by observing how SSCs marked with *lacZ* are affected by genetic or environmental perturbation.

Regulation of Germ-line Stem Cells

Internal versus External Regulatory Mechanisms

A major issue in understanding ovarian stem cell regulation has been to distinguish whether regulation occurs primarily as a result of internal or external mechanisms. In the *Drosophila* embryo, neuroblasts, progenitors of the central nervous system, clearly undergo asymmetric divisions to regulate the cell fates of both progenitors and their progeny. Interestingly, in the *Drosophila* ovary, the asymmetry of germ-line stem cell division is revealed by the asymmetric segregation of their spectrosomes, as described above. However, spectrosomes can be completely eliminated by *hts* gene mutations without affecting the maintenance and division of germ-line stem cells at all, although cyst development and the cyst asymmetry are affected. These results indicate that spectrosomes and the asymmetry associated with them are not essential for the germ-line stem cell maintenance and division, but instead, set the asymmetry for later germ cell development.

There are two major issues concerning external mechanisms that act on stem cells. The first is to identify the signals and their cellular sources that act on stem cells. During development, cells frequently send and receive information from neighboring cells that critically affect their subsequent behavior. Signals may initiate steps on differentiation pathways that change the cells' internal properties and subsequent responsiveness to additional signals. Moreover, as noted above, many signals regulate cell proliferation rates. Consequently, understanding the signals that affect stem cell differentiation and growth lies at the heart of understanding stem cell function. A number of genes have been identified that are strong candidates for encoding products that directly or indirectly control signals sent from somatic cells and received by GSC. At present, the best evidence exists for Dpp, and in addition, mutations in at least four genes outside the known *dpp* pathway also seem to act via changes in external signaling. However, there are still some uncertainties regarding how each of these genes acts on GSC.

A second issue regarding external signals provided by surrounding somatic cells concerns their stability in the absence of a stem cell target. In mammalian systems, stromal cells have been hypothesized to form special microenvironments or niches for stem cells. Some recent evidence from diverse organisms and tissue types supports the existence of stem cell niches. However, the difficulty in studying and manipulating these stem cells and their surroundings in vivo has so far made it impossible to define the structure and function of any specific stem cell niches. *Drosophila* ovarian stem cells have recently been shown to reside in a functional stem cell niche. Investigations into how it operates are beginning to reveal interesting regulatory principles that may pertain more widely.

Extracellular Regulation of GSC: The Existence of a Stem Niche

The cells surrounding ovarian stem cells are organized into an asymmetric structure. In principle, this asymmetry would allow the two daughters of a stem cell to find themselves in different micro-environments whose differing signals would lead them to adopt different cell fates. The anterior tip of the germarium contains three differentiated somatic cell types: terminal filament cells, cap cells, and inner germarial sheath cells. GSCs directly contact cap cells anteriorly and inner sheath cells laterally. In contrast, cystoblasts are closer to inner sheath cells than to cap cells. Somatic stem cells also experience an asymmetric environment. SSCs reside in region 2a, near the boundary with region 2b, where they touch inner sheath cells anteriorly. However, the progeny of SSC division move posteriorly where they are no longer in direct contact with IGS cells. The asymmetry in the organization of the cells surrounding ovarian stem cells may ensure that two stem cell daughters interact with different sets of cells, and potentially receive different signals, which makes the existence of the stem cell niches possible. In the *Drosophila* testis, germ-line stem cells and somatic stem cells are anchored to the terminally differentiated somatic hub cells, and their differentiated progeny also move away from the hub cells, suggesting that both ovaries and testes use similar strategies to regulate their stem cells.

Some or all of these cell types appear to produce signals that are required for the normal functioning of the GSCs. Presently, the BMP pathway is the only characterized signaling pathway that is required to maintain GSCs. Genetic studies of downstream components in the pathway show that a BMP signal must be received in the GSC itself. This signal is most likely encoded by *decapentaplegic* (*dpp*), a *Drosophila* homolog of human *bmp 2* and *4*, and is directly received by germ-line stem cells. Dpp and BMP2 and 4 are transforming growth factor-β (TGFβ)-like secreted growth factors.

Overexpression of *dpp* in somatic cells prevents germ cell differentiation and results in the formation of stem cell-like tumors, whereas reductions in *dpp* expression cause germ-line stem cells to differentiate and exit the germarium. Interestingly, BMP4 is also implicated in the proliferation and formation of primordial germ cells in the mouse, suggesting that the ability of TGFβ-like growth factors to regulate germ-line stem cells has been conserved. Dpp mRNA is expressed in cap cells, IGS cells, and the SSC lineage, and levels appear to be relatively uniform in the neighborhood of the GSCs. Unfortunately, the lack of a suitable anti-Dpp antibody means that it remains uncertain whether Dpp protein levels are also uniform. Finally, until a genetic function can be demonstrated in somatic clones of cup cells or IGS cells, it cannot be completely ruled out that a low level of *dpp* expression takes place in germ cells themselves and that it is this expression which is essential for stem cell maintenance.

Two other growth factors, *wingless* (*wg*) and *hedgehog* (*hh*), often function together with *dpp* synergistically or antagonistically to regulate cell growth and differentiation during embryonic and larval development.

Both Wg and Hh are also produced in the germarium. *wg* is highly expressed in cap cells, and *hh* is expressed highly in terminal filament and cap cells. The presence of Hh, for example, a known inducer of Dpp and/or Wg signaling molecules, suggests that some of the same pathways that signal embryonic and larval cell growth might be used in the ovary. At least some of the interactions among these players appear to differ between the embryo and the ovariole. For example, increasing Hh up-regulates Ptc, but does increase *wg* expression, at least as measured by a *wg-lacZ* reporter. Roles for *wg* and *hh* in GSC regulation remain to be determined.

Two other genes, *fs(1)Yb* (*Yb*) and *piwi,* also contribute to GSC maintenance. Unlike *dpp*, the biochemical function of these proteins is unclear. Whereas the structure of Yb is novel, Piwi is part of a highly conserved protein group found in plants, worms, *Drosophila*, mice, and humans. Both genes act in somatic cells, because mutations in either gene cause GSC loss, but clones of germ cells lacking either Yb or Piwi can complete differentiation. Piwi is highly expressed in the somatic cells including cap cells that adjoin GSCs, and may influence maintenance signals (including possibly Dpp) from these cells. When *piwi* was overexpressed in somatic cells using a hs-GAL4 driver and an EP-piwi responder, additional GSC-like cells accumulated in region 1 of the germarium, similar to the effects of somatically over-expressing *dpp*. In addition, *piwi* is expressed at a high level in GSCs and at lower levels in cystoblasts and dividing cysts. GSCs lacking *piwi* are maintained but divide slowly. Piwi is localized to the nucleoplasm, and to explain these effects, Piwi is hypothesized to influence the production of gene transcripts that have different targets in somatic cells and GSCs.

The differentiated somatic cells surrounding the GSCs send important signals that help maintain and regulate these cells. However, if the surrounding somatic cells form a true GSC niche, a functional niche should exist without resident stem cells, and an empty niche should allow other undifferentiated cells to occupy it and become stem cells. This prediction has been directly tested by generating marked GSCs mutant for *dpp* downstream components in the ovary. Removal of *dpp* downstream components increases germ-line stem cell loss rates and generates more empty stem cell niches. Interestingly, these empty stem cell niches are efficiently repopulated. Wild-type GSCs have a half-life of 4–5 weeks, indicating that stem cells also turn over naturally. Because stem cell replacement also happens naturally, stem cell loss is much slower than is predicted by its half-life of 4–5 weeks. In this way, the functional life of the ovary has been prolonged, and the number of stem cells can be stable for an extended period of time.

The way germ-line stem cells are replaced is also very interesting. Normally, a germ-line stem cell divides along the anterior–posterior axis of the germarium so that one daughter remains in contact with cap cells and the other daughter moves away and differentiates. After a GSC is lost, one of its neighboring stem cells changes its division plane by about 90°, and the usually differentiating daughter occupies the empty stem cell niche and both stem cell daughters gain stem cell identity. This change in the division plane is reminiscent of the cleavage orientation change of neuronal stem cells in the mammalian brain. Vertical cleavages produce identical daughters that resemble stem cells, which may serve to maintain the stem cell pool. In contrast, horizontally dividing cells produce basal daughters that behave like differentiated neurons and apical daughters that remain as stem cells. Therefore, the existence of stem cell niches not only explains how stem cell self-renewal versus differentiation is regulated, but also how a stable stem cell number is maintained.

Different Roles for Different Somatic Cell Types at the Ovariole Tip

The germarium also provides an excellent system to study the structure and function of niches. Developmental studies implicate the cap cells as one key component of the GSC niche. Normally, approximately 6 cap cells are organized into a cap-like structure at the germarial tip, some of which form special junctions with GSCs. In addition, cell junctions are known to be important for cell–cell communication. Cap cells may serve to anchor GSCs and/or directly signal to them. Interestingly, the number of stem cells in the germarium is significantly correlated with the number of cap cells. Germaria with one GSC have an average of 4.2 cap cells, germaria with two GSCs have an average of 5.3 cap cells, germaria with three GSCs have an average of 6.6 cap cells. Thus, cap cells most likely determine niche size and stem cell number. The genes *dpp*, *piwi,* and *Yb,* which are known to be essential for germ-line stem cell maintenance, are expressed in cap cells, supporting an important role of cap cells in the regulation of germ-line stem cells.

Both *inner germarial sheath cells* (IGSs) and *terminal filament cells* (TFs) could also be an integral part of the germ-line stem cell niche. In young females, approximately 9 terminal filament cells are tightly packaged linear disc-shaped cells in which the most posterior one makes extensive contact with a cap cell, whereas *inner germarial sheath cells* (IGSs) consist of approximately 35 stretched somatic cells covering the regions from region 1 to region 2a. When females age, TFs start to detach from the cap cells and form ball-like structures, and their number dramatically reduces, while the number of IGS cells also decreases with age. No good correlations between TFs and GSCs or between IGSs and GSCs have been detected in aging females. Therefore, the role of TFs and IGSs in the GSC niche could be very complex, and merits further study in the future.

Inner germarial sheath cells are a very interesting group of somatic cells in terms of their regulation. IGS cells do not exist in a germarium derived from an embryonic gonad lacking primordial germ cells, suggesting that germ cells are required for the formation or maintenance of IGSs. Furthermore, complete elimination of germ-line cells in the adult ovary by overexpression of Bam protein in GSCs also causes the complete loss of IGS cells, indicating an essential role of germ-line cells in the maintenance of IGS cells. There is also a strong correlation between the number of IGSs and the number of germ-line cysts in aging females, further supporting the notion that germ-line cells maintain IGS cells. Because all germ cells are generated by GSCs, the mitotic activity of GSCs directly regulates the number of IGS cells by undefined signals. Therefore, the integrity of the GSC niche structure is likely controlled by both functional GSCs and niche cells.

Intracellular Mechanisms to Regulate Ovarian Stem Cells

The identification of intrinsic factors will greatly help in understanding how extrinsic signals from the niche are interpreted in germ-line stem cells, and also how asymmetric signaling is translated into intrinsic asymmetry, which regulates stem cell self-renewal and differentiation. Screens for female-sterile mutations have identified two major classes of genes with opposite effects on germ-line stem cells. Mutations in stem cell maintenance genes, such as *pumilio* (*pum*), *nanos* (*nos*), and *vasa* (*vas*), cause the differentiation of germ-line stem cells and result in stem cell loss. They seem to be also required for the differentiation of early germ-line cysts since these mutant ovaries accumulate ill-differentiated early germ cells. Nos functions cooperatively with Pum as a translational repressor in posterior patterning of the early embryo and in pole cell migration. Vas is a functional homolog of human eIF-4A, a major translation initiation factor, and is expressed throughout cell development of the germ line, including in GSCs.

Therefore, Nos, Pum, and Vas must function in germ-line stem cells and early germ cells to regulate gene expression at the level of translation, but their targets remain to be identified. *piwi* has been reported to regulate germ-line stem cell division but not maintenance in a cell-autonomous manner, besides its non-cell-autonomous role in the regulation of germ-line stem cell self-renewal. Its gene product is present in the nuclei of germ-line stem cells, but its function in the nucleus remains unclear.

Mutations in early germ cell differentiation genes, such as *bag-of-marbles* (*bam*), *benign germ cell neoplasia* (*bgcn*), *orb*, and *Sex-lethal* (*Sxl*), cause the accumulation of many GSC-like germ cells or early germ-line cysts. The *bam* and *bgcn* mutant ovaries accumulate a population of single germ cells that resemble GSCs, suggesting that both genes function in the same pathway to promote cystoblast differentiation. Bam is a novel protein with two different functional forms, fusome-associated and cytoplasmic. It can also interact with one of the vesicle transport proteins, Ter94, an AAA ATPase protein, suggesting that Bam may be involved in membrane trafficking. The cytoplasmic form of Bam is present in cystoblasts and is required for their differentiation. Forced expression of *bam* in GSCs results in their loss. Ectopic expression of *bam* in *bgcn* mutants will not cause germ-line stem cells to differentiate, suggesting that *bgcn* functions downstream of *bam*. Fusome-associated Bam but not its mRNA is affected in *bgcn* mutants, indicating that *bgcn* must regulate *bam* function at posttranslational levels.

It remains to be determined how *bgcn* regulates *bam* molecularly. Mutations in *orb* and *Sxl* cause the excessive accumulation of mixed early germ cells that fail to form 16-cell cysts. Orb and Sxl are expressed in both germ-line stem cells and developing cysts. Orb is structurally related to vertebrate *cytoplasmic poly-adenylation element binding protein* (CPEB). CPEB regulates the translation of stored mRNAs after fertilization in *Xenopus*. Sxl is localized in the cytoplasm of germ-line stem cells, cystoblasts, and two-cell and four-cell cysts. The cytoplasmic form of Sxl can bind to U-rich sequences in the untranslated regions of mRNAs and regulates their translation. Likely, Sxl is also involved in translational regulation of some mRNAs in early developing germ cells in the ovary.

As mentioned earlier, *dpp* directly signals to germ-line stem cells and regulates their maintenance and normal cell division. Many known signaling components in the *dpp* pathways have been shown to be required in germ-line stem cells to transduce the signal. Germ-line stem cells mutant for *punt*, *thick veins* (*tkv*), *saxophone* (*sax*), *mothers against dpp* (*mad*), *Medea* (*Med*), and *schnurri* (*shn*) have significantly shorter life spans and divide significantly more slowly than wild-type cells. Mutations in *Daughters against dpp* (*Dad*), a negative regulator of the *dpp* pathway, can prolong the life span of germ-line stem cells.

Mutations in these genes do not affect the normal development of cystoblasts and early developing cysts, indicating that the *dpp* pathway is very specific to germ-line stem cells. *punt*, *tkv,* and *sax* encode transmembrane serine-threonine kinase receptors, whereas *mad*, *Med*, and *shn* encode DNA-binding proteins. Dad is related to Mad and Med, and prevents Mad phosphorylation by receptors and further nuclear translocation.

Likely, the *dpp* pathway transcriptionally regulates the expression of other genes, which are important for germ-line stem cell maintenance and division. Identifying these *dpp* target genes, and revealing the relationships between *dpp* signaling and the two classes of genes discussed above will greatly aid in understanding how germ-line stem cell maintenance and division are controlled intracellularly.

SSC Regulation

Less is known about the intrinsic factors for somatic stem cells. One of the important signals controlling follicle cell production is the *hh* pathway. *patched* (*ptc*)-mutant somatic follicle cells overproliferate as in *hh*-overexpressing ovaries because *ptc,* encoding a seven-transmembrane protein, is a negative regulator of the *hh* signaling pathway. Thus, SSCs or early follicle progenitors may directly receive the IGS-produced *hh* signal. *Notch* (*N*) is known to regulate differentiation of somatic follicle cells. Mutations in *N* cause the fusion of egg chambers because of ill-differentiated follicle cells. N starts its expression at the region 2a/2b junction, suggesting that it could also function in somatic stem cells. Overexpression of activated N and its ligand Delta prevents the proper differentiation of somatic stalk cells and causes the accumulation of excessive somatic cells in stalks. Notch family members have been known to regulate stem cell proliferation in different organisms. It is still unclear whether the mitotic activity and number of SSCs are affected in loss-of-function and gain-of-function *N* mutants.

Little is known about the structure and function of the SSC niche. However, circumstantial evidence suggests that posterior IGS cells form the SSC niche. Hh is expressed at low levels in IGSs, which is revealed by one enhancer trap line. Interestingly, overexpression of *hh* or removal of a negative regulator of the *hh* pathway, *patched* (*ptc*), from SSCs and their progeny can lead to overproduction of somatic follicle cells. This overproliferation takes place near the 2a/2b junction where SSCs and their immediate daughters are located, suggesting that it could be caused either by increasing stem cell number or activity. In order to study SSCs and their niche effectively, SSC molecular markers and the IGSs surrounding SSCs need to be better defined in the future.

Defining Stem Cells Independently from the Behavior of their Progeny

In this chapter, we have defined stem cells by their ability to produce offspring indefinitely while remaining in a stable physical location and cellular state. What the ultimate fate of these progeny cells is, or their manner of differentiation, has no real importance to the mechanisms that govern the stem cell itself. For example, even if each progeny cell simply moved away from its site of birth, failed to undergo any changes in its state of differentiation, and finally died, the stem cell would still be a stem cell in this sense. Although such a pointless situation is unlikely to exist normally, mutations such as *bam* and *bgcn* cause just such a change; in our view, such mutations are not stem cell mutations but mutations regulating progeny differentiation.

The term *stem cell* is sometimes used in a quite different sense, however, to mean the founder of a complex lineage of multiple daughter cell types. In this view, it is the complexity of its progeny cells, and in particular the regular production of at least two different types of progeny, that distinguish a cell as a stem cell. By this definition, stem cells are almost ubiquitous in embryos. More than 70% of the cell divisions during *Caenorhabditis elegans* embryogenesis are asymmetric, indicating that the progeny have different fates. All these cells would be stem cells.

Most organisms lack the low cell number and invariant cell lineage of *C. elegans*; whether a cell's progeny differentiate into multiple cell types or not is only determined later, as a result of cell–cell interactions. We believe that this alternative definition of a stem cell has lost its usefulness now that the predominant role of intercellular signaling in controlling development and differentiation has been established.

Future Issues

We have seen that GSCs, and probably SSCs, are controlled by complex signals orchestrated by surrounding somatic cells that form a niche. The structure of the niche likely gives it the ability to maintain GSCs located within it and to specify a cystoblast fate for more posteriorly located progeny cells. Ovariole tips with apparently normal morphology form in the absence of germ cells. A future goal will be to understand how the GSC niche is specified during ovarian development and how its structure is maintained. Are there commonalities in the design and specification of anatomical units that can maintain proliferative signals indefinitely in a fixed morphological framework?

Our current knowledge of the molecular regulation of stem cell behavior is imperfect but suggests that certain issues are likely to be of key importance. Hh, Wg, and Dpp signals are associated with proliferating epithelial cells at many stages of development, and they appear to be a main engine of ovarian stem cell proliferation as well. How is Hh expression maintained in the terminal filament and cap cells, and Wg expression in the cap cells? How do these potentially antagonistic signals interact to maintain a signaling center that is critical for SSC divisions and possibly for GSC activity as well? Second, what is the nature of the junction that holds GSCs to cap cells?

Finally, more needs to be learned about the requirements for a cell to become a GSC. Is a highly undifferentiated state necessary? Is it sufficient? Since GSC maintenance genes such as *nanos* and *pumilio* appear to act as repressors, one may well answer yes. This would explain why GSC daughter cells destined to become cystoblasts could be easily shifted to become GSCs. If so, converting other cells into GSCs might require that the cells' current state of differentiation be erased.

Concluding Remark

The *Drosophila* ovary provides excellent opportunities for studying the relationship between adult stem cells and their niches. Recent identification of intrinsic factors and signals from surrounding cells has provided significant insight into molecular mechanisms regulating the maintenance and division of ovarian stem cells, particularly germ-line stem cells. The surrounding differentiated somatic cells have been shown to function as a niche for germ-line stem cells. This will be likely true for somatic stem cells in the *Drosophila* ovary and adult stem cells in other organisms. In the future, the identification of additional signals and intrinsic factors by genetic and molecular approaches will help reveal how multiple signals from the niche are integrated to regulate the self-renewal, division, and differentiation of ovarian stem cells. Further functional and structural studies on *Drosophila* ovarian stem cell niches will reveal how niches are built during development, and how long they are maintained during the aging process. Finally, the ability of the GSC niche to reprogram other cells toward a GSC fate will be more fully assessed.

7

GERM-LINE STEM CELLS

Spermatogenesis is a classic stem cell system. Continuous production of highly differentiated, short-lived sperm is maintained throughout reproductive life by a small, dedicated population of male *germ-line stem cells* (GSCs). Male GSCs are unipotent, devoted solely to the generation of sperm, much like epidermal stem cells that give rise only to keratinocytes. As precursors of the spermatogonial lineage, male GSCs must maintain a balance between the production of mature sperm and the self-renewal of stem cell potential.

Male germ-line stem cells are defined by their function as persistent, clonogenic founders of differentiating germ cells. They can be identified by multiple criteria, including anatomical position within a niche in the testis and distinct behavioral and molecular phenotypes. Male GSCs exhibit many similarities to other stem cell systems. Spermatogonial stem cells are a rare, relatively quiescent population that lies in a protected region in the testis among support cells, which may regulate their behavior. Like all stem cells, male GSCs are the most resistant cells to irradiation or chemical damage. As with hematopoietic stem cells, spermatogonial stem cells in mammals are transplantable, with an ability to both expand the stem cell pool and to regenerate an entire depleted spermatogenic lineage. Spermatogonial stem cells also exhibit signature molecular features, such as high expression of β-1 integrin, much like epidermal and hematopoietic stem cells.

The function of male GSCs has broad implications for development, disease, and evolution. Spermatogenesis is fundamental for the propagation of species. Spermatogenic defects can result in infertility or disease, such as testicular germ cell cancer. The ability to identify, isolate, culture, and alter adult male GSCs will allow powerful new approaches in animal transgenics and human gene therapy relating to infertility and disease. The male germ line also offers a powerful experimental system to study fundamental questions in stem cell biology. Spermatogenesis is a well-studied and defined process with many useful tools that can be applied to stem cell research. Current advances such as functional identification of male GSCs, identification of somatic support cells, and tools to characterize their respective roles have established an important foundation for future work.

Spermatogenesis and male fertility, as dependent on the continual production of sperm, have been studied in many organisms ranging from invertebrates to vertebrates. The early germ cell stages of spermatogenesis have been investigated in arthropods, moths, teleosts, sharks, reptiles and amphibians, birds, rodents, and primates. Interesting parallels and contrasts in male GSC behavior have emerged from observations on different organisms. For example, the anatomical location of male GSCs and their intimate association with somatic cells are conserved in many species. One striking difference is variation in the spermatogenic cycle, such as seasonal regulation in species that undergo seasonal breeding compared to continual spermatogenesis throughout reproductive life of non-seasonal breeders.

This chapter focuses on male germ-line stem cell biology in the fruit fly *Drosophila melanogaster* and in mammals (primarily rodents). Much of the information on male GSC identification and function has resulted from the helpful tools and methods available in these systems. There are striking similarities in male GSC location, behavior, and function between *Drosophila* and mammals, suggesting that complementary approaches in the two systems will help illuminate the mechanisms of male GSC regulation.

Drosophila provides a genetic system for rapid identification of genes required for normal male GSC behavior based on unbiased mutagenesis screens. In addition, tools for mitotic recombination and genetic marking can be used in lineage analysis experiments to identify and follow the behavior of male GSCs, either in the wild type or in mutants. Similar mitotic recombination tools allow tests of whether a gene required for normal stem cell behavior functions cell-autonomously (in the germ line) or non-autonomously (in the soma). Finally, the transparency and simple organization of the *Drosophila* testis allow in vivo identification of spermatogonial stem cells and phenotypic tration of mutational effects without disrupting the local stem croenvironment.

n mammals, early spermatogonial cells, including the GSC lation, can be physically manipulated. Donor male GSCs can be splanted into recipient host testes to assay stem cell function. The nsplant assay can be used for lineage analysis or to test whether quirement for a gene function is cell-autonomous or non-cell-utonomous. As in the fly, the ability to dissect and section the mammalian testis allows phenotypic analysis of spermatogonial cells in situ. In addition, the ability to make targeted gene disruptions in mice can be used to test the genetic requirements for male GSC function.

Many of the questions that arise about male GSC biology could be asked of stem cells in any system. What regulates male GSC specification from precursor populations in the embryo? What regulates initiation of the spermatogonial stem cell divisions? How are male GSCs defined and identified? Where are GSCs located in the testis, and what cells make up their microenvironment? Do somatic cells form a niche that regulates GSC behavior? Are stem cell divisions asymmetric or symmetric? What are the molecular mechanisms that maintain a balance between self-renewal of stem cell identity and initiation of differentiation when stem cells divide? Are these critical mechanisms under intrinsic or extrinsic control?

We review these questions below, each with an introductory discussion on male GSCs in general, followed by more detailed information that specifically relates to *Drosophila* or mammalian male GSCs.

Specification of Male Germ-line and Initiation of Stem Cell Divisions

The male germ line proceeds through several developmental steps prior to establishment and initiation of spermatogonial stem cell divisions in the testis. *Primordial germ cells* (PGCs) are specified as distinct from somatic cell lineages at one of the earliest stages in embryogenesis. The PGCs proliferate and migrate from their site of origin to the future position of the gonad, where they associate with somatic gonadal precursor cells to form the gonad. Once within the gonad, the PGCs differentiate in a sex-specific manner, including a distinct program of proliferation and quiescence. PGCs in the testis become male germ-line stem cells and initiate the first round of spermatogenesis that will produce sperm at the onset of reproductive age. Finally, male

germ-line stem cell divisions are regulated so that appropriate n
of stem cells are maintained.

When does the germ line first acquire stem cell potential? Evi
in mammals suggests that the earliest PGCs derived in the em
already exhibit aspects of stem cell potential under certain experime
conditions. Unlike spermatogonial stem cells, however, PGCs a
pluripotent, giving rise to highly differentiated cells of multiple somat
and germ cell lineages. In both flies and mammals, PGCs display additional characteristic differences from germ-line stem cells established in the adult male. PGCs can proliferate expansively with no differentiation. In contrast, under normal conditions, germ-line stem cells maintain their numbers and give rise to differentiating gametes. PGCs migrate throughout the embryo, whereas germ-line stem cells normally stay within a well-defined niche. Finally, PGCs are similar in appearance and behavior in both sexes, whereas germ-line stem cells are normally restricted to either male or female gametogenesis, the products of which exhibit extreme sexual dimorphism.

Environmental cues may direct the transition from PGC to male germ-line stem cell after populating the embryonic gonad. Somatic gonadal cells are required for proper development of male germ cells, including expression of sex- and stage-specific genes. The somatic gonad also appears to direct the sex-specific program and the timing of onset of gametogenesis. However, the somatic gonadal cells do not require the presence of germ cells for testis formation.

The establishment and long-term maintenance of male germ-line stem cell divisions appear to be regulated independently. Only a subpopulation of the PGCs is maintained as male germ-line stem cells through adulthood. One hypothesis is that only some PGCs acquire and/or retain stem cell capacity, whereas other PGCs directly differentiate without self-renewing divisions. Although the mechanisms are not understood, it is thought that signals from the somatic gonad direct which cells from an initially uniform PGC population are retained and which cells differentiate.

Fly

Germ-line specification and migration in the early *Drosophila* embryo have been studied extensively. Germ cells are the first cells formed when pole cells bud off at the posterior end of the syncytial embryo. The pole cells proliferate to yield a final population of up to 40 germ cells. To reach the developing gonad, the pole cells are first passively moved with the posterior midgut invagination, then actively e across the midgut to contact the mesoderm. The pole cells into two groups, migrate into bilateral clusters, and interact with ell-characterized population of somatic gonadal precursor cells. proximately 15 PGCs and 30 somatic gonadal precursor cells coalesce form the embryonic male gonad. After gonad formation, the pole cells are called primordial germ cells in *Drosophila*, whereas in mammals germ cells are usually referred to as gonocytes at this stage.

Much less is known about the transition from primordial germ cell fate into male germ-line stem cell identity in *Drosophila*. Early germ cells exhibit sex-specific differences once inside the embryonic gonad. For example, a certain germ cell marker expressed in the male embryonic gonad is not expressed in females. Gametogenesis is also initiated at distinct times in males versus females. Male germ cell proliferation and initiation of spermatogenesis must take place prior to hatching of the embryo, because differentiating spermatogonia are found in first-instar larval testes. In contrast, female germ cells do not initiate oogenesis until days later, at the time of the larval–pupal transition. Since male GSCs are active throughout larval and adult life, stem cell activity can be assayed in larval testes, independent of adult male viability or fertility. This is helpful when studying genes required for both adult viability and stem cell behavior. From the initial ~15 PGCs in the embryonic testis, only 5–9 male germ-line stem cells are maintained in the adult testis.

Mammals

Primordial germ cells are induced in the embryonic ectoderm of the mammalian epiblast. In the mouse embryo at 7 days, approximately 100 PGCs are detected in the extraembryonic mesoderm based on alkaline phosphatase activity, a marker of germ cell identity. The PGCs then migrate from the allantois along the hindgut to the genital ridges. PGCs proliferate during their migration, with more than 10,000 PGCs eventually populating the genital ridges.

In males, PGCs are enclosed in seminiferous cords to form the gonads. The somatic environment in the seminiferous tubules triggers the male-specific differentiation of PGCs into gonocytes. Gonocytes are morphologically distinct from either the PGCs or differentiating female germ cells. Gonocyte identity is marked by transition into a nonmitotic state, growth in cell size, and the onset of distinct gene expression. It is hypothesized that somatic Sertoli cells in the seminiferous tubules produce an inhibitory signal that prevents differentiation of PGCs into female germ cells and allows progression along the male pathway. Cessation of PGC proliferation and go. survival are also dependent on the somatic cells. Whereas l survived in vitro when cocultured with one of any multiple som cell types, gonocytes survived only when cocultured in the presence Sertoli cells.

Gonocytes relocate to the basal lamina during morphogenesis oi seminiferous somatic cells. Gonocytes undergo subsequent proliferative expansion, producing up to 50,000 cells in the mouse, then mitotically arrest in the G_1/G_0 phase until after birth. Stem cells may develop directly from gonocytes or may arise indirectly via prospermatogonia that differentiate at the time of the gonocyte mitotic arrest. In the newborn mouse, spermatogonial stem cells are likely to arise when proliferation resumes. Finally, onset of puberty triggers initiation of differentiation along the spermatogenic lineage. In mammals, as in the fly, it is difficult to conclude precisely when male germ-line stem cell identity is established. Some interconnected gonocytes have been found in the embryonic gonad, which may reflect initiation of spermatogenesis as early as in the gonocyte stage. In rat and hamster, kinetic studies indicate that some gonocytes may differentiate directly into spermatogonia, suggesting that only a subpopulation of gonocytes retain stem cell status.

Role of Germ-line Stem Cells in Spermatogenesis

Male germ-line stem cells must self-renew as well as produce progeny that initiate differentiation. To study this crucial aspect of stem cell behavior and to unambiguously identify GSCs, we need to understand the relationship of spermatogonial stem cells to the more differentiated germ cells that comprise the spermatogenic lineage.

As in other stem cell systems, differentiation along the spermatogenic lineage progresses through three distinct compartments. Male GSCs maintain the lineage throughout reproductive age by balancing stem cell self-renewal with differentiation. This balance can be obtained either by asymmetric or symmetric stem cell divisions. Stem cell daughter cells that initiate differentiation undergo several rounds of mitotic amplification divisions. These mitotic divisions are specialized in that cytokinesis is incomplete, resulting in cysts of synchronously dividing, interconnected germ cells. Due to the amplification divisions, a single daughter cell committed to differentiation eventually gives rise to large numbers of sperm. In this regard, germ cells in the amplification divisions are comparable to progenitor cells in the hematopoietic lineage, which maintain a limited proliferative capacity. Finally, the germ cells exit from mitotic division and initiate terminal differentiation, including progression through the stages of spermatocyte growth, the meiotic program, and morphogenesis into spermatozoa. The entire process of spermatogenesis takes approximately 10 days in *D. melanogaster*, 35 days in mice, and 61 days in rats.

Spermatogenesis is a complex process that is regulated at multiple stages. Thus, infertility is not necessarily due to a defect in the stem cell compartment. However, it has been suggested for mammals that perturbations in later stages may affect or alter stem cell behavior, indicating feedback control on the GSC population.

Fly

In *Drosophila*, male GSC divisions normally have an asymmetric outcome. One daughter cell self-renews stem cell identity and one daughter cell initiates differentiation as a gonialblast. The gonialblast is the founder spermatogonial cell that initiates the spermatogonial amplification divisions, much as the cystoblast in the *Drosophila* female germ line initiates the amplification divisions that produce interconnected germ cells of the developing egg chamber. In *D. melanogaster*, precisely four rounds of synchronous amplification division invariably result in 16 interconnected spermatogonia. The number of amplification divisions varies among different *Drosophila* species, suggesting that germ cell proliferation is restricted at a distinct regulatory point that is under genetic control.

Stem cells divide asynchronously and continuously throughout larval and adult life. Continuous stem cell activity ensures ongoing replenishment of gonialblasts to initiate differentiation, with the effect that all stages of spermatogenesis are present in the testis at any one time.

Mammals

In mammals, all premeiotic male germ cells, including stem cells, are called spermatogonia. The different stages of spermatogonia are roughly classified by temporal order as either undifferentiated A-type spermatogonia or differentiating spermatogonia. GSCs are among the single undifferentiated A spermatogonia, termed A_{single} (A_s) spermatogonia. A_s spermatogonia self-renew stem cell identity and give rise to other undifferentiated Atype spermatogonia that eventually differentiate. The successive stages of undifferentiated spermatogonia proliferate synchronously as A_{paired} (A_{pr}) spermatogonia in clusters of 2 germ cells and $A_{aligned}$ (A_{al}) spermatogonia in clusters of 4, 8, or 16 germ cells. Undifferentiated A spermatogonia develop into differentiating spermatogonia, which continue through six additional rounds of synchronous interconnected division as A_1, A_2, A_3, A_4, Intermediate, and finally, B spermatogonia. Within a given region of the seminiferous tubules, mammalian spermatogenesis is cyclic, with synchronous bursts in stem cell activity alternating with periods of relative inactivity.

Location of Male Germ-line Stem Cells in the Testis

Male germ-line stem cells are localized anatomically to a defined compartment within the testis. The ability to identify stem cells in situ is at present relatively unique to the male germ line, in contrast to the difficulty of unambiguously identifying most other stem cell types in vivo (e.g., hematopoietic stem cells within the bone marrow). Spermatogenesis proceeds in a spatial gradient within the testis, allowing GSC activity to be easily visualized as the continual production of differentiating germ cells.

Fly

The *Drosophila* adult testis is a coiled tube closed at the apical end and opening at the base into the seminal vesicle. The progression of spermatogenesis transits the length of the tube. The GSCs reside in the germinal proliferation center at the apical testis tip. Just distal to the stem cells lie the gonialblasts and interconnected spermatogonia. Cysts of 16 growing primary spermatocytes are displaced down the testis tube, completing meiosis approximately one-third of the way toward the base. Elongating spermatids extend back up the length of the testis lumen then exit the basal opening. A similar spatial gradient is observed in larval testes, with the difference that the larval testis is ovoid and normally contains only premeiotic germ cells and spermatocytes. The germ cells at the apical testis tip can be visualized by phase microscopy of live dissected testes, although the stem cells and spermatogonia are difficult to distinguish without the aid of molecular markers.

Mammals

Mammalian spermatogenesis takes place in the seminiferous epithelium, inside the multiple seminiferous tubules that compose the testis. Male GSCs lie at the tubule periphery next to the basement membrane. Germ cells move radially inward as spermatogenesis proceeds until the sperm are released into the central lumen. Spermatogenesis occurs in successive waves along the length of the tubules. Each wave contains a discrete cohort of germ cell stages, categorized as either 12 epithelial stages in the mouse (I–XII) or 14 stages in the rat (I–XIV). In the mouse, for example, undifferentiated spermatogonia normally divide during stages X–III, but rarely divide during the other stages. The dynamics of the epithelial cycle imply that stem cell activity is selectively reinitiated and/or inhibited within groups of staged spermatogonia.

Stem Cell Niche: The Relationship Between Germ-line Stem Cells and the Soma

Male germ-line stem cells are closely associated with somatic cells. Throughout the testis, the local microenvironment varies along with the different spermatogenic stages. Male GSCs lie within a protected region of the testis, surrounded by somatic cells that isolate them from differentiating germ cells. The close association between GSCs and specific somatic cell types or stages suggests a role for somatic support cells in regulating GSC behavior.

Fly

Drosophila male germ-line stem cells are associated with two somatic cell types within the germinal proliferation center. GSCs lie in a ring closely apposed to a cluster of somatic apical cells that form a compact structure called the *hub*. The apical cells of the hub exhibit epithelial-like characteristics, such as localization of fasciclin III and E-cadherin at the cell membrane. A pair of somatic stem cells, the cyst progenitor cells, flanks each germ-line stem cell. The cyst progenitor cells contact the hub with cytoplasmic extensions, thus enclosing the GSCs. The development of hub and cyst cells does not require the presence of the germ line, supporting their somatic origin. Like the GSCs, cyst progenitor cells divide asymmetrically to self-renew a cyst progenitor cell and to produce a cyst cell that initiates differentiation. Two somatic cyst cells enclose one germ-line-derived gonialblast to form a unit called a *germ-line cyst*. The two cyst cells do not divide again, but continue to enclose the growing clone of differentiating germ cells. Germ cells and somatic cyst cells are easily distinguished from one another on the basis of cell morphology and marker expression.

The testis is enclosed in a sheath made up of an inner layer of muscle cells and an outer layer of squamous, pigmented cells. The muscle layer of the sheath is open only over the apical cells of the hub. The germinal proliferation center, however, is firmly attached to the testis apex by a thick, convoluted layer of basal lamina that overlays the hub. If the sheath is torn open from a dissected testis, cysts of differentiating germ cells spill freely away from the testis wall, while the germ cells in the germinal proliferation center remain tightly associated with the testis apex.

The number of GSCs appears to be developmentally regulated. Morphological studies by reconstruction from serial electron micrographs indicated that the number of GSCs decreases from 16–18 cells in third-instar larval testes to 5–9 cells in 3-day-old adult testes. Similarly, cyst progenitor cell numbers also decrease from 19–20 cells in larval stages to 9–17 cells in the adult, with two cyst progenitor cells associated with every GSC. Concomitantly, the size and shape of the hub is compacted from a sheet of 16–18 cells into a densely packed, dome-shaped structure of 8–16 cells in the adult. It is possible that only the germ cells that maintain contact with the developing somatic hub retain GSC identity.

Mammals

Mammalian spermatogonial stem cells are neighbors to several somatic cell types along the basement membrane of the seminiferous tubule. The most closely apposed somatic cells are the large Sertoli cells. Sertoli cells contact all stages of developing germ cells, extending processes from the basement membrane to within the central lumen of the epithelium. The Sertoli cells create distinct microenvironments along their length via directional secretion of multiple regulatory factors. Sertoli cells form a tight junction barrier that isolates the spermatogonia into an exterior or basal compartment of the epithelium, separating them from the more differentiated germ cell stages within the interior or lumenal compartment. This barrier may serve to expose and/or shield stem cells from specific regulatory signals. GSCs may also form junctions directly with the Sertoli cells. Desmosomes formed between spermatogonia and Sertoli cells have been described in vivo.

Peritubular myoid cells and the lymphatic endothelium together form a somatic boundary layer that encloses the basement membrane surrounding the spermatogonial stem cells. The myoid cells, along with Sertoli cells, secrete the underlying basement membrane as well as regulatory factors that affect Sertoli cell behavior. Although the somatic Leydig cells in the intertubular spaces do not directly contact the spermatogonial stem cells, they do secrete steroids and peptides with potential regulatory roles in spermatogenesis.

Identification of Male Germ-line Stem Cells by Phenotype and Function

Male germ-line stem cells can be distinguished from other early germ cells by distinct behavioral and molecular phenotypes. Generally, male GSCs exhibit an asynchronous pattern of mitotic division and a slower cell cycle time, making them more resistant to damage from low doses of radiation than the later undifferentiated or differentiated spermatogonia. Since the location of male germ-line stem cells within the testis is known, specific gene products can be assayed for expression in either the GSCs or their differentiated progeny. Investigation of the function of genes that are expressed in stem cells but not in more differentiated spermatogonia, or vice versa, may help uncover intrinsic mechanisms that specify stem cell self-renewal versus differentiation. In addition, such cell-type-specific expression markers greatly facilitate the analysis and interpretation of mutant phenotypes.

Stem cells are most stringently defined by their functional capacity. To be classified as a stem cell by this definition, a single, adult, male germ cell must demonstrate the ability to self-renew stem cell identity and continually produce cells that differentiate into sperm. Methods to identify male GSCs functionally have been developed for both the fly and mammalian systems.

Fly

In *Drosophila*, a combination of expression markers, characteristic subcellular structures, and division behavior are used to distinguish stem cells from later germ cell stages. In the wild-type testis, all cells undergoing mitotic division are restricted to the region of the apical tip. Stem cells and gonialblasts are distinguished from spermatogonia as the single cells that divide asynchronously close to the hub. In contrast, interconnected spermatogonia divide in synchrony and normally lie at a greater distance from the hub. A stem cell division was calculated to take place once every 10 hours at the apical tip of an adult testis. Only one GSC in mitosis was observed in 50 GSCs examined morphologically.

Drosophila germ cells have a characteristic subcellular structure called the *fusome*, which is composed of multiple membranous and cytoskeletal components, including α-spectrin. The fusome takes on a different shape in single germ cells and in interconnected germ cells. In GSCs and gonialblasts, this subcellular structure forms a ball-shaped spectrosome. Spectrosomes are normally found in a rosette of cells only one or two cells deep from the apical hub. In contrast, the fusome is extended and branched, running through the ring canals that connect mitotically related spermatogonia or spermatocytes within a cyst.

Distinct gene expression profiles can further define stem cell versus spermatogonial identities. A collection of enhancer trap lines was isolated based on distinct expression patterns of the *lacZ* marker gene in testes. For example, the enhancer trap markers *M34a* and *S1-33* drive expression of β-galactosidase activity in the GSCs and gonialblasts at the most apical tip, but not in later stages. Other lines mark different stages of differentiating germ cells or the accompanying somatic cyst cells. Antibodies to specific proteins also detect subpopulations of germ cells. The cytoplasmic *bag-of-marbles* protein (Bam-C) is detected in spermatogonia from the 2- to 16- cell stages, but not in the ring of germ-line stem cells around the hub.

Clonal analysis as a test of Drosophila male germ-line stem cell function

Male germ-line stem cells are genetically identified in the *Drosophila* testis by clonal analysis. The existence of GSCs was first proposed in 1929 to explain the size and distribution of clusters of identical mutations in brooding studies. Analysis of the progeny from irradiated males demonstrated that large groups of identical mutant progeny must have derived from the division of one original, mutagenized germ cell. On the basis of these studies, it was proposed that a few indefinitely reproducing germ cells at each division produced one germ cell like themselves and one cell with a limited potential, allowing one germ cell to generate many sperm over an extended time. Similar genetic analysis of brood patterns using a sex-linked mutable trait extended the observation, notably, that cluster size was larger when the mutation occurred in a GSC rather than at a later spermatogenic stage. The results from this study accurately predicted the existence of 7–10 GSCs per adult testis and indicated that the number of germ-line stem cells stays relatively constant regardless of the number of sperm produced. The distribution of brood clusters also suggested that the GSCs divide asynchronously, with quiescent periods of 24–48 hours.

Lineage-tracing experiments using the FLP/FRT site-specific recombination system conclusively verified both the function and location of GSCs in wild-type testes. The FLP recombinase of yeast can be used to induce site-specific recombination at *cis*-acting FRT sites incorporated into *Drosophila* chromosomes. The FLP/FRT system can be used to permanently mark cells by inducing mitotic recombination between differently marked homologous chromosomes or by inducing excision of a marker gene from within a single chromosome. Both germ-line stem cells and spermatogonia can be marked in this system. However, spermatogonia are present only transiently, as they eventually clear from the proliferation center within 2 days and differentiate. In contrast, stem cells persist by self-renewal of stem cell identity, continually producing waves of marked progeny undergoing spermatogenesis. Such clonal marking studies produced single, persistent, marked germ cells adjacent to the hub at the apical tip in ~50% of the induced testes examined, confirming the identity and location of germ-line stem cells deduced from ultrastructural studies. The function and location of the somatic cyst progenitor stem cells were demonstrated by the same method.

The FLP/FRT lineage-tracing system can also be used in mutant genetic backgrounds to identify and follow stem cell behavior in males carrying mutations that cause stem cell defects. In addition, to test whether a gene required for stem cell function acts cell-autonomously or non-cell-autonomously, the same technology can be used to make mosaic animals in which the induced clones are homozygous for a mutation in an otherwise heterozygous background.

Mammals

In mammals, several phenotypic markers are used to distinguish the As spermatogonial cells from later spermatogonial stages. GSCs undergoing asynchronous mitosis or apoptosis can be detected as single cells interspersed along the periphery of the seminiferous tubule. In contrast, clusters of interconnected spermatogonia undergo mitosis and enter apoptosis synchronously within each epithelial stage. Due to the characteristic wave-like distribution of spermatogenic stages along the length of the seminiferous tubule, certain regions show a higher frequency of A_s stem cell proliferation, whereas other epithelial stages contain relatively inactive A_s spermatogonia. The number of A_s spermatogonia, however, remains constant throughout the epithelial cycle. In rat and hamster, male GSCs and undifferentiated spermatogonia generally have a longer cell cycle time than do differentiating spermatogonia. In rat testes, A_s stem cells are estimated to divide every 60 hours, the A_{pr} and A_{al} undifferentiated spermatogonia every 55 hours, and the differentiating spermatogonia every 42 hours.

A-type spermatogonia, including spermatogonial stem cells, exhibit higher levels of telomerase activity and expression of the EE2 protein than later germ cell stages. Undifferentiated A spermatogonia can be further distinguished from differentiating spermatogonia on the basis of low nuclear heterochromatin and low levels of c-Kit expression.

Transplantation as a test of mammalian germ-line stem cell function

Mammalian germ-line stem cells are able to maintain stem cell function upon transplantation to a host testis. Methods for germ cell transplants have been established in the mouse testis. Although a mixed population of testicular cells is used for the transplant injections, only the spermatogonial stem cells are thought to establish donor-derived spermatogenesis in the host. Donor spermatogonial stem cells can provide long-term reconstitution to produce functional sperm and restore fertility in a sterile host testis, with colonization rates of >80% repopulated tubules in at least 71% of the recipient mice. The best colonization occurred when donor cells were injected into hosts depleted of germ cells, obtained either by cytotoxic treatment with busulfan or by genetic mutation with alleles of the *Dominant white spotting* (*W*) or *Steel* (*Sl*) loci. These results suggest competition between transplanted and endogenous germ-line stem cells for access to limited sites capable of supporting stem cell function, in support of the niche hypothesis.

The frequency of colonization upon transplantation indicates that germ-line stem cells are rare within the testis, consistent with the estimated two GSCs per every 10^4 germ cells. This frequency compares with the measured frequency of approximately 1 hematopoietic stem cell in every 10^4 bone marrow cells. Individual colonization events upon transplantation can be distinct, allowing the activity of a single stem cell to be followed. The ability to visualize individual stem cells verified stem cell localization at the periphery of the tubule lumen. Newly transplanted spermatogonia were shown to migrate to the basement membrane, where Sertoli cells extended processes to surround the germ cells.

The transplantation assay has been used to identify purification methods that enrich for male GSCs within populations of spermatogonia. Positive selection for germ cells with high expression of the specific β-1 and α-6 integrins provided a 3- to 5-fold enrichment in transplantable GSCs from mouse testes. β-1 integrin expression also co-segregates with both epidermal and hematopoietic stem cell populations. Similarly, selection for binding to laminin also enriched for germ cells capable of stem cell transplantation. A negative selection method utilizing cryptorchid mice, which have a high concentration of undifferentiated spermatogonia, as the transplant donors resulted in a dramatic 50-fold enrichment of transplantable stem cells over controls. Cryptorchidism induced by retention of the testis in the body cavity leads to reversible depletion of differentiating germ cells, apparently without detriment to the stem cell population. The ability to transplant mammalian germ-line stem cells coupled with the ability to persistently transfect GSCs with retroviral constructs raises the possibility of introducing genetic modifications into the male germ line for either animal transgenics or human gene therapy.

Maintaining a Balance Between Two Stem Cell Fates: Asymmetric versus Symmetric Division

Stem cell function is twofold: Stem cells must both maintain the stem cell population and give rise to differentiating cells. The mechanisms that maintain the balance between these two daughter cell fates are the crucial question in stem cell biology. Two prevailing models have been advanced to explain how stem cells give rise to progeny with two different cell fates. The first model proposes that stem cell divisions are asymmetric, ensuring that both cell populations are evenly maintained at every division. An asymmetric outcome upon stem cell division results in one daughter cell that self-renews stem cell identity and another cell that commits to differentiation. This model was first advanced nearly 100 years ago on the basis of cytomorphological evidence in grasshopper spermatogenesis, then developed by further studies in the fruit fly. The second model suggests that stem cells normally undergo symmetric divisions. Symmetric stem cell divisions must alternate between proliferative divisions, producing two stem cells, and differentiating divisions, resulting in two cells that initiate differentiation. For this discussion, these models focus on the *outcome* of stem cell division and do not specify whether the mechanisms for stem cell fate decisions are intrinsically determined or extrinsically controlled.

At present, the consensus on male germ-line stem cell division pattern is split between the fly and mammalian systems. However, there is not enough evidence to rule out the possibility in either system that male germ-line stem cells normally undergo both asymmetric and/or symmetric divisions.

Fly

Male germ-line stem cells in *Drosophila* usually divide with an asymmetric outcome. As described above, asymmetric stem cell divisions were first proposed to explain the appearance of large, clonal populations obtained from male brood experiments. More recently, the asymmetric outcome of GSC divisions was confirmed by lineage tracing: Single, marked germ cells persisted at the testis apical tip and continually gave rise to differentiated progeny. By this criterion, the somatic stem cells also undergo asymmetric divisions. Although it has not yet been analyzed in the male, symmetric proliferative divisions may be possible under conditions that induce stem cell expansion to replace lost stem cells, as was described in the *Drosophila* female germ line.

Both the germ-line stem cells and cyst progenitor cells divide radially within an inherently asymmetric microenvironment. The cell fates adopted by the two stem cell daughters correlate with their physical location with respect to the hub. Upon germ-line stem cell division, the daughter cell directly adjacent to the hub and enclosed by the cyst progenitor cells retains stem cell identity and self-renewal capacity. The daughter cell displaced away from the hub becomes enclosed in cyst cells and initiates differentiation. A similar spatial relationship between the germ cells and somatic apical cells has been described for the germinal proliferation center in other insects. The conserved organization and the correlation between physical position and fate suggest that extrinsic cues from surrounding support cells may be important for the asymmetric cell fate decisions made by the two daughters of each stem cell division.

Mammals

It is not yet clear whether mammalian stem cell divisions are normally asymmetric. Historically, three different models have been advanced to explain how mammalian male germ line stem cells maintain steady state. All models agree that the germ-line stem cells are rare, only 0.03% of the total germ cells. In addition, all models assume that stem cells maintain a steady state so that, despite continual cell division, stem cell numbers do not increase. The A_s model is currently the most widely accepted. According to this model, A_s spermatogonial stem cells usually undergo symmetric divisions, resulting in either two self-renewing A_s spermatogonia or two interconnected A_{pr} spermatogonia that initiate differentiation. It is possible that the cell which gives rise to A_{pr} spermatogonia is analogous to the gonialblast in *Drosophila* rather than a true stem cell. The observed delays in spermatogenesis both after irradiation and after transplantation indicate that stem cell expansion precedes initiation of differentiation, suggesting that male GSCs can divide symmetrically to produce two stem cells under these conditions.

Molecular Mechanisms of Stem Cell Regulation: Roles for Intrinsic versus Extrinsic Control

Mutational analysis can be used to identify genes that mediate normal stem cell behavior. Genes required for stem cell specification can be identified in screens for mutations that disrupt stem cell formation. Genes required for the cell fate decisions that specify stem cell self-renewal can be identified in screens for mutations that lead to premature stem cell loss. Likewise, genes required for the initiation of differentiation instead of stem cell self-renewal can be identified in screens for mutations that lead to unrestricted proliferation of stem cells.

According to the stem cell niche hypothesis, the microenvironment plays a crucial role in regulating stem cell behavior. If so, then genes that are expressed in surrounding somatic support cells that extrinsically regulate germ-line stem cell behavior and genes that act intrinsically within the GSCs are both likely to be required for normal stem cell function. A somatic role in germ-line maintenance was elegantly demonstrated by ablation experiments in the nematode. Subsequent genetic analysis revealed the underlying molecular mechanism. Maintenance of gametogenesis in the nematode syncytial gonad requires expression of the Delta-related ligand *lag-2* in the somatic distal tip cell, which signals via the Notch-like receptor *glp-1* expressed in the germ line to maintain mitotic proliferation of germ cell nuclei. Without the signal from the distal tip cell, the germ cell nuclei cease proliferation and enter meiosis and terminal differentiation.

In both flies and mammals, somatic gonadal cells have been shown to provide important signals for post-stem-cell stages of spermatogenesis. Recent evidence demonstrates that signals from somatic cells also regulate the choice of GSC fates. The best current examples from each system suggest that counteracting signals are likely to balance the choice between stem cell self-renewal and differentiation.

Fly

In *Drosophila*, specification of somatic gonadal precursor cells under control of a genetic hierarchy of pattern formation genes is necessary for gonad coalescence and subsequent PGC development into germ-line stem cells. Screens for zygotic genes required for germ cell migration and gonad formation have also identified additional genes that act in the soma. Although the somatic gonad can form in the absence of germ-line cells, the behavior of somatic cells is altered in agametic testes. In testis lacking germ line, cyst cells continue to proliferate and can take on an altered identity. This suggests that cross-regulation between germ cells and somatic cells may coordinate normal germinal proliferation center function. Little is known about the mechanisms that regulate the transition from primordial germ cell to stem cell identity once the germ line is established in the gonad.

A small number of genes specifically required for function of *Drosophila* male germ-line stem cells have been identified from screens for viable, male-sterile mutants. However, since spermatogenic stem cell activity initiates at the end of embryogenesis, genes essential for embryonic viability cannot readily be assessed for effects on male GSC regulation if loss-of-function mutations are homozygous lethal. Chemical mutagenesis can produce specific alleles that affect only the male germ line, weak alleles, or conditional alleles (such as temperature-sensitive alleles) of essential genes to address their roles in spermatogonial stem cell regulation. In addition, production of homozygous mutant clones in otherwise heterozygous males by mitotic recombination can be used to make gonadal mosaics to test the role of essential genes. Furthermore, the ability to make marked mutant clones in the testis can be used to test whether genes required for GSC regulation act cell-autonomously in the germ line or non-cell-autonomously in surrounding somatic support cells.

Mutations in genes required for survival or maintenance of GSCs cause loss of germ-line renewal. This phenotype can be easily identified by phase microscopy of whole testes. Typically, mutations in which germ-line stem cells are initially active but are not maintained or become quiescent lead to testes with some differentiating spermatids but loss of the developmental gradient of early germ cells at the apical tip. Phenotypes can range from few to many sperm, depending on whether a mutation affects establishment of stem cell identity or continual stem cell self-renewal upon division. Additional phenotypic analysis using molecular markers can then distinguish whether loss of early germ cell differentiation is due to loss of germ-line stem cells, stem cell quiescence, or cell death. Mutants representing all of these classes have been isolated in mutagenesis screens and are currently being characterized.

Mutations in the *Drosophila* gene *piwi* cause loss of renewing germ line in both male and female flies. Males homozygous for *piwi* mutations exhibit tiny testes with a nearly complete absence of spermatogenesis. Piwi protein is normally expressed in the early male germ cells as well as somatic hub and cyst cells. Although the male phenotype has not been examined in detail, *piwi* has been shown to be required for both the maintenance and division kinetics of *Drosophila* female GSCs. Wild-type *piwi* function appears to be required both in somatic cells and in the germ line for normal female germ-line stem cell behavior. Proteins homologous to Piwi have been found in a wide range of organisms from plants to humans. The plant homologs, *zwille* and *argonaute*, both play similar roles in maintenance of the plant meristem. Mutations in the *Drosophila* gene *escargot* also disrupt maintenance of spermatogenesis; however, a substantial number of sperm are produced before male germ-line stem cells are lost. Expression of *escargot* mRNA in the testis is restricted to early germ cells and the somatic hub, consistent with either an intrinsic or extrinsic role in male germ-line stem cell maintenance.

An opposite phenotype results from mutation of genes required to specify the gonialblast fate instead of germ-line stem cell self-renewal or to specify further steps of differentiation instead of spermatogonial proliferation. Unrestricted GSC self-renewal or spermatogonial mitotic amplification at the expense of differentiation can be detected by staining for DNA, which reveals expansion of compact, brightly staining mitotic cells at the apical tip. Overproliferation of stem cells or gonialblasts can be easily distinguished from overproliferation of interconnected spermatogonia using the gene expression, subcellular structure, and cell division behavior markers described above.

Genetic analysis has identified several *Drosophila* genes (*bam, bgcn, punt, schnurri*) that control the decision to exit the spermatogonial divisions at the 16-cell stage. Function of two genes (*punt* and *schnurri*) is required in the somatic cyst cells to restrict spermatogonial proliferation, indicating extrinsic regulation of germ cell amplification divisions. Similar extrinsic mechanisms may limit the proliferative capacity of amplifying progenitors in other lineages, such as the amplifying keratinocytes in the skin or the myeloid progenitors in the bone marrow.

Recent evidence suggests that a signal(s) from somatic cyst cells also restricts GSC self-renewal and allows differentiation of the gonialblast, thus ensuring that male germ-line stem cell divisions have an asymmetric outcome. Mutations in two different genes suggest that wild-type function of the epidermal growth factor receptor pathway is required in somatic cyst cells for the normal choice of GSC fates. Conditional loss of function of either the *Egfr* or *raf* genes results in a massive increase in the number of early germ cells, including the

GSCs and spermatogonia. In both mutants, many of the accumulated GSCs maintain expression of GSC markers outside of their normal niche alongside the hub or cyst progenitor cells. Accumulation of GSCs and spermatogonia is associated with a block in their differentiation, as no new spermatocytes are observed. Germ-line clones of either *Egfr* or *raf* null alleles demonstrated that wild-type function of these genes was not required in the germ line itself to allow germ-line differentiation. Cyst cell clones homozygous for mutant *raf* resulted in unrestricted proliferation of the encysted germ cells, which still contained a wild-type copy of the *raf* gene. These results suggest that cyst cells play a guardian role to ensure that upon GSC division, one daughter cell down-regulates the ability to self-renew stem cell identity and adopts a gonialblast fate.

Mammals

In mammals, mutations that cause germ cell depletion or early arrest of spermatogenesis are characterized initially by a "Sertoli-cell only" phenotype—atrophic tubules with few germ cells. The absence of renewing germ line can indicate a failure in germ cell specification, migration, gonad formation, or maintenance of the stem cells. Several genes are known to act extrinsically for primordial germ cell survival; for example, the growth factors LIF and OncM. The *Dominant white spotting* (W) and Steel (Sl) receptor–ligand pair are both needed for proper primordial germ cell proliferation and migration. Mutations in either gene result in reduced numbers of primordial germ cells in the embryonic gonad. Interestingly, the *W* and *Sl* mutations also affect the proliferation and migration of erythropoietic and melanocytic precursors. Chimera experiments testing where c-Kit and Sl function for normal germ cell development demonstrated that the receptor is only required to act in germ cells, whereas the ligand functions in the soma. Although their molecular identity or mechanism of action is not yet known, the Hertwig's anemia (*an*) and atrichosis (*at*) genes are also required for primordial germ cell survival and gonocyte development. Mice mutant for either the *an* or *at* genes have few germ cells due to germ cell death in the embryonic gonad. Finally, disruption of the growth factor BMP8b gene, which is normally expressed in early germ cells, results in a reduction in the number of spermatogonial stem cells established in the embryonic gonad, causing a delay in initiation of spermatogenesis.

In other examples, loss of renewing germ line follows relatively normal germ cell specification, gonad formation, and occasionally an initial wave of spermatogenesis. However, the presently known mouse mutations that cause depletion of early germ cells all appear to affect a similar stage just downstream of the stem cells, and not stem cell self-renewal directly. Five distinct situations appear to cause developmental arrest at the switch from undifferentiated A-type spermatogonia (A_{al}) into differentiating-type spermatogonia, resulting in proliferation but not accumulation of undifferentiated A-type spermatogonia in mouse or rat testes. These situations include (1) mutations in the mouse genes *Steel* or *W* (Sl^{17H}, Sl^{d}, and W^{f} alleles); (2) mutation of the mouse gene *juvenile spermatogonial depletion* (*jsd*); (3) conditions of vitamin A deficiency; (4) cryptorchidism; and (5) inhibition by intratesticular testosterone after radiation or other toxicant exposures. Colonization of recipient testes with donor GSCs transplanted from *Sl* mutant or cryptorchid mice demonstrated that functional spermatogonial stem cells are in fact retained in these conditions where later stages of differentiating germ cells are depleted.

Somatic cells in the testis play an important regulatory role in spermatogenesis, although most hormones and cytokines tested exhibit either pre- or post-spermatogonial effects. Less is known about the possible role of somatic cells in regulating spermatogonial stem cell behavior. Recently, the level of *glial-cell-line-derived neurotrophic factor* (GDNF) produced in Sertoli cells was implicated in regulation of spermatogonial stem cell fate decisions. Reduction in GDNF function disrupted stem cell maintenance, causing stem cell loss due to differentiation. Conversely, overexpression of GDNF from a transgene construct blocked early germ cell differentiation, resulting in unrestricted proliferation of stem cells and spermatogonia.

Transplantation experiments can be used in mammals to address whether genes that regulate germ-line stem cell behavior are required in the germ line or the somatic lineages. Transplantation of male germ cells between sterile W/W^{v} and Sl/Sl^{d} mutant mice demonstrated W is required in germ cells and Sl is required in the soma, in agreement with previous evidence from chimeric mice. Sl/Sl^{d} mutant germ cells lacking functional ligand were successfully transplanted into sterile W/W^{v} mutant recipient testes lacking functional receptor, giving rise to Sl/Sl^{d}-derived fertile sperm.

The cyclic nature of mammalian spermatogenesis in a given region of seminiferous tubules suggests that reentry into or exit from GSC proliferation must be regulated at distinct stages of the epithelial cycle. An inhibitory activity in testicular extracts capable of blocking the stem cell proliferation that normally reinitiates during mouse epithelial stages II and III has been reported, possibly derived from the differentiating spermatogonia. Interestingly, the inhibitory factor was tissue- but not species-specific, suggesting that a conserved, testis-specific factor may regulate the timing of GSC divisions. More dramatic regulation of spermatogonial proliferation is seen in certain other seasonally breeding vertebrates, such as the shark.

Aberrations leading to germ cell neoplasia reveal important points of normal germ-line regulation in humans. Gonadoblastomas occur when PGCs fail to populate the embryonic gonad, such as in XY females, demonstrating the importance of somatic–germ cell interactions in restricting proliferation of GSC precursors. Adult male germ cell tumors may derive from latent gonocytes that transformed into *carcinomas in situ* (CIS), suggesting that the developmental transition from a gonocyte to an established male GSC is a critical regulatory step.

The male germ line offers a powerful system in which to study central questions in stem cell biology. Male germ-line stem cells have been identified. in situ, allowing analysis of the role of surrounding somatic support cells in regulation of stem cell behavior. Analysis of male germ-line stem cell behavior in *Drosophila* and mammals has revealed striking parallels. The ability to combine genetic screens with well-developed descriptive analysis in both these organisms promises functional identification of genes and regulatory pathways that regulate critical aspects of stem cell biology. The ability to construct mosaic animals by germ cell transplantation in mammals or mosaic analysis in *Drosophila* allows tests of whether crucial genes control stem cell behavior by intrinsic or extrinsic mechanisms. In *Drosophila*, powerful genetic tools and the availability of the full genome sequence allow rapid identification of molecules controlling stem cell behavior. In mice, the ability to construct knockout and conditional knockout mutations allows tests of the role of candidate genes in stem cell function. These tools, and the possibility that genes identified by forward genetics in *Drosophila* may have functional homologs that can be tested in mammals, promise to reveal fundamental principles and underlying molecular pathways that may govern stem cell specification, self-renewal, and differentiation in a variety of stem cell systems.

Primordial Germ Cells as Stem Cells

Germ cells are the precursors of the mature gametes, making their status as stem cells apparently unassailable. The fusion of the gametes to produce a totipotent zygote initiates the whole program of embryonic development, leading to the formation of the stem cells of all adult tissues as well as the next generation of germ cells.

The germ cell lineage usually originates as a very small founding population that is segregated from somatic cells early in development, at least in organisms where the overall body plan is also established early. Perhaps the physical separation of germ cells from organizing centers helps to protect them from the influence of potent signaling factors and morphogenetic movements. In vertebrates and *Drosophila*, there is considerable proliferation of the founding population as it moves from its site of origin to the gonads. The term *primordial germ cells* (PGCs) is strictly applied to the diploid germ cell precursors that transiently exist in the embryo before they enter into close association with the somatic cells of the gonad and become irreversibly committed as germ cells.

Male and female PGCs are indistinguishable, and in mammals both will finally stop dividing and enter into meiosis when associated with the somatic cells of the ovary, or even with tissues such as the adrenal gland outside the gonads. However, in the testis, PGCs behave differently, since they come under the influence of the XY gonadal cells that produce a short-range, diffusible, meiosis-inhibiting factor. Male PGCs therefore normally undergo mitotic arrest in G_1 as prospermatogonial stem cells that do not divide again until puberty. Some limited proliferation of spermatogonial stem cells can be obtained

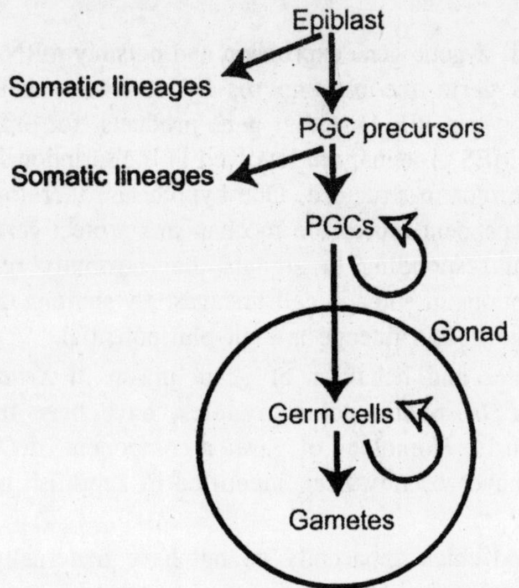

Fig. 7.1. Germ cell lineage in mammals.

in culture. Therefore, the mammalian germ line consists of two distinct stem cell populations, the transient population of PGCs outside the gonad and the spermatogonial stem cells within, that self-renew and differentiate into sperm throughout the fertile life of the adult male.

Origin of Primordial Germ Cells: Inheritance of Cytoplasmic Determinants versus Induction by Extrinsic Factors

In most organisms studied, but with several exceptions including mammals and birds, the segregation of pluripotent germ cells from somatic cells involves maternal factors or determinants. These are deposited in the cytoplasm of the egg and during cleavage are asymmetrically segregated into a small number of blastomeres that subsequently differentiate into PGCs.

Germ cell determinants are complexes of RNA and protein that have been best characterized in *Caenorhabditis elegans* (where they associate into organelles called *P granules*) and in *Drosophila* (where they constitute the polar granules or pole plasm). For example, in *Drosophila*, components include oskar, nanos, vasa, and tudor. Ectopic expression of oskar in the *Drosophila* blastula is sufficient to initiate the formation of ectopic germ cells. However, in *C. elegans*, P granule components are necessary but not sufficient for the specification of germ cells. Very little is known about the way in which germ plasm components regulate gene expression and cell behavior in germ cell precursors.

In *C. elegans*, zygotic gene expression and possibly mRNA stability are repressed in germ-line blastomeres by at least one P granule component (the protein PIE-1). Other gene products, for example, the polycomb group MES proteins, are involved in transcriptional silencing at the level of chromatin structure. One hypothesis, therefore, is that a number of independent repression mechanisms protect germ cells in *C. elegans* from responding to signals that normally restrict the developmental options of somatic cell lineages. By shutting down gene expression, the germ cell lineage is kept pluripotential.

The properties and behavior of germ plasm in *Xenopus*, and similarities with *Drosophila* polar granules, have been thoroughly discussed previously. Homologs of vasa, a component of *Drosophila* polar granules, have recently been identified in zebrafish primordial germ cells.

Mammals and chick apparently do not have maternally derived germ cell determinants. Mouse genes that encode homologs of *Drosophila* polar granule components, for example, vasa (*Mvh*) and *germ cell-less* (*Gcl*), are not expressed in PGCs but in adult male germ cells. If maternally encoded germ plasm is absent from mammals, what regulates PGC formation? The current idea is that PGC precursors are induced in the embryo by secreted signaling factors produced by adjacent extraembryonic cells. It is still possible that the localized production of these inducing factors is under the control of maternal determinants segregated to extraembryonic cells, but this hypothesis has not yet been tested.

To enable critical evaluation of the induction of PGCs, a brief description of early mouse development is in order. At around the time of implantation (~4.0–4.5 days post coitum, dpc) the blastocyst consists of two populations of cells; an outer epithelial layer of trophoblast cells that surrounds a tightly packed cluster of undifferentiated *inner cell mass* (ICM) cells. The ICM subsequently differentiates into an inner epiblast or embryonic ectoderm population and an outer primitive or extraembryonic visceral endoderm. After implantation, all cell types proliferate rapidly and the trophoblast forms a knob-like mass of cells known as the *extraembryonic ectoderm*, and the *epiblast* cells become organized into a cup-shaped epithelium.

There is a clear morphological demarcation or junction between the epiblast and the extraembryonic ectoderm, and lineage analysis strongly suggests there is no mixing of cells between the two tissues after about 4.0 dpc. Around 6.0 dpc, the embryo begins to undergo gastrulation. Proximal epiblast cells move posteriorly, lose their epithelial organization, and give rise to unpolarized mesodermal cells. The first cells to delaminate from the epiblast give rise to extra-embryonic mesoderm, whereas cells that drop out later give rise to embryonic mesoderm in the primitive streak. As described in detail below, the first time PGCs can be clearly distinguished from somatic cells in the mouse embryo is around 7.5 dpc. A cluster of about 45–50 cells that express high levels of the genes encoding *tissue nonspecific alkaline phosphatase* (TNAP) and the OCT4 transcription factor can be identified posterior to the primitive streak at the base of the allantois. They are surrounded by somatic mesoderm cells that express much lower levels of these markers.

Several lines of evidence show that PGCs arise from cells in the epiblast, although the precise time at which the PGC progenitors are committed to their fate has not been determined. The first evidence comes from the elegant lineage analysis experiments of Kirstie Lawson. She injected single cells in the epiblast at the pre-streak (pregastrulation) and early streak stages with a fluorescent lineage marker and then cultured the embryos for 40 hours and determined the location of labeled descendants. Analysis showed that PGCs (as judged by alkaline phosphatase staining) were derived from cells originally located in the proximal region of the prestreak epiblast, dispersed within about three cell diameters of the junction with the extraembryonic ectoderm. A crucial finding was that no injected epiblast cells gave rise to PGCs alone.

Cells that generated labeled PGCs also gave rise to labeled cells in the extraembryonic mesoderm (most frequently allantois, but also amnion and extraembryonic mesoderm of the yolk sac). This finding showed that the so-called PGC precursor cell population must generate descendants committed to either the extraembryonic mesoderm or the germ cell lineages. It is thought that this allocation takes place when the precursors are posterior to the primitive streak at around 7.5 dpc, but nothing is known about the mechanisms involved.

For example, the process may be stochastic and cell-autonomous or influenced by extrinsic factors. It may involve lateral inhibition, or asymmetric cell division and the localization of zygotic (rather than maternal) gene products to the PGC lineage. One limitation of the cell lineage analysis described above is that it does not say anything about the time at which the PGC precursors are first set aside. It only tells us that they already exist at 6.0 dpc in the prestreak epiblast. The process leading to the generation of the precursors could have been initiated significantly.earlier.

The fact that PGC precursors are located in the epiblast close to the junction with the extraembryonic ectoderm suggests that this environment contains factors inducing PGC precursor fate. To test this hypothesis, Tam and Zhou (1996) carried out an embryonic grafting experiment. They isolated clumps of 5–20 distal epiblast cells from early streak stage embryos of a reporter transgenic line that expresses β-galactosidase constitutively in all cells. The pieces were then grafted close to the junction with the extraembryonic ectoderm of wild-type, early streak stage (6.5 dpc) embryos, and the chimeric embryos were cultured in vitro. Analysis of these embryos showed many lacZ-positive cells in the extraembryonic and posterior mesoderm. In a very small number of embryos a few PGCs could be found that coexpressed alkaline phosphatase and β-galactosidase.

This important experiment indicated that distal embryonic ectoderm cells, which would normally have developed into anterior ectoderm or neurectoderm, can change in response to exogenous signals and acquire more ventral posterior cell fates, including that of PGCs. However, it does not reveal the whole window of time when induction of PGC precursors normally takes place; it only tells us that the inducing activity is still available at the early streak stage. In fact, it may be that induction occurs over an extended period of time, with some epiblast cells receiving sufficient inducing signal early, even at the blastocyst stage, and others not until 6.5 dpc.

If mouse or human blastocysts are grown in vitro, the ICM can give rise to pluripotential *embryonic stem* (ES) cell lines that can both self-renew indefinitely and give rise to multiple cell types in culture. When mouse ES cells are injected into a blastocyst, they mix with the ICM cells and contribute to all the tissues of the embryo except for the trophoblast and extraembryonic ectoderm or extraembryonic visceral endoderm. Some of the ES cells are able to differentiate along the germ cell lineage. Since the ES cells have undergone extensive proliferation in culture, this finding argues strongly against maternally inherited factors in the epiblast or ICM playing a role in germ cell determination in mammals. However, the results again shed no light on when or how the induction of the germ cell precursors first takes place.

Evidence that Factors Produced by the Trophoblast or Extraembryonic Ectoderm Play a Role in Mammalian Germ-line Development

The experiments described above suggest that germ-line-inducing factors are present near the junction between the trophoblast and ICM or extraembryonic ectoderm and epiblast. The first evidence for the nature of the factors came from the observation that embryos homozygous null for the gene encoding the transforming growth factor β (TGFβ)-related growth factor, bone morphogenetic protein 4 (BMP4), completely lack both an allantois (assessed morphologically) and PGCs (assessed by staining for either alkaline phosphatase or the carbohydrate antigen, SSEA-1). *Bmp4* is first expressed at high levels in the extraembryonic ectoderm and only later in the extraembryonic mesoderm cells in the allantois and posterior primitive streak surrounding the PGCs. *Bmp4* is not expressed in the founding population of PGCs.

The early expression pattern of *Bmp4* suggests that the protein secreted by the extraembryonic ectoderm induces cells in the proximal epiblast to assume the fate of PGC/allantois precursors. The finding that wild-type ES cells cannot rescue the mutant phenotype, even when they contribute more than 90% of the cells in chimeric embryos, supports this hypothesis.

Finally, *Bmp4* heterozygous embryos have a significantly smaller founding population of PGCs than normal, even though, once formed, the cells proliferate at the same rate as wild-type PGCs. This is consistent with a model in which BMP4 produced by the extraembryonic ectoderm acts in a dose-dependent manner to control the fate of the pluripotent proximal epiblast cells. According to this model, epiblast cells that receive the highest dose of BMP4 over the longest period have a high probability of becoming PGC precursors, whereas cells receiving a lower dose are more likely to give rise to extraembryonic or lateral mesoderm.

Further experiments are needed to distinguish between this model and alternatives. For example, rather than acting instructively, BMP4 may function simply as a permissive factor, maintaining the survival of PGC precursors segregated by a different mechanism. One prediction of the instructive or morphogen model is that in *Bmp4* homozygous mutants (and in mutants of genes encoding receptors or components of downstream signaling pathways), the fate of cells in the proximal epiblast, including those that normally give rise to PGC precursors, is changed to more dorsal/anterior cell types.

Another prediction is that BMP protein should induce PGC precursors in isolated epiblasts. However, it is possible that BMP4 is necessary but not sufficient for inducing PGC precursors, and that additional factors, including ones made by the visceral endoderm, are required. Recent data suggest that at least one factor secreted by the extraembryonic ectoderm functions in collaboration with BMP4 to control PGC development. This is the related protein, BMP8b, made exclusively by the extraembryonic ectoderm at this stage of development. BMP8b may act independently as a homodimer or possibly form biologically active heterodimers with BMP4, although the data do not support the hypothesis that such heterodimers are obligatory for PGC precursor formation. Finally, it is not yet known whether BMP4 and BMP8b act directly on the epiblast, or indirectly through the extraembryonic endoderm.

Further analysis of the role of cell–cell interactions in early PGC development would be greatly facilitated by the development of specific molecular markers both for PGC precursors before they have moved into the posterior primitive streak and for the descendants of these precursors that have differentiated along the PGC lineage.

Characteristics of Mammalian PGCs before they Reach the Gonad and Maintenance of the Pluripotent State

The founding population of PGCs of the 7.5-dpc mouse embryo undergoes two important processes en route to their final resting place in the gonads. The cells proliferate, and they migrate along the endoderm of the hind gut, through the mesentery and into the genital ridges. Most PGCs have reached the ridges at 11.5 dpc, and proliferation ceases by about 13.5 dpc. Migration is common to germ cells of several organisms; e.g., *Drosophila*, *Xenopus*, zebrafish, and chick. However, since it is not obviously relevant to the stem cell status of PGCs, it will not be considered further here. It should be noted, however, that chick PGCs migrate to the gonads via the bloodstream, but no intravascular PGCs have been seen in mammals, even though their migration route carries them near major blood vessels.

Proliferation increases the number of PGCs in the embryo from around 150 at 8.5 dpc to about 25,000 by 13.5 dpc, giving a population doubling time of about 16 hours. It is not known whether the number of cell divisions undergone by each PGC in vivo is invariant.

Unlike hematopoietic stem cells, there are still very few molecular markers characteristic of mammalian PGCs. Indeed, no gene is yet known that is exclusively expressed in PGCs and PGC precursors. The markers that are most frequently used to distinguish PGCs after about 7.5–8.5 dpc are TNAP, stage-specific embryonic antigen-1 (SSEA1, a complex surface carbohydrate), and OCT4 (a POU-domain transcription factor).

OCT4 (encoded by the *Pou5f1* gene in mice and also known as OCT3/4) is of particular interest because it appears to be a key regulator of the pluripotential phenotype. It is expressed in all the cells of the cleavage-stage embryo and late-stage morula, but switched off in the trophoblast, and remains active in the embryonic ectoderm and primitive endoderm until gastrulation. It is then gradually down-regulated in the derivatives of the embryonic ectoderm and endoderm, and by 8.5 dpc is only expressed in the PGCs. It is finally extinguished in the germ line when the PGCs begin to differentiate in the gonad, only to be reactivated as the gametes reach maturity. This pattern of expression led to the hypothesis that OCT4 is a guardian of the pluripotential phenotype and prevents cells from becoming restricted in their developmental potential. This idea is supported by the observation that *Pou5f1* null embryos lack an inner cell mass and consist entirely of trophoblast cells. More recent studies have suggested that the precise level of *Pou5f1* expression in undifferentiated ES cells regulates their differentiation in vitro. Intermediate levels of OCT4 protein appear to favor the pluripotential, undifferentiated phenotype, whereas low levels promote the differentiation of ES cells into trophoblast, and high levels into endoderm and mesoderm.

It is thought that OCT4 maintains the undifferentiated state by regulating gene transcription in collaboration with coactivators such as SOX2 or ROX-1. High levels of these cofactors may be maintained by signaling through the IL-6/LIF (*leukemia inhibitory factor*) receptor subunit, gp130, mediated by STAT-3 activation. LIF is a cytokine that was recognized for its ability to maintain the undifferentiated state of ES cells in vitro. However, homozygous null *Lif* mutant embryos develop normally, so that if gp130 signaling plays a role in vivo, it must be activated by LIF-related cytokines. Obviously, an important goal is to identify genes up- or down-regulated by OCT4 in pluripotential epiblast cells and PGCs in vivo. One candidate is *Fgf4*; another is *Kit*. It is noteworthy that OCT4 appears to regulate gene expression in mammalian PGCs rather differently from pole plasm determinants in *C. elegans*, which apparently function by generally repressing gene expression in the PGCs.

Mammalian PGCs are distinguished from somatic cells by a number of genome-wide modifications. Normally, during preimplantation development, zygotic DNA is demethylated, except at sites associated with allele-specific parental imprinting. Remethylation occurs in somatic cells before gastrulation, but the PGCs (and presumably their precursors) do not undergo this epigenetic modification. In addition, PGCs go one step further and remove the methylation of parentally imprinted loci that persists in somatic cells. This reprogramming appears to occur gradually, as the PGCs migrate to the genital ridges, and is completed by 13.5 dpc.

New imprints are subsequently added during germ cell maturation. Erasure of parental imprinting in PGCs has two consequences. First, PGCs that have not yet begun their differentiation into mature germ cells are unique in having no modification of their genome at all, at least at the level of DNA methylation. This may be necessary to erase the epigenetic influences or modifications of the parents and to restore the totipotency of the germ line. Second, PGCs late in the migratory pathway or just arrived in the gonad have a different phenotype from PGCs at 13.5 dpc. It could therefore be argued that the proliferation phase of PGCs does not strictly speaking involve a self-renewal, but rather the rapid amplification of a transitional precursor population.

Another difference in genomic modification between PGCs and somatic cells is the fact that female PGCs avoid random inactivation of their X chromosomes, at least during the early stages of their existence. Both X chromosomes are active in the epiblast until shortly before gastrulation, when a wave of random X inactivation goes through the population. However, at the earliest time they can be recognized, which is posterior to the primitive streak and in the hindgut endoderm, most PGCs still have two active X chromosomes. By the time they have reached the gonad, however, most have asynchronously undergone X inactivation. The X chromosomes are then reactivated before the onset of meiosis. Understanding how the PGCs initially avoid X inactivation will provide important information about the mechanism of germ-line specification at the genomic level.

Proliferation of PGCs

At least three different extracellular ligand/receptor signaling systems have been identified that promote the survival and proliferation of PGCs. These are (1) stem cell factor and its tyrosine kinase receptor, (2) bFGF and FGF receptors, and (3) cytokines of the interleukin/LIF family and their receptors that signal through a common gp130 subunit.

Stem cell factor (SCF, also known as Steel factor and mast cell growth factor) encoded by the *Mgf* (formerly *Steel*) locus, and its transmembrane tyrosine kinase receptor, c-KIT, encoded by the *Kit* (formerly *W*) gene, were first identified as growth factors for PGCs from genetic analysis in the mouse. *Mgf* is expressed in the somatic cells through which the PGCs migrate, whereas *Kit* is expressed by the PGCs themselves, at least until a few days after arrival in the genital ridge when it is down-regulated.

A role for FGFs and receptors (e.g., FGFR1 and 2) in promoting PGC proliferation was first suggested from experiments in which purified bFGF was added to cells in culture. Whether FGFs play a role in vivo is not known, but *Fgf 3,4, 5,* and *8* genes are variously expressed in the epiblast, posterior primitive streak, and mesoderm along the migration route of the PGCs.

Other factors that promote the survival and proliferation of PGCs in vitro are *leukemia inhibitory factor* (LIF), *oncostatin M* (OSM), interleukin-6 (IL-6), and *ciliary neurotrophic factor* (CNTF), all members of the IL-6/LIF cytokine family. These factors function through a dimeric transmembrane receptor expressed in PGCs. One subunit of the receptor (e.g., LIF receptor-β) binds specific ligands. The other is a common, non-ligand-binding subunit called glycoprotein 130 (gp130) that acts as a signal transducer by activating STAT-3. Antibodies to gp130 block the activity of LIF on PGCs. LIF was first tested for its effect on PGCs in vitro because it promotes the undifferentiated, pluripotent phenotype of mouse ES cells in culture; in the absence of LIF and feeder cells, ES cells rapidly differentiate.

Recent studies suggest that LIF functions in combination with a specific level of OCT4 to maintain the undifferentiated phenotype. Signaling through gp130 is therefore likely to play a key role in maintaining the pluripotency and self-renewal ability of PGCs. However, despite the key role apparently played by LIF/gp130 in controlling the PGC phenotype, it is still unclear which member(s) of the ligand family functions in vivo, since mice lacking LIF, LIFRβ, and IL-6 all have normal numbers of PGCs. Likewise, mutation of genes encoding IL-4 and IL-2R has no effect on PGC number even though IL-4 promotes the survival of PGCs in vitro. The most likely explanation is that several interleukins regulate PGC proliferation and survival and can compensate for each other in vivo.

Finally, it is very likely that as-yet-unidentified growth factors influence PGC proliferation because optimal growth of the cells in vitro requires fibroblast cell feeder layers. The possible identity of some of these factors has been discussed previously.

Although SCF, LIF, and FGF alone have some activity on PGC survival and proliferation in vitro, in combination they have a dramatic effect on the behavior of cells isolated before about 13.5 dpc. Rather than showing a finite number of cell doublings in vitro, the PGCs continue to proliferate indefinitely. Moreover, the PGCs change their phenotype to resemble pluripotent ES cells that are derived from the inner cell mass cells of the blastocyst. Precisely how this "*trans-differentiation*" from a PGC to ES cell phenotype is brought about is not known. Like ES cells, embryonic germ-cell-derived cell lines (known as *EG cells*) can differentiate extensively in culture and can contribute to all the tissues of the embryo, including the germ line, when injected into a host blastocyst.

However, many undifferentiated EG cells have differences in the methylation of imprinted loci compared with ES cells. This reflects the fact, discussed in the previous section, that PGCs remove allele-specific parental imprints as they migrate toward, and enter, the gonads. This phenotype is dominant, because if EG cells are fused with somatic cells (thymic lymphocyte), there is demethylation of several imprinted and non-imprinted genes in the somatic nuclei. It is not yet known whether human EG cell lines show differences in the methylation of imprinted loci.

A process similar to the trans-differentiation of PGCs to EG cells in vitro may occur during the rare in vivo development of teratocarcinomas in the testis of some strains of mice, e.g., 129/Sv. The frequency of testicular teratomas can be increased from about 1% to 95% in the 129/Sv strain by the introduction of the homozygous *Ter* mutation, but the identity of this modifier is not yet known. Teratomas can also be induced experimentally in vivo in mice by transplanting genital ridges to ectopic sites, or in some rodents from extraembryonic endoderm cells of the early yolk sac. Evidence suggests that this reflects the trans-differentiation of endoderm cells, rather than proliferation of yolk sac cells that have remained undifferentiated.

What Changes When PGCs Enter the Gonad and Come into Close Association with Somatic Cells? The End of the Road for PGCs

When PGCs enter the genital ridge, they come into close association with somatic gonadal cells derived from the intermediate mesoderm, and after continuing proliferation for a few days, they differentiate into germ cells. By about 13.5 dpc, female PGCs are entering into the prophase of meiosis, while male germ cells go into mitotic arrest and do not resume mitosis until the onset of puberty.

In the mouse, there is a down-regulation of c-KIT receptors in germ cells in the gonad. This presumably plays a role in making the germ cells unresponsive to stem cell factor after they have entered the gonad. Since the formation of teratomas is rare, a whole variety of additional mechanisms probably operate normally to protect intragonadal PGCs from the influence of other mitogenic factors. These mechanisms also appear to operate in PGCs that fail to reach the gonad, since extragonadal teratomas, which are presumed without any direct evidence to be derived from PGCs, are also rare.

As discussed earlier, PGCs that come to lie in the fetal adrenal gland cease proliferating and enter meiosis in the mouse. However, experiments in which *Xenopus* PGCs were isolated from the mesentery, labeled in vitro, and then transplanted into the blastoceol cavity of host embryos provided evidence that under these conditions the cells could become incorporated into various tissues and differentiate into somatic cells such as muscle and notochord. Analogous experiments in which in-vitro-labeled PGCs from 10.5-dpc mouse embryos were injected into blastocysts failed to show incorporation into either somatic tissues or the germ line. The ability of individual PGCs to change their fate in ectopic sites in vivo needs to be explored in more detail using robust genetic lineage markers.

In conclusion, PGCs constitute a stem cell population that plays important, evolutionarily conserved roles in germ-line development. The advantages of this population for the organism are that it expands the initially very small pool of germ cell precursors, moves them from extraembryonic regions to the gonads, and helps to ensure that the cells are protected from influences driving them down somatic lineages. The PGCs thus remain pluripotential until they come to the end of the road and differentiate into germ cells.

8

EMBRYONIC STEM CELLS

Embryonic stem (ES) cells are pluripotent stem cell lines derived directly from early mouse embryos without use of immortalizing or transforming agents. They can be propagated as homogeneous stem cell cultures and expanded without apparent limit. Unusually among established cell lines, ES cells maintain a stable euploid karyotype. Yet more remarkably, ES cells retain the character of embryo founder cells, even after prolonged culture and extensive manipulation. Thus, they are able to reintegrate fully into embryogenesis when returned to the early embryo. Chimeric mice can be produced in which ES cell descendants are represented among all cell types, including functional gametes. ES cells are also readily amenable to sophisticated genome engineering, in particular via homologous recombination. These properties are widely exploited to introduce gene knock-outs and other precise genetic modifications into the mouse germ line.

The ability to propagate pluripotent ES cells presents unique opportunities for experimental analysis of gene regulation and function during self-renewal, cell commitment, and differentiation. The combination of intrinsic and extrinsic factors that maintain developmental identity and potency is beginning to be defined. Progress is also being made toward understanding and controlling lineage- and/or cell-type-specific differentiation of ES cells in vitro. When harnessed effectively, ES cell differentiation can provide defined cell populations for pharmacological testing and cellular transplantation. Generation of ES cell equivalents from other species, most particularly human, is now anticipated for realization of the full power of ES cell technologies both as research tools and, ultimately, as cell therapy reagents.

Origins of Mouse Embryonic Stem Cells

Pluripotent Stem Cells in the Early Embryo

The mammalian fetus develops from a founder population of cells that are present before and shortly after implantation. These cells are pluripotent, meaning that they are individually capable of giving rise to derivatives of each of the three primary germ layers and to germ cells. Initially defined as the entire internal cell component of the blastocyst, the *inner cell mass* (ICM), pluripotent cells are segregated to a subcompartment, the epiblast, prior to implantation. After implantation the epiblast expands rapidly to generate the cellular substrate for gastrulation and formation of the embyro proper. Once gastrulation commences, the epiblast cells (often termed primitive or embryonic ectoderm at this stage) progressively differentiate into definitive mesoderm, endoderm, and ectoderm. Pluripotent cells are thus succeeded by lineage-committed precursors in the fetus.

Although the possibility that rare pluripotent cells may persist cryptically within differentiated tissues has not been definitively excluded, the epiblast per se is clearly transient and does not appear to meet one of the conventional criteria for a stem cell population, persistence throughout the lifetime of the organism. However, a range of experimental interventions in mouse embryos have revealed that epiblast cells are highly plastic and that their self-renewal and differentiation are regulated according to the embryonic context. Thus, the ICM and epiblast can adjust to either the removal or addition of significant numbers of cells and still give rise to a normal fetus. Indeed, this capacity to accommodate extra cells provides the foundation on which chimeric fetuses are produced.

The destined tissue contribution of epiblast cells is dictated by their position in the egg cylinder at the onset of gastrulation, but they can adopt new fates if grafted heterotopically. Most strikingly, if the early mouse embryo is removed from the uterus and the epiblast cells are grafted to a permissive ectopic site, such as the testis or kidney capsule of a syngeneic or immunocompromised mouse, they will generate large multi-differentiated tumors known as *teratocarcinomas*. Teratocarcinomas contain differentiated cell types of all germ layers and, in addition, an undifferentiated, proliferative component that can be maintained on serial transplantation. Teratocarcinomas can be produced at a high frequency from single epiblasts, but not at all from post-gastrulation embryos. The persistence and expansion of undifferentiated stem cells in embryo-derived teratocarcinomas indicates that mouse epiblast cells do in fact have an intrinsic potential for prolonged self-renewal.

Embryonal Carcinoma Cells

The undifferentiated component of teratocarcinoma is described as embryonal carcinoma due to its resemblance to early embryonic tissue. The propagation of this population can be maintained in explant culture, and continuous embryonal carcinoma, or EC, cell lines may be derived. EC cells have several distinctive features. In particular, they are often capable of multilineage differentiation. Crucially, this capacity is retained by clonal isolates, formally establishing the presence of pluripotent stem cells. Along with evidence of similarities in immunophenotype and protein expression profile to the ICM/epiblast, this gave rise to the concept that EC cells are counterparts of normal pluripotent embryo cells. The finding that EC cells could participate in embryonic development and contribute to chimeric fetuses and, in some cases, even live offspring, substantiated this notion. This ability demonstrates that the capacity of EC cells for extended self-renewal in teratocarcinomas or in culture does not represent an oncogenic transformation. In this regard, it is noteworthy that the epiblast does not express G_1 cyclins and is therefore unlikely to be subject to normal cell cycle control mechanisms. The corollary of extrauterine teratocarcinoma development is that extended self-renewal of epiblast within the embryo is actively suppressed.

However, the derivation of EC cells via expansion in tumors, usually involving serial transfers, compromises their genetic constitution. Almost all have an aneuploid karyotype, and more subtle changes are also likely due to the pressure for selective growth advantage. Indeed, most EC cells show some restrictions in differentiation potential and in their ability to integrate normally into embryogenesis. Consequently, their value as a developmental model is undermined, and they do not provide a suitable system for germ-line transgenesis.

Derivation of Embryonic Stem Cells

Studies with EC cells laid the intellectual and experimental groundwork for the establishment of "true" embryo stem cell cultures. A seminal point was the realization that pluripotency was best sustained in a coculture system. Martin and Evans observed that in primary cultures of teratocarcinoma, EC cells tended to thrive in proximity to differentiated cell types but to expand poorly in isolation. This prompted investigation of the potential of established cell lines to support EC cell propagation. Coculture with mitotically inactivated embryonic fibroblasts was found not only to allow the efficient establishment of EC cultures, but also to result in stem cells with high differentiation capacity. It was reasoned that the fibroblasts were providing some critical nutrient or trophic factor support, hence they were described as "*feeder*" cells.

Feeders were employed in renewed efforts to establish cell cultures directly from cultured embryos. In 1981, the derivation of pluripotent cell lines from mouse blastocysts was reported. The protocols for ES cell derivation are relatively simple. Embyros at the expanded blastocyst stage are plated, either intact or following immunosurgical isolation of the ICM, onto a feeder layer. Conventional tissue culture medium is supplemented with 2-mercaptoethanol and 10–20% fetal calf serum. After several days of culture, epiblast outgrowths are disaggregated and replated onto fresh feeders. Various types of differentiated colonies arise along with colonies of undifferentiated morphology. The latter are individually dissociated and replated. If secondary colonies of undifferentiated cells arise, these can generally be expanded further, and continuous ES cell lines can be established.

Factors Influencing ES Cell Derivation

Establishing an ES cell culture entails the liberation of pluripotent epiblast cells from their fated differentiation. Prior induction of diapause (implantation delay) appears to enhance the efficiency of ES cell generation. This may be attributable to an increase in epiblast cell numbers during diapause, although this is relatively modest. Perhaps more likely is that the arrest of normal development pre-configures the epiblast cells for continued self-renewal by activating dependency on cytokine signaling for maintenance of pluripotency. The process by which a state of continuous self-renewal is arrived at is not automatic, however, and is poorly understood. The standard protocol can reproducibly yield ES cell lines from inbred 129 strains and somewhat less efficiently from C57BL/6 strains. However, usually only a minority of embryos give rise to ES cells, suggesting that some epigenetic event is rate-limiting. Furthermore, the isolation of ES cell lines from other strains of mice has generally proven very problematic. Thus, there is a strong genetic component to ES cell derivation. Interestingly, this is not reflected in the propensity of embryos to give rise to teratocarcinomas, which does not exhibit significant strain dependency.

Isolation of epiblast cells from inductive influences of adjacent hypoblast is reported to enhance the efficiency of ES cell derivation. Brook and Gardner even demonstrate the derivation of multiple ES cell lines from separate cells of the same epiblast by this approach, although it remains to be shown formally that all cells of the epiblast are equally competent to produce ES cells. Removal of differentiated lineages has been applied to circumvent non-permissiveness in CBA mice, but whether this will hold for other strains has not been reported.

It is evident that ES cells originate from the epiblast, that is, after differentiation of the hypoblast, but the point of embryo development at which capacity to generate ES cells is lost is not clear. To date no success has been reported with egg cylinder stages, even though these give rise to teratocarcinomas very efficiently. Possibly the epithelial organization of the egg cylinder imposes constraints on epiblast cells that may be disrupted on ectopic grafting but are not readily erased in primary culture.

Embryonic Germ Cells

In addition to experimental induction from explanted embryos, teratocarcinomas can originate spontaneously from germ cells. Testicular teratocarcinoma is particularly prevalent in strain 129 mice. The evidence that undifferentiated germ cells can give rise to embryonal carcinoma remained something of a curiosity in the absence of methods for propagating germ cells in vitro. Following molecular cloning of the *Steel* growth factor and the cytokine leukemia inhibitory factor, limited expansion of germ cells became possible. Building on this, it was found that on additional inclusion of basic fibroblast growth factor (FGF-2) in the cultures, mouse primordial germ cells converted after several days in culture into cells resembling ES cells that could then be maintained indefinitely. These are termed *embryonic germ* (EG) cells to denote their origin. In most respects, they are indistinguishable from blastocyst-derived ES cells, including pluripotency and even germ-line competence. However, irregularities in imprinting arising from their germ cell origin can compromise full developmental potential so that EG cells may colonize chimeras less effectively than ES cells. Evidence has also been presented that EG cells retain the unique capacity of germ cells to erase imprints, a property that has not been shown in ES cells.

Pluripotency of Embryonic Stem Cells

Teratocarcinoma Formation

ES cells closely resemble EC cells in morphology, growth behavior, and marker expression. This relationship extends to the capacity to give rise to multi-differentiated teratomas and teratocarcinomas. ES cells readily produce tumors containing well-differentiated mesodermal, ectodermal, and endodermal tissue and cell types. The representation of undifferentiated stem cells in the tumors tends to be less than in EC cell-generated teratocarcinomas, most likely reflecting the latter's history of tumor selection. The ability clonally to give rise to teratocarcinomas is a defining feature of pluripotent embryo cells, shared by ES, EG, and EC cells.

Integration into the Developing Embryo

The most extraordinary feature of ES cells is that, even after extended propagation on tissue culture plastic in synthetic media, they remain capable of participating in normal embryogenesis. Several techniques can be used to introduce ES cells into the preimplantation mouse embryo, but regardless of method of delivery, the ES cells can colonize all fetal lineages plus yolk sac mesoderm. Consistent with their epiblast origin, ES cells contribute poorly to extraembryonic endoderm and rarely, if ever, to trophoblast. In contrast to EC cells, ES cells behave relatively consistently in their ability to integrate into the embryo and produce viable chimeras. ES cells produce functional differentiated progeny in all tissues and organs. Genetic coat-color markers therefore provide a simple and fairly reliable means of monitoring overall chimeric contribution.

Incorporation into embryogenesis not only confirms that ES cells are pluripotent, but also demonstrates that they can respond appropriately to developmental cues for proliferation, differentiation, migration, and patterning. ES cells thus retain in full the identity and capacity of resident epiblast cells.

Germ-line Transmission

A key property of ES cells is that they maintain a euploid karyotype. This is crucial because a balanced diploid chromosome complement is permissive for meiosis. Thus, unlike EC cells, if ES cells colonize the germ-cell lineage in a chimera, they are capable of progression to functional gametes. The landmark of deriving mice from cultured stem cells was reported by the Evans laboratory in 1984. In the early days of ES cell culture, germ-line transmission was often elusive. This became more frequent as the skills required for maintaining the diploid pluripotent phenotype were disseminated. Retention of germ-line competence depends absolutely on adherence to a rigorous tissue culture regime, with avoidance of any untoward selective pressures such as overgrowth or nutrient deprivation. Of course, random mutational events will always occur in the culture and epigenetic modifications may also arise; for example, alterations in imprinting status, so it is advisable to use low-passage stocks and/or to isolate new subclones periodically for transgenic work.

The great majority of ES cell lines are 40XY. The implication that two active X chromosomes may somehow be disadvantageous for ES cell propagation is consistent with the high incidence of spontaneous X chromosome deletions found in established XX ES lines. In any case, the XY genotype confers particular advantages for establishing germ-line transmission. Not only can male chimeras produce more offspring than females, but XY cells can convert the undifferentiated genital ridge of an XX recipient into testicular development. Since XX germ cells do not survive in a male gonad, this phenomenon of sex conversion results in chimeric males in which all the mature germ cells are of ES cell origin. In addition, it has been observed that the extent of contribution of strain 129 ES cells to chimeras is strongly influenced by the genotype of the recipient embryo. In particular, microinjection into C57BL/6 blastocysts results in very high ES cell contributions and a greatly increased frequency of germ-line transmission.

ES Cell-derived Fetuses

ES cells are not in themselves capable of generating a blastocyst and should therefore not be described as totipotent. The issue of whether ES cells are self-sufficient for generation of the fetal component of the conceptus has been addressed by Nagy et al. (1991, 1993), who introduced ES cells into tetraploid recipient embryos. In tetraploid embryos, extraembryonic lineages are produced normally but fetal lineages develop poorly. Consequently, in chimeras between tetraploid and diploid embryos, the fetus becomes almost exclusively colonized by the diploid cells. ES cells show a similar propensity to dominate the tetraploid contribution to the fetus, and such fetuses can develop to term. Thus, it can be argued that ES cells alone are competent to generate the entire fetus. However, although there may be few or possibly no tetraploid cells persisting in the animal at birth, a resident tetraploid ICM compartment is present initially.

To date, fetal development has not been reported following microsurgical replacement of the ICM with ES cells. Therefore, requirement for a "*normalizing*" signal from the host ICM to induce ES cells to reenter into an embryonic differentiation program cannot be discounted. Furthermore, although liveborn offspring may be obtained from ES cell–tetraploid chimeras, many embryos die in utero, and those that do persist usually die shortly after birth, in contrast to the situation with ICM chimeras. This is likely attributable to cryptic epigenetic or possibly mutational changes that have arisen during derivation or propagation of the ES cells. Such changes may be masked in diploid chimeras. Consequently, ES cells that give good somatic and germ-line colonization in diploid chimeras vary greatly in performance in the tetraploid setting. Thus, a note of caution is required in any assertion that an ES cell is unaltered from an epiblast cell in situ.

Genome Manipulation in ES Cells

Insertional Mutagenesis and Gene Trapping

DNA can be introduced into ES cells by conventional infection or transfection protocols. Their capacity for clonogenic expansion then allows independent integrants to be expanded and transgenic mice to be generated. Random insertion of viral vectors into the ES cell genome has been employed to mutate and tag genes in phenotype-driven screens. Gene trapping is a refinement of this approach that facilitates isolation of a disrupted gene and can allow a degree of pre-selection for desired categories of target gene based on expression pattern or subcellular localization of the gene product. This technique has been widely used as a gene discovery tool in mice and further pursued as a method for annotated mutagenesis of the entire mouse genome.

Targeted Gene Modification

The major use of ES cell genetic modification to date, however, has been for the directed modification of nominated genes, known as gene targeting. Pioneering work in the mid-1980s established that transfected DNA could be integrated into designated loci in the ES cell genome via homologous recombination. In 1989 the first incidence of germ-line transmission of a targeted allele was reported, demonstrating that the manipulations and drug selections involved in isolating homologous recombinant clones did not in themselves compromise ES cell pluripotency.

There are now well-established procedures for introducing a range of different types of modifications, such as deletion, point mutation, reporter insertion, or coding sequence replacement, into the mouse genome. Conditional mutations can be created by incorporation of site-specific recombinase technology. In such cases, short recognition sequences for a recombinase such as Cre or Flp are targeted by homologous recombination to flank the gene segment of interest. This interval can then be deleted in a stage- or tissue-specific fashion by appropriate transgenic expression of the recombinase.

Chromosome Engineering

The use of site-specific recombination can be extended to the engineering of long-range modifications in the ES cell and thence the mouse genome. Deletions, inversions, duplications, or translocations can be generated according to the respective orientation and *cis* or *trans* localization of the recombinase recognition sequences. This is a powerful method for interrogating the genome, increasingly so with the amassing of sequence information and gene localization data.

Autonomous chromosomal elements have also been introduced into ES cells via cell fusion. These minichromosomes can be maintained stably in ES cells and chimeras, and in some cases, can be transmitted through the germ line. This creates the foundations of a system for genetic dissection of centromere function in mammalian mitosis and meiosis. Minichromosome vectors may also find applications in biotechnology; for example, the creation of humanized antibodies.

Maintenance of ES Cells Pluripotency

Symmetrical Self-renewal

ES cells multiply by symmetrical cell division. They can routinely be expanded to give relatively homogeneous and undifferentiated populations, judged by morphology, marker expression, efficient generation of equipotent subclones, and reproducibly broad colonization of chimeras from a few cells. This expansion can be continued over several weeks, and very large (10^9–10^{10}) populations of substantially pure stem cells can be generated. In fact, ES cells appear to be immortal and show no evidence of either crisis or senescence, in contrast to other primary cultures.

The symmetric amplification of ES cells contrasts with most other stem cells ex vivo and, in conjunction with the facility for genetic manipulation, provides a tractable system for experimental characterization of self-renewal.

Oct-3/4: Governor of Transcription and Fate in Pluripotent Cells

Oct-3/4 is a POU family transcriptional regulator restricted to early embryos, germ-line cells, and undifferentiated EC, EG, and ES cells. In vivo, zygotic expression of Oct-3/4 is essential for the initial development of pluripotential capacity in the ICM. In ES cells, continuous function of Oct-3/4 is necessary to maintain pluripotency. If Oct-3/4 expression is acutely eliminated in ES cells, self-renewal ceases and an unorthodox differentiation process is triggered. Instead of forming the normal ES cell derivatives, endoderm and mesoderm, the cells differentiate.into trophoblast. In the presence of FGF-4 and feeders, it is even possible to isolate *trophoblast stem* (TS) cells.

The interest of this observation is that the differentiation of trophectoderm and ICM in the mouse blastocyst is considered to be associated with a segregation of developmental capacity such that the former can generate only trophoblast lineages and the latter only yolk sac and fetal tissues. Consistent with this, ES cells do not normally form trophoblast either in vitro, in teratomas, or in chimeras. It appears that this developmental restriction may be necessary for manifestation of pluripotency and is imposed directly by Oct-3/4. In other words, Oct-3/4 acts in part as a lock that prevents default differentiation into trophoblast.

Oct-3/4 also contributes positively to pluripotency by directing expression of multiple target genes. This is achieved via interaction with several coactivators, and probably also corepressors. The complexity of Oct-3/4 function is indicated by the finding that marginally increased expression in ES cells provokes differentiation, but in this case into endoderm and mesoderm. Although Oct-3/4 is normally down-regulated during pluripotent cell differentiation, this is a consequence rather than a cause of germ-layer commitment. Indeed, in the ICM, Oct-3/4 levels transiently increase in nascent hypoblast. One hypothesis is that differential lineage commitment may be determined by altered interaction of Oct-3/4 with specific partners, expression or activity of which may be regulated by inductive signals.

Finally, although Oct-3/4 seems to be a pivotal player in the determination of pluripotent cell fate, maintenance of Oct-3/4 expression is not in itself sufficient to sustain the pluripotent phenotype. Significantly, an extrinsic signal is also needed.

Cytokine Stimulation of Self-renewal

In monoculture using media supplemented with serum alone, ES cells can neither be derived nor maintained. As discussed above, ES cells were originally isolated by coculture with a feeder layer. Subsequently it was discovered that the feeders can be substituted by conditioned medium preparations, indicating that their key function is to provide trophic stimulation. In fact, a purified cytokine, *leukemia inhibitory factor* (LIF), is sufficient to sustain ES cell self-renewal. This effect is exclusive to LIF and a small group of related cytokines that act via the gp130 receptor. LIF is expressed by feeder cells, and this expression is elevated in the presence of ES cells. LIF does not act via inducing expression of Oct-3/4 because transgenic expression of Oct-3/4 does not remove the requirement for LIF.

On withdrawal of LIF (or feeders), proliferation continues, but differentiation is induced and ES cells do not persist beyond a few days. It has been suggested that self-renewal equates to the inhibition of differentiation. However, this is only true provided survival and division are constitutive. In the case of ES cells, replication does indeed appear to be autonomous. The cell cycle is uncoupled from cdk/cyclin checkpoints in G_1, and no method has been reported for producing quiescence in ES cells. Apoptosis can be induced in ES cells, however, which implies that an anti-apoptotic pathway could be a crucial element of the self-renewal signal. Although ES cells remain viable in defined media lacking LIF, serum, or added growth factors, this is heavily dependent on cell density, suggesting a requirement for autocrine survival signaling.

Two major signal transduction pathways are recruited downstream of ligand-induced dimerization of gp130 receptors. The latent transcription factor STAT3 is activated by tyrosine phosphorylation mediated by JAK kinases, and engagement of the adapters SHP2 and Gab1 lead to stimulation of the Ras-Erk mitogen-activated protein kinase cascade. Analysis of modified receptors indicated that recruitment of STAT3 is essential for ES cell propagation. This conclusion was substantiated by demonstration that expression of the dominant interfering STAT3F molecule induced differentiation in the presence of LIF. In contrast, suppression of the SHP2/Erk signaling arm actually enhanced ES cell self-renewal. Finally, studies with directly activatable or constitutively activated variants of STAT3 have provided strong evidence that this transcription factor alone can provide the self-renewal signal in serum-supplemented medium.

Cytokines of the LIF family are not dedicated to stem cell regulation, but have diverse effects on a variety of cell types. Interestingly, most of these actions are to promote differentiation, for example of myeloid cells or astrocyte precursors, or to induce expression of differentiated functions, such as acute phase protein synthesis by hepatocytes. STAT3 is the major mediator of these responses. Therefore, ES cell self-renewal is stimulated by a conventional signal transduction pathway, but the output of this signal, inhibition of differentiation, is peculiar to the stem cell context. A key task is to define the level of interaction, direct or indirect, with

Oct-3/4 and to identify the important target genes in ES cells. These are likely to include repressed genes, expression of which could direct commitment and differentiation.

Intracellular Signaling Network in ES Cells

The antagonistic effect of SHP2 and Erk activation on self-renewal can partly be explained by a negative regulation of JAK-STAT signaling by SHP2 tyrosine phosphatase activity. In addition, blockade of the Erk activating enzyme MEK1 with the inhibitor PD05809 reduces ES cell differentiation both in monolayer and aggregate culture. Continued cell proliferation without stimulation of Erk activation is unusual but may be accounted for by the lack of restriction on entry into S phase in ES cells. Promotion of self-renewal by PD05809 further implies that there is likely to be a direct pro-differentiative effect of Erk activation. This relates not only to coupling downstream of gp130, but to growth factors and other inductive stimuli that signal through the Ras-Raf-MEK-Erk cascade, and probably also to aspects of integrin signaling that involve Erk activation.

As discussed above, self-renewal may require suppression of apoptosis. It remains to be determined whether the anti-apoptotic function that has been ascribed to STAT3 in other cell systems is operative in ES cells, or whether activation of the PI3-kinase/Akt pathway may fulfill such a role.

The requirement to achieve and maintain a high level of STAT3 activation and low level of Erk activity may underlie some of the difficulty experienced in ES cell derivation. It is significant, therefore, that application of PD05809 can increase the efficiency of ES cell establishment by promoting expansion of primary stem cell colonies. It is also noteworthy that, although ES cells established on feeders can be adapted to grow on gelatin in the presence of LIF, this is generally preceded by extensive differentiation, and the stem cells that persist and regenerate the cultures are often compromised and show reduced contribution to chimeras. This may indicate that the signaling network is tuned slightly differently during culture on feeders and must be readjusted for growth in LIF alone.

Although ES cells employ classical signal transduction mechanisms, the likely existence of stem-cell-specific signaling adapters should not be overlooked. For example, Erk activation in response to various stimuli appears attenuated in ES cells relative to other cell types, despite the presence of comparable levels of Erk proteins. This probably results in part from expression of an altered form of the Gab1 adapter protein in ES cells that suppresses linkage of certain receptors to the Ras-Erk cascade. Such redirection of primary signal transduction so as to minimize pro-differentiative outputs may turn out to be a cardinal aspect of stem cell propagation in general.

Alternative Pathways of Self-renewal

The dependency of ES cells on gp130 signaling presents a paradox, however. In vivo, a critical role for the gp130 pathway in the epiblast is evident only on induction of embryonic diapause. In uninterrupted embryogenesis, there is no apparent requirement for LIF, gp130, or STAT3 prior to gastrulation. This implies that normal expansion of the epiblast is either autonomous or under the direction of a separate signaling pathway.

The regulative properties of the epiblast argue against the former. For example, giant blastocysts made by aggregation of two or more cleavage embryos produce normal-sized egg cylinders after implantation on the same time schedule as ordinary blastocysts. Conversely, small blastocysts created by splitting 2-cell embryos "*catch up*" after implantation and produce full-sized egg cylinders. Regulation of the epiblast population occurs within 36 hours of implantation. This adaptive response may be governed via juxtacrine signaling within the epiblast compartment or by paracrine stimulation from neighboring tissue.

ES cells provide both an assay for, and a potential source of, epiblast regulatory signals. Like epiblast, ES cells can be sustained independently of gp130 and STAT3, at least transiently. Such a process operates in addition to LIF signaling during culture of ES cells on mouse embryo fibroblast feeders. The factor(s) and signaling mechanisms that mediate this effect have yet to be characterized at the molecular level, however. Evidence has also been presented that pluripotent cells may be maintained in a slightly altered state using conditioned medium from the HepG2 human hepatocyte cell line.

Under such conditions, cell morphology, expression of certain genes, differentiation behavior, and ability to colonize the embryo are altered. Significantly, this transition appears reversible with regard to all these features. Therefore, it either reflects an inherent plasticity of the ES cell phenotype or marks two distinct but continuous stages of normal epiblast progression. It will be of interest to resolve whether integrin signaling may contribute to ES cell regulation, given its significance for hematopoietic, keratinocyte, and other stem cell populations, and requirement for egg cylinder development.

In vitro Differentiation of ES Cells

A major aspiration at the outset of EC and ES cell research was to elucidate the decision-making processes in lineage commitment and cell type differentiation of pluripotent cells. This issue is now reemerging to the fore with increasing interest in the application of ES cell systems for efficient in vitro analysis of gene function and pharmacological screening, and for the potential development of cell therapy. From their differentiation in teratomas and chimeras, ES cells clearly have the capacity to produce every type of fetal and adult cell. Understanding and controlling cell fate determination remains a major challenge, however.

Differentiation in Embryoid Bodies

Our abilities to direct pluripotent cells into specific pathways and then to support the viability and maturation of individual differentiated phenotypes in vitro are currently limited, and the approaches are rather unsophisticated. The principal method used to trigger differentiation of ES cells into defined cell types is cell aggregation in suspension culture. This technique, originally developed with EC cells, leads to formation of multi-differentiated structures called embryoid bodies. In these structures, the developmental program of ICM/epiblast cells is reactivated in the ES cells. Cellular differentiation proceeds in a similar fashion to that which occurs in the embryo, albeit in the absence of proper axial organization or elaboration of a body plan. Each embryoid body develops multiple different cell types. A range of differentiated products can readily be obtained, including yolk sac endoderm, cardiomyocytes, embryonic and definitive hematopoietic cells, endothelial cells, skeletal myocytes, adipocytes, neurons, and glia. It is possible to bias the differentiation for or against certain cell types by addition of retinoic acid. However, the final cultures are always a heterogeneous mixture of various cell types.

In the absence of knowledge of how to instruct ES cells into a lineage of choice, which in any case may never be completely effective in the context of complex multicellular interactions that occur in an embryoid body, an alternative approach is to isolate cells of interest from the mixture of differentiation products. For hematopoietic cells, this can readily be done by selective culture in semi-solid media in the presence of hematopoietic growth factors. A complementary strategy is to purify lineage-specific precursors or terminal differentiated phenotypes based on marker gene expression. This can be achieved by immunopurification where suitable cell-surface markers are available, or more generally by introduction of a transgene marker conferring drug resistance and/or cell-sorting capacity.

Differentiation in Monolayer Culture

ES cells differentiate readily in monolayer culture when deprived of LIF or feeder support. Various differentiated morphologies emerge, and markers of mesoderm and endoderm become expressed. However, the identities of the major cell types produced under such conditions have not been carefully defined. It is possible that many of the cells may not represent bona fide embryonic or fetal phenotypes but rather could be aberrant products arising from mis-regulated or scrambled differentiation programs. Nonetheless, in a very elegant study, Nishikawa has shown that distinct mesodermal subsets can be produced during monolayer differentiation, and in particular that clonogenic endothelial and hematopoietic progenitors can be isolated by *fluorescence-activated cell sorting* (FACS).

These progenitors can even proceed to form vasculature. This is a very significant result because it establishes that "true" differentiation can be uncoupled from morphogenesis and does not require complex multicellular interactions. Furthermore, the development of some lineages is actually suppressed by such interactions and is therefore enhanced by purifying the precursors. This finding also has implications on a practical level because monolayer culture is much more amenable than aggregation for experimental dissection and manipulation. For example, the application of candidate inductive signaling molecules such as Wnts and BMPs is likely to have more profound and interpretable consequences in homogeneous monolayer cultures than on embryoid bodies. Use of defined media will likely be required to realize the full potential of this approach.

Mechanism of Differentiation

Do ES cells undergo asymmetric division during differentiation? In the absence of LIF, undifferentiated ES cells are rapidly depleted from the cultures. This must occur either by symmetrical division leading to differentiation of both daughter cells, or by a limited number of asymmetric divisions followed by selective death of the stem cells. The absence of any overt polarity in ES cells in monolayer culture might suggest that there is no foundation for development of asymmetry. However, this issue should be investigated directly by time lapse recordings. It would also be instructive to determine whether in embryoid bodies the initial differentiation of an outer layer of extraembryonic endoderm is an asymmetric event or is directed solely by external position. Although there is also selective activation of a small number of specialized genes, lineage commitment is in essence a restriction of global gene expression potential. This entails the heritable repression of the majority of non-house-keeping genes. How such epigenetic mechanisms operate and how they may be erased during nuclear transfer or dedifferentiation has yet to be determined. Important insights could be obtained, however, by studying how chromatin architecture is modified during ES cell differentiation.

The chromatin organization in a pluripotent cell nucleus must be permissive for activation of lineage-specific gene transcription. A noteworthy observation from gene trapping studies is that many developmentally regulated genes are already transcriptionally active at low levels in undifferentiated ES cells. It is also possible to detect allegedly tissue-restricted transcripts in ES cells by reverse transcription PCR. This is reminiscent of the "*lineage priming*" concept proposed for hematopoietic stem cells; but if stem cells express lineage-specific genes, how is the undifferentiated pluripotent state maintained? There are at least three possible and nonexclusive explanations:

1. The level of expression of differentiation genes may not be functionally significant, but may simply reflect random transcription occurring through open chromatin.
2. Commitment may require coordinated expression of a battery of genes, individual expression of which has no consequence.
3. Stem-cell-specific transcriptional determinants may specifically antagonize the action of lineage commitment genes.

Self-renewal of ES cells appears to rely on an interplay of conflicting intracellular signals and transcriptional determinants. This is so finely balanced that the alternative outcome of differentiation can readily be triggered. Thus, some level of "*spontaneous*" differentiation is usually evident in ES cell cultures. It will be interesting to discover whether other types of stem cells are regulated in a similar manner such that they are constantly "*poised*" to differentiate.

Pluripotent Embryo Cells from other Species

Derivation of permanent stem cell lines that fulfill the criteria of epiblast origin, sustained symmetrical self-renewal, pluripotency, integration into fetal development, and germ-line colonization has to date only been validated in mice. Germ-line colonization from cultured cells has been reported in chickens and medaka fish, but only after short-term culture. Chimeras have been reported in rabbits, pigs, and cattle, but in no case has germ-line colonization been corroborated. These data may suggest that the situation in the mouse is the exception rather than the rule.

The ES cell phenotype represents a ground state for mouse epiblast or primordial germ cells in teratocarcinomas or ex vivo. However, although diploid stem cell cultures can readily be derived from rat ICMs, rather than exhibiting multi-lineage differentiation, these cells appear restricted to extraembryonic development. It may be significant in this regard that ectopically grafted rat embryos do not produce teratocarcinomas and that rat epiblast appears to retain the ability to produce hypoblast even into egg cylinder stages. Therefore, the possibility should be considered that the ES cell phenomenon is specific to inbred laboratory mice and that the ground state in other species may differ and perhaps even be a more primitive "*pre-pluripotent*" cell.

Stem cell cultures have also been established from human blastocysts. These cells can generate teratomas in immunocompromised mice and show some capacity for multilineage differentiation in vitro. Therefore, they could represent human equivalents of ES cells. However, the critical functional tests of chimera contribution and gamete production obviously should not be undertaken for ethical reasons. It is noteworthy that these human cells differentiate into trophoblast, indicating that they may not represent exactly the same developmental stage as mouse ES cells. Furthermore, they are difficult to expand and seemingly do not respond to LIF. Intriguingly, human EG-like cells derived from fetal primordial germ cells, in contrast, appear to be dependent on LIF for continued propagation. The molecular characterization of these human cells and comparison against mouse ES and EG cells is now a pressing issue, particularly in light of the desire to develop human pluripotent cells for regenerative therapies.

Trophoblast Stem Cells

Trophoblast Lineage and Placental Development

In the mammalian embryo, the formation of the extraembryonic lineages, the trophectoderm and primitive endoderm, precedes the differentiation of cells that will contribute to the fetus itself. The precocious differentiation of these lineages in mammals is related to their essential roles in promoting survival of the embryo in the uterine environment. Cells of the primitive endoderm give rise to the endoderm layers of the yolk sacs, whereas descendants of the trophectoderm will form the trophoblast portion of the placenta. The development of both extraembryonic lineages has been reviewed previously. The scope of the following discussion is to provide an overview of trophectoderm development in the preimplantation embryo and of the trophoblast lineage and placenta in post-implantation stages, as the basis for the study of trophoblast stem cells.

The trophectoderm is the first cell type to differentiate in the mammalian embryo. In the mouse, it is morphologically distinguishable by the blastocyst stage (3.5 days post coitum [dpc]), where it forms an outer monolayer of cells surrounding the blastocoelic cavity and the *inner cell mass* (ICM). The trophectoderm cells are characterized by their flattened epitheloid-like appearance, with apical tight junction complexes. In addition, trophectoderm cells harbor sodium pumps (Na^+, K^+-ATPase) whose activity leads to accumulation of fluid within the blastocoel cavity, resulting in blastocoel expansion. This process is thought to facilitate hatching from the zona pellucida, a thick protein coat surrounding the embryo.

The distinction of the trophectoderm from the ICM may be initiated during the events of compaction at the eight-cell stage, when cells become polarized. During successive divisions, cells that inherit an apical region will end up in the trophectoderm, while apolar cells contribute to the ICM. However, irreversible commitment of individual cells to the trophoblast lineage does not occur until the blastocyst stage.

In the blastocyst, trophectoderm cells make up approximately 75% of the total cell number, underscoring the fact that most of the cells in the embryo at this stage are set aside for the role of establishing the maternal-embryonic contact. By 4.5 dpc, at the peri-implantation stage, the formation of different trophectodermal subtypes is initiated. While the cells overlying the ICM continue to divide and form the polar trophectoderm, the cells away from the ICM, the mural trophectoderm, cease dividing, but continue to replicate their DNA. This process, known as *endoreduplication*, gives rise to cells with multiple copies of their genome. They will form the primary trophoblast giant cells that mediate implantation of the embryo into the uterine epithelium and later take on endocrine functions.

Following implantation, the polar trophectoderm continues to divide and grows initially toward the blastocoelic cavity, to form the *extraembryonic ectoderm* (ExE), and then outward to form the *ectoplacental cone* (EPC). This typical directional growth may occur due to mechanical constraints placed on the implanting blastocyst by the uterine wall. Secondary trophoblast giant cells arise at the periphery of the EPC and eventually come to surround the entire embryo and its membranes. These cells provide direct contact with the maternal environment during the early part of pregnancy. These polyploid cells appear phenotypically indistinguishable from primary giant cells of the blastocyst.

At 8.5 dpc, formation of the chorioallantoic placenta is initiated by fusion of the diploid trophoblast of the EPC and chorionic plate with the allantois, a structure derived from extraembryonic mesoderm. Subsequently, by 10.0 dpc, three distinctive trophoblast cell layers are formed, which persist for the rest of gestation. The innermost layer, the labyrinth, is formed following the attachment of the allantois to the chorion, and eventually will be the site for exchange of nutrients and gases with the maternal blood. This function is achieved via the large surface area of the labyrinthine layer, formed through extensive folding and branching that occur as it develops. The intermediate compact layer of spongiotrophoblast cells, through which the maternal blood cells pass, is formed as a consequence of the expansion and flattening of the EPC after 7.5 dpc. The third layer, the outer layer of giant cells, is formed from the post-mitotic trophoblast cells lying at the periphery of the EPC, and practically covers the entire placenta. The trophoblast cells perform many other functions connected with survival in the uterus, including hormone and cytokine production and immune protection of the fetus.

Evidence for the Existence of Trophoblast Stem Cells In Vivo

A basic requirement for cells within a specific lineage to serve as stem cells is their ability for self-renewal, and the capacity to give rise to all descendants of that lineage. The existence of such cells in the trophoblast lineage of the mouse has been implied by numerous studies, as summarized below.

Studies on the trophoblast lineage

As discussed earlier, following implantation the mural trophoblast cells of the blastocyst stop dividing and transform into primary giant cells, while the polar trophoblast cells continue to divide. In culture, mural trophectoderm fragments were shown to form trophectoderm vesicles that induce decidual reactions when transferred to uteri of pseudopregnant recipients. Although these vesicles implanted at equal efficiencies as control blastocysts, only a limited number of giant cells were recognized at the sites of implantation of the trophectoderm. This suggested that the mural trophectoderm by itself is limited in its ability to contribute additional giant cells due to its inability to proliferate. Indeed, a later study showed that further increase in the number of primary giant cells relies on cell division originating in the polar trophectoderm followed by cell migration to the mural trophectoderm. Thus, these experiments demonstrated that trophectoderm proliferation relies on cell division to be maintained in the polar trophectoderm.

Ectopic transplantation of EPC led to the formation of trophoblast giant cells at the transplantation sites. This suggested that post-implantation trophoblast still required some positive signal to maintain its proliferation. In addition, it was demonstrated that transplanting or culturing pieces of ExE isolated from mouse embryos up to 8.5 dpc also resulted in cells with the morphological characteristics of trophoblast giant cells, emphasizing the similarity between the properties of the ExE and EPC.

The ExE and EPC are not identical, however, although they are both diploid trophectoderm derivatives. They respond differently to different culture conditions: EPC cells gave rise to giant cells under all conditions tested, whereas ExE exhibited a period of sustained proliferation in some conditions. The two tissues also differed in their 2D protein synthetic profiles. During in vitro culture, ExE cells pass through a brief stage of diploid maintenance and synthesis of EPC-like proteins before transforming into giant cells. Under the same conditions, EPC cells rapidly commenced giant cell transformation. In addition, when 5.5 dpc and 6.5 dpc ExE were injected into 3.5 dpc blastocysts, they contributed to ExE as well as the EPC and trophoblast giant cells in resulting chimeras. Taken together, all these results suggest that the diploid ExE may act as a stem cell pool for the post-implantation trophoblast, giving rise to EPC cells as well as secondary giant cells.

ICM and its derivatives in trophoblast proliferation

Giant cell transformation is apparently the normal path of differentiation of all trophoblast cells of the blastocyst not in contact with the ICM. This process begins at the abembryonic pole of the mouse blastocyst and eventually involves the whole trophectoderm except those cells in the region immediately overlaying the ICM. Reconstitution experiments involving the injection of ICM cells into mural trophoblastic vesicles resulted in the development of normal embryos, thus suggesting the ICM is an active source of signals promoting trophoblast proliferation. Continued dependence of trophoblast proliferation on the ICM and its derivatives could be important in ensuring that the development of the trophoblast is coordinated with that of the embryo it supports. Indeed, it was suggested that cell division in the diploid ExE trophoblast cells may be maintained by their proximity to the embryo proper during post-implantation stages, whereas EPC cells farther away from the ICM derivatives will transform into the secondary giant cell population. Maintaining post-implantation trophoblast cells in close contact with one another in the absence of ICM derivatives was insufficient to maintain them in a diploid state. However, when pieces of ExE were inserted into embryonic tissue, grafted ectopically, and scored for the presence of trophoblast giant cells, none was observed. In control grafts, where ExE was not enclosed in embryonic tissue, trophoblast giant cells were readily found. Thus, it appears that a signal emanating from the embryonic tissues was able to maintain the ExE cells in a diploid state. In contrast, EPC cells analyzed in similar conditions differentiated into trophoblast giant cells, further suggesting that the ExE is likely the source of stem cells for the trophoblast lineage.

Model for the postimplantation trophoblast cell lineage

From all the studies described above, it is possible to present a model for the early trophoblast cell lineage incorporating the role of the ICM and its derivatives in the induction and maintenance of trophoblast proliferation. The model presents a pathway of differentiation in which the diploid polar trophectoderm of the blastocyst gives rise to the ExE. A diploid cell population is maintained in the ExE, but it is also capable of producing diploid EPC cells. Self-renewal in the diploid EPC cells is limited, and some go on to form secondary giant cells. Maintenance of a diploid cell population within the polar trophectoderm or the ExE is dependent on interaction with the ICM or the embryonic ectoderm, respectively. In the mature chorioallantoic placenta, the spongiotrophoblast, components of the labyrinth, and the giant cells are direct descendants of the trophoblast lineage.

Evidence for FGF Signaling in the Preimplantation Embryo and its Importance for Trophoblast Proliferation

As described earlier, the proliferative capacity of diploid cells within the trophoblast lineage depends on their interaction with the ICM or its derivatives. Recent expression and genetic studies point to a critical role of *fibroblast growth factor* (FGF) signaling in regulation of proliferation and development in the trophoblast lineage.

The FGF family consists of at least 22 FGF ligands that function in various processes throughout embryonic development. FGF4 is the only ligand known to be expressed in the preimplantation embryo. *Fgf4* transcripts can be detected at the 1-cell stage through to the blastocyst stage, where its expression becomes confined to the ICM. During peri- and postimplantation stages (4.5–6.0 dpc), it continues to be expressed solely in the epiblast in a uniform pattern. Embryos homozygous for an *Fgf4* null mutation die shortly after implantation (5.5 dpc). Their postimplantation phenotype was characterized by impaired growth of the embryonic component. This defect is suggested to occur after implantation, since *Fgf4*$^{-/-}$ blastocysts appeared normal when compared to wild-type littermates. Nonetheless, the presence of maternal *Fgf4* mRNA or protein may partially compensate for the loss of zygotic transcripts. This in turn could allow additional cell divisions, resulting in extended survival of the embryos until the postimplantation stage.

In vitro outgrowths from mutant blastocysts revealed the lack of ICM and endoderm, which could be partially rescued by exogenous FGF4. Exogenous administration of FGF4 to wild-type cultured ICM cells resulted in proliferation of these cells and promoted their differentiation into parietal and/or primitive endoderm. Thus, it appeared that FGF4 serves as a critical component for the survival and development of the ICM in the early postimplantation phase of mouse embryogenesis. However, *Fgf4*$^{-/-}$ *embryonic stem* (ES) cells can be generated, and they proliferate normally in vitro, although growth and survival of the differentiated cell types that arise from null ES cells are affected. This discrepancy between the effects of loss of FGF4 in ES cells and embryos could be explained by an additional requirement for FGF4 in signaling to the trophectoderm lineage. Some data to support this were observations that administration of FGF4 to blastocyst outgrowths led to an increase in the number of outgrowing trophectoderm cells. Furthermore, addition of FGF4 to *Oct4*$^{-/-}$ embryos, which only make trophectoderm, also promoted its proliferation.

FGF signaling is mediated through the cooperative interaction of high-affinity tyrosine kinase receptors and heparin sulfate proteoglycans. The group of high-affinity FGF receptors consists of four members, FGFR1–FGFR4. There is some cross-reactivity between the four receptors and ligands, FGF1–FGF9, and some alternatively spliced forms of FGFR1–FGFR3 have ligand specificity. Isoforms of all four FGF receptors bind FGF4, and expression of all four receptors has been detected at the blastocyst stage. Both *Fgfr3* and *Fgfr4* transcripts are detectable in all early blastocyst cells, with the latter being more pronounced in the trophectoderm. However, these receptors are unlikely to play a critical role in trophoblast development, since embryos mutant for *Fgfr3* only, or *Fgfr3* and *Fgfr4*, display late skeletal and lung defects, respectively.

A recent analysis of *Fgfr2* expression pattern in the preimplantation embryo reveals an early onset of expression of two alternatively spliced forms of this receptor (termed IIIb and IIIc) that differ in an extracellular Ig domain. The *Fgfr2 IIIc* transcript is present in the oocyte, whereas the *FgfR2 IIIb* isoform is first detected in the two-cell stage. Both alternatively spliced transcripts are detected through the compacted morula stage. Whole-mount in situ and immunohistochemistry analyses detect *Fgfr2* expression specifically in the outer cell layer of the morula. These cells later give rise to the trophectoderm, where the expression of *Fgfr2* persists in the expanded blastocyst. Postimplantation expression of *Fgfr2* in the trophoblast lineage continues in the ExE through the late-streak stage.

The generation of two different *Fgfr2* mutations led to embryonic lethality at different stages. The more severe *Fgfr2* phenotype resulted in embryos that implanted randomly with respect to the mesometrial–antimesometrial axis of the uterus. These embryos died at 4.5–5.5 dpc. Since the trophectoderm mediates implantation, these observations suggested that the lethality of these mutants was due to trophoblast defects. Analysis of the other allele of *Fgfr2* revealed that mutant embryos die at 10.0–11.0 dpc due to placental failure. The lethal defects are characterized by failure in chorioallantoic fusion in about one-third of the mutant embryos, and others lack the labyrinthine portion of the placenta. Thus, *Fgfr2* appears to be important also for later stages of placental development. Since different regions of the *Fgfr2* locus were targeted, it is possible that only one is a null and the other is a new allele. The early, severe phenotype could be the result of a dominant-negative allele that inhibited all FGF signaling or the later, less severe phenotype could be due to a hypomorphic allele that allowed partial FGFR2 signaling to occur. The discrepancy between these two phenotypes has yet to be resolved. However, genetic background cannot account for the differences, since both mutants were analyzed on a 129 background.

Evidence for a critical role for FGF signaling in the proliferation and development of the trophectoderm was further demonstrated by the analysis of mouse embryos that express a dominant negative FGF receptor (dnFGFR). In mosaic embryos, cell division ceased at the fifth cell division in all cells that expressed the mutant receptor, resulting in an overall lower number of cells in these mosaic blastocysts. Moreover, no mitotic trophoblast cells adjacent to the ICM were found to express dnFGFR, whereas expression of dnFGFR was detected in postmitotic trophoblast cells farther away from the ICM. Thus, it seems that functional FGF signaling, probably via the FGF4 ligand and FGFR2 receptor, is essential for trophoblast proliferation in vivo, and in culture, as discussed below.

Derivation of Trophoblast Stem Cell Lines

Several attempts have been made to culture trophoblast cells ex vivo. Invariably, the cultures differentiated into postmitotic giant cells, and consequently, cell lines were not established. However, the recent evidence implicating FGF signaling in trophoblast proliferation and development warranted a renewed attempt at trophoblast cultures. Initially, the ExE of 6.5 dpc embryos was isolated and disaggregated into a near-single-cell suspension with trypsin. These ExE cells were plated in the presence of FGF4 and its required cofactor, heparin. In these conditions, the ExE cells eventually differentiated into giant cells, indicating that FGF4 alone was not sufficient to maintain trophoblast cells in a proliferative state. However, when the ExE cells were plated on a feeder layer of *mouse embryonic fibroblasts* (MEFs) in the same conditions, the ExE cells produced epithelial colonies that could be passaged without differentiation into giant cells. Upon removal of FGF4 or the MEFs, the epithelial cells ceased to divide and differentiated into giant cells.

As described below, gene expression studies and chimeric analysis determined these cells to be of trophoblast origin, and they retained the capacity to contribute to all trophoblast cell types in vivo. Thus, they were termed *trophoblast stem* (TS) cells. First, the general properties of TS cells are described, including their derivation, maintenance in culture, differentiation, and their developmental potential in chimeras. This is followed by gene expression studies of TS cells in their proliferative state and as they differentiate.

TS cell lines can be derived from several different stages of development. In addition to the ExE of 6.5 dpc embryos, TS cell lines have been derived from 3.5 dpc blastocysts and the chorionic ectoderm of 7.5 dpc embryos. However, TS cell lines could not be derived from the embryonic ectoderm or the EPC of 6.5 dpc embryos. This suggests that TS cells exist in vivo for at least a 4-day window within the trophoblast lineage. TS cell lines derived from different stages of development exhibited similar properties in culture, and their gene expression profiles and developmental potential were indistinguishable. The efficiency of TS cell derivation is very high, approaching 100% of explanted embryos. Furthermore, the ability to generate lines is not strain-dependent, as observed for ES cell line derivation, and male and female lines can be generated equally well.

TS cell lines are dependent on both FGF4 and MEFs. If either component is removed, or if the cofactor for FGF4, heparin, is removed, the cultures differentiate into giant cells. However, FGF4 may be replaced with either FGF1 or FGF2 without loss of stem cell maintenance. This is not surprising, since FGF1 and FGF2 can activate the same receptors as FGF4. More interestingly, the MEFs may be replaced with *MEF-conditioned medium* (MEF-CM). This CM must be present at a critical concentration to be effective, suggesting that there is a soluble factor or factors in the MEF-CM required for TS cell maintenance. The identification of the critical MEF-CM component(s) should shed more light on the requirements of trophoblast stem cells in vivo. The identity of the required factor in the MEF-CM has remained elusive. Several candidates, such as the ligands LIF, EGF, BMP2, and BMP7, have been unable to replace the MEF-CM in TS cell cultures. Even the nature of the stem-cell-promoting effect of the feeder cells is still unknown. Biochemical characterization of the MEF-CM, such as by protease treatments and fractionations, is under way.

As mentioned above, TS cells were observed to differentiate into cells that morphologically resembled trophoblast giant cells. *Fluorescence-activated cell scan* (FACS) analysis of propidium iodide-stained cells confirmed that they were indeed increasing their DNA content, as do giant cells in vivo. These FACS studies also supported the microscopy observations that TS cell lines are considerably heterogeneous. The FACS profiles of TS cells grown in stem cell conditions indicated that ~10% of the cells exhibited an increased ploidy, which is indicative of differentiation. In an attempt to reduce the heterogeneity of the cultures, TS cells were FACS-sorted on the basis of DNA content using the vital dye, Hoechst 33342. TS cells with a 2N DNA content were selectively FACS-sorted from higher ploidy cells and cultured. These 2N-sorted TS cells were considerably less heterogeneous in culture, and their FACS profiles confirmed these observations.

The most convincing results that speak to the true stem cell nature of TS cell lines were their specific and exclusive contributions to trophoblast lineages in chimera studies. When TS cell lines were injected into 3.5 dpc blastocysts, contributions were only observed in trophoblast lineages and never in embryonic lineages or extraembryonic tissues derived from mesodermal and endodermal lineages, such as the yolk sac. Chimeras were analyzed from 6.5 dpc conceptuses through to term placenta. Contributions were observed in all trophoblast subtypes, but some preferences were observed. In early chimeras, TS cells were found to contribute most frequently to the ExE and less frequently to the EPC. Contributions to the giant cell layer were rare, and when they were observed, the clones were small.

At later stages of development, TS cells contributed well to the labyrinth and the spongio-trophoblast, but to a lesser extent to the secondary giant cell layer. In fact, large clones in the trophoblast were observed right up to the edge of the placenta where the secondary giant cells begin to form, but the TS cells rarely contributed to the secondary giant cell layer itself. These restrictions may reflect the mode of generation of chimeras via blastocyst injection, where much of the trophectoderm has already formed. If technical difficulties with aggregation of TS cells with morulae can be overcome, more extensive colonization of the trophoblast can be expected. Another interesting observation was the coherent nature of most of the TS cell clones. This was reminiscent of clones in the primitive endoderm lineage and in direct contrast to the very mosaic contributions observed in the embryo proper.

Gene Expression Studies

The genes expressed by TS cells cultured in stem cell conditions, and as they differentiate, closely resemble the gene expression profile of the trophoblast lineage in vivo. The receptor, *Fgfr2*, and the transcription factors, *Cdx2*, *mEomesodermin* (*mEomes*), and *Err2* are all expressed in early diploid trophoblast cells, the ExE. In agreement with these in vivo observations, these genes are strongly expressed in TS cells and are down-regulated as they differentiate. The secreted ligand, BMP4, is expressed in the ExE of the early mouse conceptus and is expressed only in stem cells and not in differentiated TS cells.

The *basic helix-loop-helix* (bHLH) transcription factor, Mash2, and the novel gene *Tpbp* (formerly *4311*), are expressed in the EPC and spongiotrophoblast. In TS cells, *Mash2* and *Tpbp* were not expressed in stem cells, but were induced upon differentiation. A similar pattern was observed for the transcription factor, Gcm1. *Placental lactogen*-1 (PL-1) is a hormone secreted by giant cells. Expression of this gene was only detected in differentiated TS cells. Hand1, a bHLH transcription factor, is also expressed in differentiated trophoblast lineages such as the EPC and the giant cells. However, *Hand1* was highly expressed in both stem and differentiated TS cells, which was not in agreement with the pattern in vivo.

Genes representative of other early embryo cell types were not detected in TS cells. Oct4, a transcription factor expressed in the early embryonic ectoderm, was not detected in TS cells. *Brachyury*, a mesodermal marker, and *Hnf4*, an endodermal marker, were also not expressed in TS cells.

With few exceptions, the gene expression profile of TS cells in culture is largely representative of the trophoblast in vivo. ExE-specific genes are expressed in TS cells maintained in stem cell conditions, whereas EPC and giant-cell-specific genes are induced upon differentiation.

New Model for Trophoblast Development

With the derivation of FGF4-dependent TS cell lines and recent expression and mutational data, an updated model of trophoblast development can be formulated. The most significant change from the former model is the identification of a major component of the embryo-derived signal required for the maintenance and proliferation of trophoblast stem cells in vivo. In addition, transcription factors thought to be critical for the trophoblast lineage have been added to the model. It is proposed that the growth factor, FGF4, produced and secreted by the ICM (and later by the embryonic ectoderm), signals to the overlying trophoblast cells via an FGFR, of which FGFR2 is a good candidate, maintaining them in a diploid, proliferative state. As trophoblast cells are displaced distally from the embryo proper, they cease to receive the embryo-derived signals and begin to differentiate into other trophoblast subtypes, such as EPC cells and giant cells. It is not known whether the differentiated trophoblast subtypes are default states that are manifested after removal of stem-cell-maintaining signals or whether there are additional signals that promote differentiation. The ex vivo culture system of TS cells would suggest the former, but the mechanism in vivo is not likely to be that simple.

The transcriptional regulation of *Fgf4* has been well studied in cultured cells. The findings suggest that two transcription factors, Oct4 and Sox2 (an HMG-box transcription factor), synergistically activate

Fgf4 expression in the embryo proper. The *Oct4* null mutation resulted in embryos with significantly reduced levels of *Fgf4* expression. The major putative receptor for FGF4 in the trophoblast is FGFR2. The expression data and mutant analysis make this receptor the best candidate to functionally receive the embryo-derived FGF4 signal. In fact, the more severe *Fgfr2* mutant embryos exhibit a peri-implantation lethal phenotype. Interestingly, the embryos implanted randomly and not at the abembryonic pole, implying that the trophectoderm of the blastocyst was equivalent and did not segregate into polar and mural regions. The *Fgfr2* gene is differentially spliced to encode two receptors (IIIb and IIIc) that differ in their ligand specificity.

Targeted mutation of the IIIb isoform resulted in a later phenotype that differed from both of the two *Fgfr2* mutations. The pups died at birth with major limb and lung abnormalities. When *Fgfr2* mutant embryos had their placentas rescued by tetraploid aggregations, the embryonic phenotype was strikingly similar to the IIIb isoform mutation mentioned above. These results indicated that the IIIb isoform was critical for signaling in the embryo proper and that the IIIc isoform of *Fgfr2* is likely the functional receptor in the trophoblast. This would predict that a IIIc isoform-specific mutation would phenocopy the complete null mutation.

Among the transcription factors known to be expressed in TS cells, Cdx2 and mEomes are among the most interesting because they belong to gene families that are known to be regulated by FGF signaling in other vertebrates. *Cdx2* is a member of the *caudal*-related subfamily of homeobox genes. Since the discovery of the prototypic member, *caudal*, in *Drosophila melanogaster*, numerous orthologs have been isolated from both invertebrates and vertebrates. Studies on *Xcad3*, a *Xenopus caudal* homolog, indicated that this gene was an immediate-early target of FGF signaling. Of the three *caudal*-related genes known to exist in the mouse (*Cdx1*, *Cdx2*, and *Cdx4*), only *Cdx2* was found to be expressed in trophoblast tissues. It would be of interest to determine whether *Cdx2* is a direct target of the FGF4–FGFR signal in the trophoblast. Interestingly, the *Cdx2* null phenotype is similar, if not identical, to the severe *Fgfr2* mutation, suggesting a critical role for *Cdx2* in the trophoblast and consistent with its being a putative target of FGF signaling.

The T-box gene family, of which *mEomes* is a member, defines a large group of transcription factors with a conserved DNA-binding domain known as the T-box. The prototypic T-box gene, *Brachyury*

(*T*), was shown to be an FGF immediate-early response gene in *Xenopus*. Studies in zebrafish also found two T-box genes, *no tail* (*T* homolog) and *spadetail*, to be regulated by FGF signaling. However, it was not determined whether this regulation was direct or indirect. *mEomes* is first expressed in the trophectoderm of the 3.5 dpc blastocyst and is later restricted to the ExE of the early postimplantation embryo. A targeted mutation of this gene resulted in embryonic lethality shortly after implantation due to trophectoderm defects. This phenotype was similar to the *Cdx2* null mutation and one of the *Fgfr2* mutations, suggesting that FGF signaling may also regulate *mEomes*.

Concluding Remark

The derivation and characterization of trophoblast stem cell lines has provided a new cell culture system for studying trophoblast development and placentation in the mouse. The ability of TS cells to both differentiate in culture and contribute to trophoblast tissues in vivo is unique among trophoblast or placental cell lines studied to date. When TS cell lines are compared to other rodent trophoblast cell lines, several fundamental differences can be noted. The Rcho-1 cell line from a rat choriocarcinoma has the ability to differentiate directly into giant cells in vitro, which makes it a valuable system for studying giant cell differentiation and endoreduplication. However, the starting population is tetraploid, and they do not differentiate into other trophoblast subtypes, such as labyrinthine or spongiotrophoblast cells, and their ability to contribute to chimeras has not been reported. The SM cell lines (SM9-1, SM9-2, and SM-10) were derived from mouse mid- to late-gestation placentas. These cell lines exhibit differential invasiveness in culture, but their inability to differentiate restricts their usefulness as models for broad aspects of trophoblast development.

Both ES and TS cell lines express genes that are highly representative of their tissues of origin, but several exceptions can be noted. TS and ES cells can differentiate into several cell types in culture, and both exclusively contribute to their respective lineages in chimeric conceptuses. TS and ES cells behave as classic stem cells in vitro with indefinite maintenance of a stem cell pool that generates different progeny. However, in vivo they both show a limited period of stem-cell-like activity. Their existence in vivo is relatively short (3–4 days) in comparison to the length of gestation, and they are restricted to the earliest moments of development.

Numerous interactions are thought to occur between the epiblast and trophoblast in vivo. With the availability of both TS and ES cells, these interactions may be studied ex vivo. Preliminary experiments have shown that TS and ES cells do interact to form coherent aggregates in suspension cultures. Interestingly, the cells most often segregate to form a structure where a group of ES cells is interacting with a group of TS cells at a planar surface, reminiscent of the situation in vivo. Phenomena such as primordial germ cell induction and mesoderm induction may be investigated in such TS–ES cell aggregates.

The universality of the existence of TS cells in other mammalian species has yet to be determined. However, attempts to derive TS cell lines from other species are under way. It is encouraging to note that a porcine trophoblast cell line, TE1, has been derived that shares several morphological characteristics with mouse TS cell lines. This cell line is dependent on feeder cells, but a requirement for FGF signaling has not been established.

Since the placenta is the most diverged organ among mammals, and there are considerable differences between human and mouse trophoblast development, the derivation of human TS cell lines would provide a significant model of the human placenta, which would not be redundant with mouse TS cell lines. Although blastocyst formation in the mouse and human appear quite similar, the events following implantation are quite different. The mouse polar trophectoderm exhibits rapid proliferation subsequent to implantation to give rise to the ExE, whereas many of the human trophectodermal cells undergo cell fusion to immediately form the multinucleated syncytiotrophoblast. There is a layer of diploid trophoblast cells between the syncytiotrophoblast and the epiblast, known as the cytotrophoblast. If TS cells exist in human, they would most likely arise from the cytotrophoblast layer. However, rapid proliferation of the cytotrophoblast layer does not occur until the early villous stage, which is about 7 days after implantation. These observations suggest that TS cells may more likely be found in later stages of human placental development rather than in the blastocyst. Human TS cell lines could be of potential therapeutic use, such as in the production of clinically useful hormones or cell-based therapies for patients with placental insufficiencies.

9

SCALING-UP OF CELL CULTURE

Small scale culture of cells in flasks of up to 1 litre volume (175 cm^2 surface area) is the best means of establishing new cell lines in culture, for studying cell morphology and for comparing the effects of agents on growth and metabolism. However, there are many applications in which large numbers of cells are required, for example extraction of a cellular constituent (10^9 cells can provide 7 mg DNA); to produce viruses for vaccine production (typically 5×10^{10} cells per batch) or other cell products (interferon, plasminogen activator, interleukins, hormones, enzymes, erythropoietin, and antibodies); and to produce inocula for even larger cultures. Animal cell culture is a widely used production process in biotechnology, with systems in operation at scales over 10000 litres. This has been achieved by graduating from multiples of small cultures, an approach which is tedious, labour-intensive, and expensive, to the use of large '*unit process*' systems. Although unit processes are more cost-effective and efficient, achieving the necessary scale-up has required a series of modifications to overcome limiting factors such as oxygen limitation, shear damage, and metabolite toxicity. One of the aims of this chapter is to describe these limitations, indicate at what scale they are likely to occur, and suggest possible solutions. This theme has to be applied to cells that grow in suspension and also to those that will only grow when attached to a substrate (anchorage-dependent cells). Another aspect of scale-up that will be discussed is increasing unit cell density 50- to 100-fold by the use of cell immobilization and perfusion techniques.

Free suspension culture offers the easiest means of scale-up because a 1 litre vessel is conceptually very similar to a 1000 litre vessel.

The changes concern the degree of environmental control and the means of maintaining the correct physiological conditions for cell growth, rather than significantly altering vessel design. Monolayer systems (for anchorage-dependent cells) are more difficult to scale-up in a single vessel, and consequently a wide range of diverse systems have evolved. The aim of increasing the surface area available to the cells and the total culture volume has been successfully achieved, and the most effective of these methods (microcarrier) will be described in this chapter.

General Methods and Culture Parameters

Familiarity with certain biological concepts and methods is essential when understanding scale-up of a culture system. In small scale cultures there is leeway for some error. If the culture fails it is a nuisance, but not necessarily a disaster. Large scale culture failure is not only more serious in terms of cost, but also the system demands that conditions are more critically met. This section describes the factors that need to be considered as the culture size gets larger.

Cell Quantification

Measuring total cell numbers (by haemocytometer counts of whole cells or stained nuclei) and total cell mass (by determining protein or dry weight) is easily achieved. It is far more difficult to get a reliable measure of cell viability because the methods employed either stress the cells or use a specific, and not necessarily typical, parameter of cell physiology. An additional difficulty is that in many culture systems the cells cannot be sampled (most anchorage-dependent cultures), or visually examined, and an indirect measurement has to be made.

Cell viability

The dye exclusion test is based on the concept that viable cells do not take up certain dyes, whereas dead cells are permeable to these dyes. Trypan blue (0.4%) is the most commonly used dye, but has the disadvantage of staining soluble protein. In the presence of serum, therefore, erythrocin B (0.4%) is often preferred. Cells are counted in the standard manner using a haemocytometer. Some caution should be used when interpreting results as the uptake of the dye is pH and concentration-dependent, and there are situations in which misleading results can be obtained. Two relevant examples are membrane leakiness caused by recent trypsinization and freezing and thawing in the presence of dimethyl sulfoxide. A colorimetric method using the MTT assay can be used both to measure viability after release of cytoplasmic contents into the medium from artificially lysed cells, and for microscopic visualization within the attached cell.

Indirect measurements

Indirect measurements of viability are based on metabolic activity. The most commonly used parameter is glucose utilization, but oxygen utilization, lactic or pyruvic acid production, or carbon dioxide production can also be used, as can the expression of a product, such as an enzyme. When cells are growing logarithmically, there is a very close correlation between nutrient utilization and cell numbers. However, during other growth phases, utilization rates, caused by maintenance rather than growth, can give misleading results. The measurements obtained can be expressed as a growth yield (Y) or specific utilization/respiration rate (Q):

$$\text{Growth yield (Y)} = \frac{\text{change in biomass concentration (dx)}}{\text{change in substrate concentration (ds)}}$$

$$\text{Specific utilization/respiration rate } (Q_A) = \frac{\text{change in substrate concentration (ds)}}{\text{time (dt)} \times \text{cell mass / numbers (dx)}}$$

Typical values of growth yields for glucose (10^6 cells/g) are 385 (MRC-5), 620 (Vero), and 500 (BHK).

A method which is not so influenced by growth rate fluctuations is the *lactate dehydrogenase* (LDH) assay. LDH is measured in cell-free medium at 30°C by following the oxidation of NADH by the change in absorbance at 340 nm. The reaction is initiated by the addition of pyruvate. One unit of activity is defined as 1 μmol/min NADH consumed. LDH is released by dead/dying cells and is therefore a quantitative measurement of loss of cell viability. To measure viable cells a reverse assay can be performed by controlled lysis of the cells and measuring the increase in LDH.

Equipment and Reagents

Culture vessel and growth surfaces

The standard non-disposable material for growth of animal cells is glass, although this is replaced by stainless steel in larger cultures. It is preferable to use borosilicate glass (e.g. Pyrex) because it is less alkaline than soda glass and withstands handling and autoclaving better. Cells usually attach readily to glass but, if necessary, attachment may be augmented by various surface treatments. In suspension culture, cell attachment has to be discouraged, and this is achieved by treatment of the culture vessel with a proprietary silicone preparation (siliconization). Examples are Dow Corning 1107 (which has to be baked on) or dimethyldichlorosilane which requires thorough washing of the vessel in distilled water to remove the trichloroethane solvent.

Complex systems use a combination of stainless steel and silicone tubing to connect various components of the system. Silicone tubing is very permeable to gases, and loss of dissolved carbon dioxide can be a problem. It is also liable to rapid wear when used in a peristaltic pump. Thick-walled tubing with additional strengthening (sleeve) should be used. Custom-made connectors should be used to ensure good aseptic connection during process operation.

Safe removal of samples of the culture at frequent intervals is essential. An entry with a vaccine stopper through which a hypodermic syringe can be inserted provides a simple solution, but is only suitable for small cultures. Repeated piercing of the vaccine stopper can lead to a loss of culture integrity. The use of specialized sampling devices, also available from fermenter supply companies, is recommended. These automatically enable the line to be cleared of static medium containing dead cells, thus avoiding the necessity of taking small initial samples which are then discarded, and increasing the chances of retaining sterility.

Air filters are required for the entry and exit of gases. Even if continuous gassing is not used, one filter entry is usually needed to equilibrate pressure and for forced input or withdrawal of medium. The filters should have a 0.22 μm rating and be non-wettable.

Non-nutritional medium supplements

Sodium carboxymethylcellulose (15-20 centrepoise, units of viscosity) is often added to media (at 0.1%) to help minimize mechanical damage to cells caused by shear forces generated by the stirrer impeller, forced aeration, or perfusion. This compound is more soluble than methylcellulose, and has a higher solubility at 4°C than at 37°C.

Pluronic F-68 (trade name for polyglycol) (BASF, Wyandot) is often added to media (at 0.1%) to reduce the amount of foaming that occurs in stirred and/or aerated cultures, especially when serum is present. It is also helpful in reducing cell attachment to glass by suppressing the action of serum in the attachment process. However, its most beneficial action is to protect cells from shear stress and bubble damage in stirred and sparged cultures, and it is especially effective in low serum or serum-free media.

Practical Considerations

(a) *Temperature of medium.* Always pre-warm the medium to the operating temperature (37°C) and stabilize the pH before adding cells. Shifts in pH during the initial stages of a culture create many problems, including a long lag phase and reduced yield.

(b) *Growth phase of cells.* Avoid using stationary phase cells as an inoculum since this will mean a long lag phase, or no growth at all. Ideally, cells in the late logarithmic phase should be used.

(c) *Inoculation density.* Always inoculate at a high enough cell density. There is no set rule as to the minimum inoculum level below which cells will not grow, as this varies between cell lines and depends on the complexity of the medium being used. As a guide, it may be between 5×10^4 and 2×10^5 cells/ml, or 5×10^3 and 2×10^4 cells/cm^2.

(d) *Stirring rate.* Find empirically the optimum stirring rate for a given culture vessel and cell line. This could vary between 100-500 r.p.m. for suspension cells, but is usually in the range 200-350 r.p.m., and between 20-100 r.p.m. for microcarrier cultures.

(e) *Medium and surface area.* The productivity of the system depends upon the quality and quantity of the medium and, for anchorage-dependent cells, the surface area for cell growth. A unit volume of medium is only capable of giving a finite yield of cells. Factors which affect the yield are: pH, oxygen limitation, accumulation of toxic products (e.g. NH^4), nutrient limitation (e.g. glutamine), spatial restrictions, and mechanical/shear stress. As soon as one of these factors comes into effect, the culture is finished and the remaining resources of the system are wasted. The aim is, therefore, to delay the onset of any one factor until the accumulated effect causes cessation of growth, at which point the system has been maximally utilized. Simple ways of achieving this are: a better buffering system (e.g. Hepes instead of bicarbonate), continuous gassing, generous headspace volume, enriched rather than basal media, with nutrient-sparing supplements such as non-essential amino acids or lactalbumin hydrolysate, perfusion loops through ultrafiltration membranes or dialysis tubing for detoxification and oxygenation, and attention to culture and process design.

Growth Kinetics

The standard format of a culture cycle beginning with a lag phase, proceeding through the logarithmic phase to a stationary phase and finally to the decline and death of cells, is well documented. Although cell growth usually implies increase in cell numbers, increase in cell mass can occur without any replication. The difference in mean cell mass between cell populations is considerable, as would be expected, but so is the variation within the same population.

Growth (increase in cell numbers or mass) can be defined in the following terms.

(a) Specific growth rate, μ (i.e. the rate of growth per unit amount (weight/numbers) of biomass):

$$\mu = (1/x)\,(dx/dt)h^{-1}$$

where dx is the increase in cell mass, dt is the time interval, and x is the cell mass. If the growth rate is constant (e.g. during logarithmic growth), then

$$\ln x = \ln x_0 + \mu t$$

where x_0 is the biomass at time t_0.

(b) Doubling time, t_d (i.e. the time for a population to double in number/mass):

$$t_d = \frac{\ln 2}{\mu} = \frac{0.693}{\mu}$$

(c) Degree of multiplication, n or number of doublings (i.e. the number of times the inoculum has replicated):

$$n = 3.32 \log(x/x_0)$$

Medium and Nutrients

A given concentration of nutrients can only support a certain number of cells. Alternative nutrients can often be found by a cell when one becomes exhausted, but this is bad practice because the growth rate is always reduced (e.g. while alternative enzymes are being induced). If a minimal medium, such as Eagle's *basal medium* (BME) or *minimum essential medium* (MEM) is used with serum as the only supplement, then this problem is going to be met sooner than in cultures using complete media (e.g. 199), or in media supplemented with lactalbumin hydrolysate, peptone, or BSA (which provides many of the fatty acids). Nutrients likely to be exhausted first are glutamine, partly because it spontaneously cyclizes to pyrrolidone carboxylic acid and is enzymically converted (by serum and cellular enzymes) to glutamic acid, leucine, and isoleucine. Human diploid cells are almost unique in utilizing cystine heavily. A point to remember is that nutrients become growth-limiting before they become exhausted. As the concentration of amino acids falls, the cell finds it increasingly difficult to maintain sufficient intracellular pool levels. This is exaggerated in monolayer cultures because as cells become more tightly packed together, the surface area available for nutrient uptake becomes smaller.

Glucose is often another limiting factor as it is destructively utilized by cells and, rather than adding high concentrations at the beginning, it is more beneficial to supplement after two to three days. In order to maintain a culture, some additional feeding often has to be carried out either by complete, or partial, media changes or by perfusion.

Many cell types are either totally dependent upon certain additives or can only perform optimally when they are present. For many purposes it is highly desirable, or even essential, to reduce the serum level to 1% or below. In order to achieve this without a significant reduction in cell yield, various growth factors and hormones are added to the basal medium. The most common additives are insulin (5 mg/litre), transferrin (5-35 mg/litre), ethanolamine (20 μM), and selenium (5 μg/litre).

Cell aggregation is often a problem in suspension cultures. Media lacking calcium and magnesium ions have been designed specifically for suspension cells because of the role of these ions in attachment. This problem has also been overcome by including very low levels of trypsin in the medium (2 μg/ml).

pH

Ideally pH should be near 7.4 at the initiation of a culture and not fall below a value of 7.0 during the culture, although many hybridoma lines appear to prefer a pH of 7.0 or lower. A pH below 6.8 is usually inhibitory to cell growth. Factors affecting the pH stability of the medium are buffer capacity and type, headspace, and glucose concentration.

The normal buffer system in tissue culture media is the carbon dioxide bicarbonate system analogous to that in blood. This is a weak buffer system, in that it has a pKa well below the physiological optimum. It also requires the addition of carbon dioxide to the headspace above the medium to prevent the loss of carbon dioxide and an increase in hydroxyl ions. The buffering capacity of the medium is increased by the phosphates present in the *balanced salt solution* (BSS). Medium intended to equilibrate with 5% carbon dioxide usually contain Earle's BSS (25 mM $NaHCO_3$) but an alternative is Hanks' BSS (4 mM $NaHCO_3$) for equilibration with air. Improved buffering and pH stability in media is possible by using a zwitterionic buffer, such as Hepes (10-

20 mM), either in addition to, or instead of, bicarbonate (include bicarbonate at 0.5 mM). Alternative buffer systems are provided by using specialist media such as Leibovitz's L-15. This medium utilizes the buffering capacity of free amino acids, substitutes galactose and pyruvate for glucose, and omits sodium bicarbonate. It is suitable for open cultures.

The headspace volume in a closed culture is important because, in the initial stages of the culture, 5% carbon dioxide is needed to maintain a stable pH in the medium but, as the cells grow and generate carbon dioxide, it builds up in the headspace and this prevents it diffusing out of the medium. The result is an increase in weakly dissociated bicarbonate producing an excess of hydrogen ions in the medium and a fall in pH. Thus, a large headspace is required in closed cultures, typically tenfold greater than the medium volume (this volume is also needed to supply adequate oxygen). This generous headspace is not possible as cultures are scaled-up, and an open system with a continuous flow of air, supplied through one filter and extracted through another, is required.

The metabolism of glucose by cells results in the accumulation of pyruvic and lactic acids. Glucose is metabolized at a far greater rate than it is needed. Thus, glucose should ideally never be included in media at concentrations above 2 g/litre, and it is better to supplement during the culture than to increase the initial concentration. An alternative is to substitute glucose by galactose or fructose as this significantly reduces the formation of lactic acid, but it usually results in a slower growth rate. These precautions delay the onset of a non-physiological pH and are sufficient for small cultures. As scale-up increases, headspace volume and culture surface area in relation to the medium volume decrease. Also, many systems are developed in order to increase the surface area for cell attachment and cell density per unit volume. Thus pH problems occur far earlier in the culture cycle because carbon dioxide cannot escape as readily, and more cells means higher production of lactic acid and carbon dioxide. The answer is to carry out frequent medium changes or use perfusion, or have a pH control system.

The basis of a pH control system is an autoclavable pH probe. This feeds a signal to the pH controller which is converted to give a digital or analogue display of the pH. This is a pH monitor system. Control of pH requires the defining of high and low pH values beyond which the pH should not go. These two set points on the pH scale turn on a relay to activate a pump, or solenoid valve, which allows additions of acid or alkali to be made to the culture to bring it back to within the allowable units.

It is rare to have the pH rise above the set point once the medium has equilibrated with the headspace, so the addition of acid can be disregarded. If an alkali is to be added, then sodium bicarbonate (5.5%) is recommended. Sodium hydroxide (0.2 M) can be used only with a fast stirring rate, which dilutes the alkali before localized concentrations can damage the cells, or if a perfusion loop is installed. Normally, liquid delivery pumps are supplied as part of the pH controller. Gas supply is controlled by a solenoid valve and 95% air directly. Above the set point carbon dioxide will be mixed with the air, but below this point only air will be delivered to the culture. This in itself is a controlling factor in that it helps remove carbon dioxide as well as meeting the oxygen requirements of the cultures. pH regulation is very readily adapted to computer control systems.

Oxygen

The scale-up of animal cell cultures is very dependent upon the ability to supply sufficient oxygen without causing cell damage. Oxygen is only sparingly soluble in culture media (7.6 μg/ml) and a survey of reported oxygen utilization rates by cells reveals a mean value of 6 $\mu g/10^6$ cells/h. A typical culture of 2×10^6 cells/ml would, therefore, deplete the oxygen content of the medium (7.6 μg/ml) in under 1 h. It is necessary to supply oxygen to the medium throughout the life of the culture and the ability to do this adequately depends upon the *oxygen transfer rate* (OTR) of the system:

$$\text{OTR} = Kla\ (C^* - C)$$

where OTR is the amount of oxygen transferred per unit volume in unit time, *Kl* is the oxygen transfer coefficient, and a is the area of the interface across which oxygen transfer occurs (as this can only be measured in stationary and surface-aerated cultures, the value *Kla* is used; this is the mass transfer coefficient (vol/h). C* is the concentration of dissolved oxygen when medium is saturated, and C is the actual concentration of oxygen at any given time.

The *Kla* (OTR/C when C = 0) is in units of h^{-1} hours^{-1} and is thus a measure of the time taken to oxygenate a given culture vessel completely under a particular set of conditions.

A culture can be aerated by one, or a combination, of the following methods: surface aeration, sparging, membrane diffusion, medium perfusion, increasing the partial pressure of oxygen, and increasing the atmospheric pressure.

Surface aeration: static cultures

In a closed system, such as a sealed flask, the important factors are the amount of oxygen in the system and the availability of this oxygen to the cells growing under 3-6 mm of medium. Normally a headspace/medium volume ratio of 10:1 is used in order to provide sufficient oxygen. Thus a 1 litre flask (e.g. a Roux bottle) with 900 ml of air and 100 ml of medium will initially contain 0.27 g oxygen. This amount will support 10^8 cells for 450 h and is thus clearly adequate. The second factor is whether this oxygen can be made available to the cells. The transfer rate of oxygen from the gas phase into a liquid phase has been calculated at about 17 μg/cm^2/h. Again, this is well in excess of that required by cells in a 1 litre flask. However, if the surface is assumed to be saturated with dissolved oxygen and the concentration at the cell sheet is almost zero then, applying Fick's law of diffusion, the rate at which oxygen can diffuse to the cells is about 1.5 μg/cm^2/h. At this rate there is only sufficient oxygen to support about 50×10^6 cells in a 1 litre flask, a cell density which in practice many tissue culturists take as the norm. These calculations show the importance of maintaining a large headspace volume; otherwise, oxygen limitation could become one of the growth-limiting factors in static closed cultures.

Table 9.1. Oxygen concentrations on the gas and liquid phases of a Roux bottle culture

Oxygen in 900 ml air:

$900 \times 0.21 \times 32/22400 = 0.27$ g

Oxygen in 100 ml medium:

$100 \times 7.6 \times 10^{-4} = 0.0076$ g

Notes:

0.21 = proportion of oxygen in air;

32 = molecular weight of oxygen;

22400 = gram molecular volume;

$\times 10^{-4}$ = solubility of oxygen in water at 37°C when equilibrated with air.

Sparging

This is the bubbling of gas through the culture, and is a very efficient means of effecting oxygen transfer (as proven in bacterial fermentation). However, it may be damaging to animal cells due to the effect of the high surface energy of the bubble on the cell membrane. This damaging effect can be minimized by using large air bubbles (which have lower surface energies than small bubbles), by using a very low gassing rate (e.g. 5 ml/1 min), and by adding Pluronic F-68. A specialized form of sparging, the airlift fermenter has also been used in large unit process monolayer cultures (e.g. multiple plate propagators). When sparging is used, efficiency of oxygenation is increased by using a culture vessel with a large height/diameter ratio. This creates a higher pressure at the base of the reactor, which increases oxygen solubility.

Membrane diffusion

Silicone tubing is very permeable to gases, and if long lengths of thin-walled tubing can be arranged in the culture vessel then sufficient diffusion of oxygen into the culture can be obtained. However, a lot of tubing is required (e.g. 30 m of 2.5 cm tubing for a 1000 litre culture). This method is expensive and inconvenient to use, and has the inherent problem that scale-up of the tubing required is mainly two-dimensional while that of the culture is three-dimensional. However, several commercial systems are available.

Medium perfusion

A closed-loop perfusion system continuously (or on demand) takes medium from the culture, passes it through an oxygenation chamber, and returns it to the culture. This method has many advantages if the medium can be conveniently separated from the cells for perfusion through the loop. The medium in the chamber can be vigorously sparged to ensure oxygen saturation and other additions, such as sodium hydroxide for pH control, which would damage the cells if put directly into the culture, can be made. This method is used in glass bead systems and has proved particularly effective in microcarrier systems, where specially modified spin filters can be used.

Environmental supply

The dissolved oxygen concentration can be increased by increasing the headspace pO_2 (from atmospheric 21% to any value, using oxygen and nitrogen mixtures) and by raising the pressure of the culture by 100 kPa (about 1 atm) (which increases the solubility of oxygen and its diffusion rate). These methods should be employed only when the culture is well advanced, otherwise oxygen toxicity could occur. Finally, the geometry of the stirrer blade also affects the oxygen transfer rate.

Scale-up

Oxygen limitation is usually the first factor to be overcome in culture scale-up. This becomes a problem in conventional stirred cultures at volumes above 10 litres. However, with the current use of high density cultures maintained by perfusion, oxygen limitation can occur in a 2 litre culture.

Table 9.2. Methods of oxygenating a 40 litre bioreactor (30 litre working volume with a 1.5:1 aspect ratio)

Oxygenating method	*Oxygen delivery (mg/litre/h)*	*No. cells × 10^6/ml supported*
Air (10 ml/min at 40 r.p.m.)		
Surface aeration	0.5	0.08
Direct sparging	4.6	0.76
Spin filter sparging	3.0	0.40
Perfusion (1 vol/h)	12.6	2.10
Perfusion (1 vol/h)		
+ spin filter sparging	15.9	2.65
Oxygen (10 ml/mln at 80 r.p.m.)		
Spin filter sparging	51.0	8.50
+ perfusion (1 vol/h)	92.0	15.00
(assuming oxygen utilization rate of 2-6 μg/10^6 cells/h).		

Redox potential

The *oxidation-reduction potential* (ORP), or redox potential, is a measure of the charge of the medium and is affected by the proportion of oxidative and reducing chemicals, the oxygen concentration, and the pH. When fresh medium is prepared and placed in the culture vessel it takes time for the redox potential to equilibrate, a phenomenon known as poising. The optimum level for many cell lines is +75 mV, which corresponds to a dissolved pO_2 of 8-10%. Some investigators find it beneficial to control the oxygen supply to the culture by means of redox, rather than an oxygen electrode.

Alternatively, if the redox potential is monitored by means of a redox electrode and pH meter (with mV display), then an indication of how cell growth is progressing can be obtained. This is because the redox value falls during logarithmic growth and reaches a minimum value approximately 24 h before the onset of the stationary phase. This provides a useful guide to cell growth in cultures where cell sampling is not possible. It is also useful to be able to predict the end

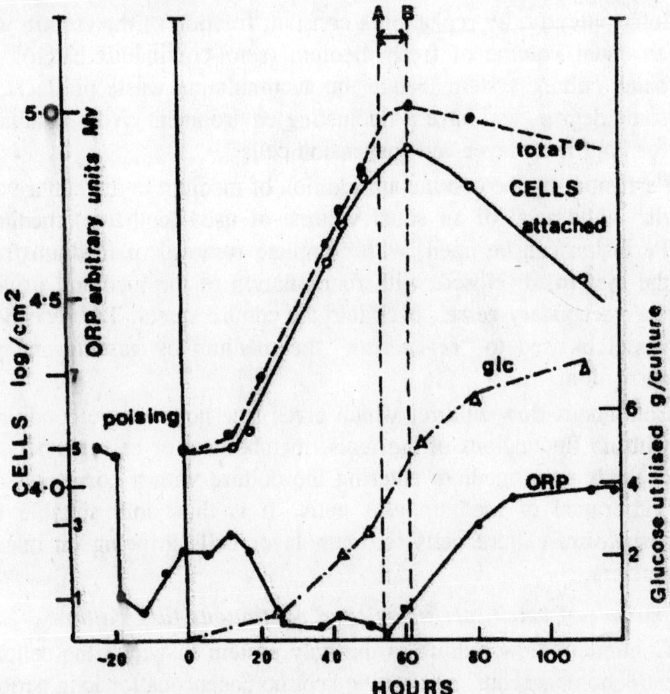

Fig. 9.1. Changes in oxygen-reduction potential (ORP) correlated to cell growth and glucose (glc) utilization.

of the logarithmic growth phase so that medium changes, addition of virus, or product promoters can be given at the optimum time.

Types of Culture Process

Batch and continuous culture

In standard culture, known as *batch culture*, cells are inoculated into a fixed volume of medium and, as they grow, nutrients are consumed and metabolites accumulate. The environment is therefore continually changing, and this in turn enforces changes to cell metabolism, often referred to as physiological differentiation. Eventually cell multiplication ceases because of exhaustion of nutrient(s), accumulation of toxic waste products, or density-dependent limitation of growth in monolayer cultures. There are means of prolonging the life of a batch culture, and thus increasing the yield, by various substrate feed methods.

(a) Gradual addition of fresh medium, so increasing the volume of the culture (fed batch).

(b) Intermittently, by replacing a constant fraction of the culture with an equal volume of fresh medium (semi-continuous batch). All batch culture systems retain the accumulating waste products, to some degree, and have a fluctuating environment. All are suitable for both monolayer and suspension cells.

(c) Perfusion, by the continuous addition of medium to the culture and the withdrawal of an equal volume of used (cell-free) medium. Perfusion can be open, with complete removal of medium from the system, or closed, with recirculation of the medium, usually via a secondary vessel, back into the culture vessel. The secondary vessel is used to "*regenerate*" the medium by gassing and pH correction.

(d) Continuous-flow culture, which gives true homeostatic conditions with no fluctuations of nutrients, metabolites, or cell numbers. It depends upon medium entering the culture with a corresponding withdrawal of medium plus cells. It is thus only suitable for suspension culture cells, or monolayer cells growing on micro-carriers.

Comparison of batch, perfusion, and continuous-flow culture

Continuous-flow culture is the only system in which the cellular content is homogeneous, and can be kept homogeneous for long periods of time (months). This can be vital for physiological studies, but may not be the most economical method for product generation. Production economics are calculated in terms of staff time, medium, equipment, and downstream processing costs. Also taken into account are the complexity and sophistication of the equipment and process, as this affects the calibre of the staff required and the reliability of the production process. Batch culture is more expensive on staff time and culture ingredients, because for every single harvest a sequence of inoculum build-up steps and then growth in the final vessel has to be carried out, and there is also downtime whilst the culture is prepared for its next run. Feeding routines for batch cultures can give repeated but smaller harvests, and the longer a culture can be maintained in a productive state then the more economical the whole process becomes.

Continuous-flow culture in the chemostat implies that cell yields are never maximal because a limiting growth factor is used to control the growth rate. If maximum yields are desired in this type of culture then the turbidostat option has to be used. Some applications, such as the production of a cytopathic virus, leave no choice other than batch culture. Maintenance of high yields, and therefore high product

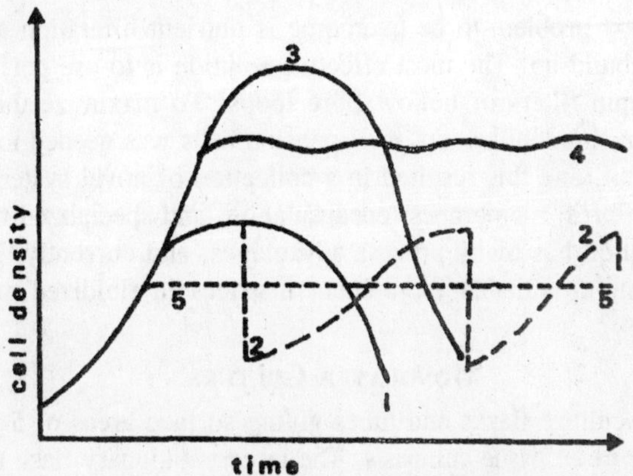

Fig. 9.2. Comparison of culture processes.

concentration, may be necessary to reduce downstream processing costs and these could outweigh medium expenses. For this purpose perfusion has to be used. Although for many processes this is more economical than batch culture, it does add to the complexity of the equipment and process, and increases the risk of a mechanical or electrical failure or microbial contamination prematurely ending the production run. There is no clear-cut answer to which type of culture process should be used—it depends upon the cell and product, the quantity of product, downstream processing problems, and product licensing regulations (batch definition of product, cell stability, and generation number). However, a relative ratio of unit costs for perfusion, continuous-flow, and batch culture in the production of monoclonal antibody is 1:2:3.5.

Summary of Factors Limiting Scale-up

In the following sections a wide variety of alternative culture systems are described. The reason for this large number, apart from scale and different growth characteristics of cells, is the range of solutions which have been used to overcome limitations to scale-up.

As already explained, oxygen is the first limiting factor encountered in scaleup. To overcome this, reactors are available with membrane or hollow fibre devices to give bubble-free aeration, often by means of external medium loops. A related factor is mixing, and low-shear options such as airlift reactors and specially designed impellers have been developed. For anchorage-dependent cells the low unit surface area has limited effective scale-up, so a range of devices using plates, spirals, ceramics and, most effective of all, microcarriers, can be used.

The next problem to be overcome is nutrient limitation and toxic metabolite build-up. The most effective solution is to use perfusion by means of spin filters or hollow fibre loops. To maximize the benefit of perfusion, immobilization of suspension cells was needed to prevent cell wash-out, and this resulted in a collection of novel systems based on hollow fibres, membranes, encapsulation, and specialized matrices. Immobilization has many process advantages, and currently the effort is on scaling-up suitable entrapment matrices in fluidized and fixed beds.

Monolayer Culture

Tissue culture flasks and tubes giving surface areas of 5-200 cm^2 are familiar to all tissue culturists. The largest stationary flask routinely used in laboratories is the Roux bottle (or disposable plastic equivalent) which gives a surface area for cell attachment of 175-200 cm^2 (depending upon type), needs 100-150 ml medium, and utilizes 750-1000 cm^2 of storage space. This vessel will yield 2×10^7 diploid cells and up to 10^8 heteroploid cells. If one has to produce, for example, 10^{10} cells, then over 100 replicate cultures are needed (i.e. manipulations have to be repeated 100 times). In addition the cubic capacity of incubator space needed is over 100 litres. Clearly there comes a time in the scale-up of cell production when one has to use a more efficient culture system. Scale-up of anchorage-dependent cells reduces the number of cultures, is more efficient in the use of staff, and increases significantly the surface area/volume ratio. In order to do this a very wide and versatile range of tissue culture vessels and systems has been developed. The methods with the most potential are those based on modifications to suspension culture systems because they allow a truly homogeneous unit process with enormous scale-up potential to be used. However, these systems should be attempted only if time and resources allow a lengthy development period.

Although suspension culture is the preferred method for increasing capacity, monolayer culture has the following advantages:

(a) It is very easy to change the medium completely and to wash the cell sheet before adding fresh medium. This is important in many applications when the growth is carried out in one set of conditions and product generation in another. A common requirement of a medium change is the transfer of cells from serum to serum-free conditions. The efficiency of medium changing in monolayer cultures is such that a total removal of the unwanted compound can be achieved.

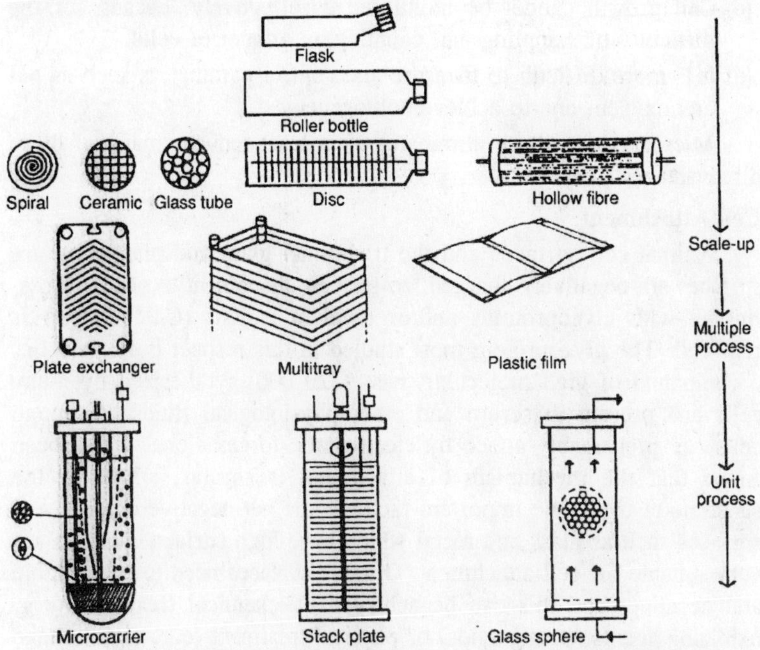

Fig. 9.3. The scaling-up of culture systems for anchorage-dependent cells.

(b) If artificially high cell densities are needed then these can be supported by using perfusion techniques. It is much easier to perfuse monolayer cultures because they are immobilized and a fine filter system (to withhold cells) is not required.

(c) Many cells will express a required product more efficiently when attached to a substrate.

(d) The same apparatus can be used with different media/cell ratios which, of course, can be easily changed during the course of an experiment.

(e) Monolayer cultures are more flexible because they can be used for all cell types. If a variety of cell types are to be used then a monolayer system might be a better investment.

It should be noted that the microcarrier system confers some of the advantages of a suspension culture system.

There are four main disadvantages of monolayer compared to suspension systems:

(a) They are difficult and expensive to scale-up.

(b) They require far more space.

(c) Cell growth cannot be monitored as effectively, because of the difficulty of sampling and counting an aliquot of cells.

(d) It is more difficult to measure and control parameters such as pH and oxygen, and to achieve homogeneity.

Microcarrier culture eliminates or at least reduces many of these disadvantages.

Cell Attachment

Animal cell surfaces and the traditional glass and plastic culture surfaces are negatively charged, so for cell attachment to occur, cross-linking with glycoproteins and/or divalent cations (Ca^{2+}, Mg^{2+}) is required. The glycoprotein most studied in this respect is fibronectin, a compound of high molecular mass (220 000) synthesized by many cells and present in serum and other physiological fluids. Although cells can presumably attach by electrostatic forces alone, it has been found that the mechanism of attachment is similar, whatever the substrate charge. The important factor is the net negative charge, and surfaces such as glass and metal which have high surface energies are very suitable for cell attachment. Organic surfaces need to be wettable and negative, and this can be achieved by chemical treatment (e.g. oxidizing agents, strong acids) or physical treatment (e.g. high voltage discharge, UV light, high energy electron bombardment). One or more of these methods are used by manufacturers of tissue culture grade plastics. The result is to increase the net negative charge of the surface (for example by forming negative carboxyl groups) for electrostatic attachment.

Surfaces may also be coated to make them suitable for cell attachment. A tissue culture grade of collagen can be purchased which saves a tedious preparation procedure using rat tails. The usefulness of collagen as a growth surface is also demonstrated by the availability of collagen-coated microcarrier beads.

Surfaces for cell attachment

Glass

Alum-borosilicate glass (e.g. Pyrex) is preferred because soda-lime glass releases alkali into the medium and needs to be detoxified (by boiling in weak acid) before use. After repeated use glassware can become less efficient for cell attachment, but efficiency can be regained by treatment with 1 mM magnesium acetate. After several hours soaking at room temperature the acetate is poured away and the glassware is rinsed with distilled water and autoclaved.

Plastics

Polystyrene is the most used plastic for cell culture, but polyethylene, polycarbonate, Perspex, PVC, Teflon, cellophane, and cellulose acetate are all suitable when pre-treated correctly.

Metals

Stainless steel and titanium are both suitable for cell growth because they are relatively chemically inert, but have a suitable high negative energy. There are many grades of stainless steel, and care has to be taken in choosing those which do not leak toxic metallic ions. The most common grade to use for culture applications is 316, but 321 and 304 may also be suitable. Stainless steel should be acid washed (10% nitric acid, 3.5% hydrofluoric-acid, 86.5% water) to remove surface impurities and inclusions acquired during cutting.

Scaling-up

Roller bottle

The aims of scaling-up are to maximize the available surface area for growth and to minimize the volume of medium and headspace, while optimizing cell numbers and productivity. Stationary cultures have only one surface available for attachment and growth, and consequently they need a large medium volume. The medium volume can be reduced by rocking the culture or, more usually, by rolling a cylindrical vessel. The roller bottle has nearly all its internal surface available for cell growth, although only 15-20% is covered by medium at any one time. Plastic disposable bottles are available in *c.* 750 cm^2 and *c.* 1500 cm^2 (1400-1750 cm^2) sizes. Rotation of the bottle subjects the cells to medium and air alternately, as compared with the near anaerobic conditions in a stationary culture. This method reduces the volume of medium required, but still requires a considerable headspace volume to maintain adequate oxygen and pH levels. The scale-up of a roller bottle requires that the diameter is kept as small as possible. The surface area can be doubled by doubling the diameter or the length. The first option increases the volume (medium and headspace) fourfold, the second option only twofold.

The only means of increasing the productivity of a roller bottle and decreasing its volume is by using a perfusion system, originally developed by Kruse *et al.*, and marketed by New Brunswick and Bellco (Auto-harvester). This is an expensive option, as an intricate revolving connection has to be made for the supply lines to pass into the bottle. However, cell yields are considerably increased and extensive multi-layering takes place.

Protocol 1. Use of standard disposable roller cultures

Equipment

- 1400 cm^2 (23 × 12 cm) plastic disposable
- Inverted microscope bottle (e.g. Costar, Bibby Sterilin)

Method

1. Add 300 ml of growth medium.
2. Add 1.5 × 10^7 cells.
3. Roll at 12 r.p.h. at 37°C for 2 h, to allow an even distribution of cells during the attachment phase.
4. Decrease the revolution rate to 5 r.p.h., and continue incubation.
5. Examine cells under an inverted microscope using an objective with a long working distance.
6. Harvest cells when visibly confluent (five to six days) by removing the medium, adding trypsin (0.25%), and rolling. Yields will be very similar to those obtained in flasks, assuming enough medium was added. This method has the advantage of allowing the medium volume/surface area ratio to be altered easily. Thus after a growth phase the medium volume can be reduced to get a higher product concentration.

Roller bottle modifications

The roller bottle system is still a multiple process, and thus inefficient in terms of staff resources and materials. To increase the surface area within the volume of a roller bottle, the following vessels have been developed.

SpiraCel

Bibby Sterilin have replaced their bulk cell culture vessel with a SpiraCel roller bottle. This is available with a spiral polystyrene cartridge in three sizes. 3000, 4500, and 6000 cm^2. It is crucial to get an even distribution of the cell inoculum throughout the spiral, otherwise very uneven growth and low yields are achieved. Cell growth can be visualized only on the outside of the spiral, and this can be misleading if the cell distribution is uneven.

Glass tubes

A small scale example is the Bellco-Corbeil Culture System (Bellco), A roller bottle is packed with a parallel cluster of small glass tubes (separated by siliconc spacer rings). Three versions are available giving surface areas of 5 × 10^3, 1 × 10^4, and 1,5 × 10^4 cm^2. Medium is perfused through the vessel from a reservoir. The method is ingenious in that it alternately rotates the bottle 360° clockwise and then 360° anti-clockwise. This avoids the use of special caps for the supply of perfused medium.

An example of its use is the production of 3.2×10^9 Vero cells ($2.3 \times 10^5/cm^2$) over six days using 6.5 litres of medium (perfused at 50 ml/min) in the 10000 cm^2 version.

Increased surface area roller bottles

In place of the smooth surface in standard roller bottles the surface is '*corrugated*', thus doubling the surface area within the same bottle dimensions; e.g. *extended surface area roller bottle* (ESRB) available from Bibby Sterilin (Corning), or the ImmobaSil surface which is a textured silicone rubber matrix surface.

Large capacity stationary cultures

Cell factory (multitray unit)

The standard cell factory unit (Nunc) comprises ten chambers, each having a surface area of 600 cm^2, fixed together vertically and supplied with interconnecting channels. This enables all operations to be carried out once only for all chambers. It can thus be thought of as a flask with a 6000 cm^2 surface area using 2 litres of medium and taking up a total volume of 12 500 ml. In practice this unit is convenient to use and produces good results, similar to plastic flasks. It is made of tissue culture grade polystyrene and is disposable. One of the disadvantages of the system can be turned to good use.

In practice it is difficult to wash out all the cells after harvesting with trypsin, etc. However, enough cells remain to inoculate a new culture when fresh medium is added. Given good aseptic technique, this disposable unit can be used repeatedly. The system is used commercially for interferon production (by linking together multiples of these units). In addition, units are available giving 1200 (2 tray) and 24 000 cm^2 (40 tray).

Costar CellCube

The CellCube has parallel polystyrene trays in a modular closed-loop perfusion system with an oxygenator, pumps, and a system controller (pH, O_2, level control). The unit is compact, with the trays being only 1 mm apart; thus the smallest unit of 21250 cm^2 is less than 5 litres total volume (1.25 litres medium). Additional units of 42500 cm^2 (2.5 litres medium) and 85000 cm^2 (5 litres medium) are available and four units can be run in parallel, giving 340 000 cm^2 growth area.

Hollow fibre culture

Bundles of synthetic hollow fibres offer a matrix analogous to the vascular system, and allow cells to grow in tissue-like densities. Hollow fibres are usually used in ultrafiltration, selectively allowing passage of macromolecules through the spongy fibre wall while allowing a continuous flow of liquid through the lumen. When these fibres are enclosed in a cartridge and encapsulated at both ends, medium can be pumped in and will then perfuse through the fibre walls, which provide a large surface area for cell attachment and growth.

Culture chambers based on this principle are available from Amicon. The capillary fibres, which are made of acrylic polymer, are 350 μm in diameter with 75 μm walls. The pores through the internal lumen lining are available with molecular mass cut-offs between 10000 and 100000. It is difficult to calculate the total surface area available for growth but units are available in various sizes and these give a very high ratio of surface area to culture volume (in the region of 30 cm^2/ml). Up to 10^8 cells/ml have been maintained in this system. These cultures are mainly used for suspension cells but are suitable for attached cells if the polysulfone type is used.

Opticell culture system

This system consists of a cylindrical ceramic cartridge (available in surface area between 0.4-12.5 m^2) with 1 mm^2 square channels running lengthways through the unit. A medium perfusion loop to a reservoir, in which environmental control is carried out, completes the system. It provides a large surface area/volume ratio (40:1) and its suitability for virus, cell surface antigen, and monoclonal antibody production is documented. Scale-up to 210 m^2 is possible with multiple cartridges arranged in parallel in a single controlled unit. Cartridges are available for both attached and suspension cells, which become entrapped in the rough porous ceramic texture.

Heli-Cel (Bibby Sterilin)

Twisted helical ribbons of polystyrene (3 mm × 5-10 mm × 100 μm) are used as packing material for the cultivation of anchorage-dependent cells. Medium is circulated through the bed by a pump, and the helical shape provides good hydrodynamic flow. The ribbons are transparent and therefore allow cell examination after removal from the bed.

Unit process systems

There are basically three systems which fit into fermentation (suspension culture) apparatus:

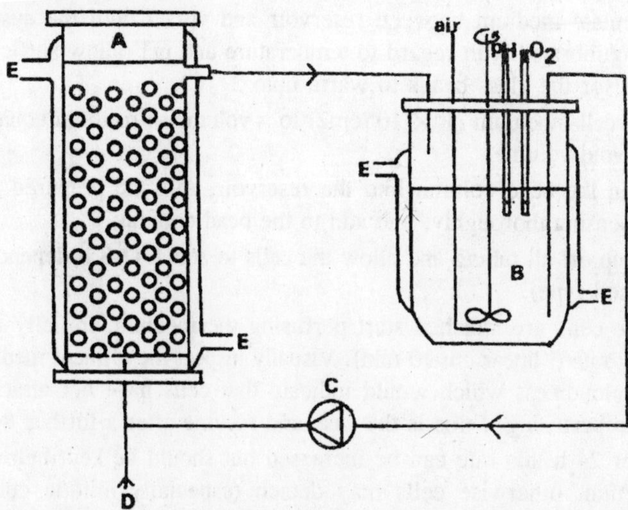

Fig. 9.4. A glass bead bioreactor: (A) Glass bead bed; (B) reservoir; (C) pump; (D) inoculation and harvest line; (E) temperature-controlled water jackets.

1. Cells stationary, medium moves (e.g. glass bead reactor)
2. Heterogeneous mixing (e.g. stack plate reactor)
3. Homogeneous mixing (e.g. microcarrier)

Bead bed reactor

The use of a packed bed of 3-5 mm glass beads, through which medium is continually perfused, has been reported by a number of investigators since 1962. The potential of the system for scale-up was demonstrated by Whiteside and Spier who used a 100 litre capacity system for the growth of BHK21 cells and systems can be commercially obtained.

Spheres of 3 mm diameter pack sufficiently tightly to prevent the packed bed from shifting, but allow sufficient flow of medium through the column so that fast flow rates, which would cause mechanical shear damage, are not needed. The physical properties of glass spheres and these data allow the necessary culture parameters to be calculated. Medium is transferred by a peristaltic pump in this example, but an airlift driven system is also suitable and gives better oxygenation. Medium can be passed either up or down the column with no apparent difference in results.

Protocol 2. Growing cells in a bead bed reactor

1. Prepare sufficient medium at the rate of 1-2 ml/10 cm^2 culture surface area.

2. Circulate medium between reservoir and vessel until the system is equilibrated with regard to temperature and pH (allow sufficient tune for the glass beads to warm upto 37°C).
3. Add cell inoculum (1 × 10^4/cm2) to a volume of medium equal to the void volume.
4. Drain the bead column into the reservoir, mix the prepared cell suspension thoroughly, and add to the bead column,
5. Clamp off all tubing and allow the cells to attach (3-8 h depending on cell type).
6. Once cells are attached start perfusing the medium, initially at a slow rate (l linear cm/10 min). Visually inspect the effluent medium for cloudiness which would indicate that cells have not attached (stop perfusing if this is the case and resume after a further 8 h).
7. After 24 h this rate can be increased but should be kept below 5 cm/min, otherwise cells may detach (especially mitotic cells). The pH readout at the exit and a comparison of glucose concentrations in the input and exit sampling points will indicate whether the flow rate is sufficient.
8. Monitor cell growth by glucose determinations. Glucose utilization rates should be found empirically for each cell and medium, but a rough indication is 2-5 × 10^8 cells produced/g glucose.
9. When the culture is confluent, drain off the medium, wash the column with buffer, add the void volume of trypsin/versene or pronase, and allow to stand for 15-30 min. Cell harvesting can be accelerated by occasionally releasing the trypsin from the column into the harvest vessel and pushing it back into the culture again. Alternatively, feed gas bubbles through the column (only very slowly, otherwise the beads will shift and damage the cells).
10. The culture vessel should be washed immediately after use, otherwise it becomes very difficult to clean. Use a detergent such as Decon, and circulate continuously through the culture system, followed by tap-water, and then several changes of distilled water. Autoclave moist if control probes (oxygen, pH) are used. Use Pyrex glass beads in preference to the standard laboratory beads which are made of soda glass.

Although 3 mm beads arc normally used in order to maximize the available surface area, the use of 5 mm spheres actually gives a higher total tell yield despite the reduced surface area per unit volume.

Heterogeneous reactor

Circular glass or stainless steel plates are fitted vertically, 5-7 mm apart, on a central shaft. This shaft may be stationary, with an airlift pump for mixing, or revolving around a vertical (6 r.p.h.) or horizontal (50-100 r.p.m.) axis. This multi-surface propagator (20) was used at sizes ranging from 7.5-200 litres, giving a surface area of up to $2 \times 10^5/cm^2$. The author's experience is solely with the horizontal stirred plate type of vessel, which is easier to use and has been successful for both heteroploid and human diploid cells. The main disadvantage with this type of culture is the high ratio of medium volume to surface area (1 ml to 1-2 cm^2). This cannot be altered with the horizontal types, although it can be halved with the vertically revolving discs.

Homogeneous systems (microcarrier)

When cells are grown on small spherical carriers they can be treated as a suspension culture, and advanced fermentation technology processes and apparatus can be utilized. The method was initiated by Van Wezel, who used dextran beads (Sephadex A-50). These were not entirely satisfactory, because the charge of the beads was unsuitable, and also possibly due to toxic effects. However, much developmental work has since resulted in many suitable microcarriers being commercially available.

The author's experience has largely been with dextran-based microcarriers and much of the following discussion is based on these products. The choice of this microcarrier was based on a preference for a dried product which could be accurately weighed and then prepared *in situ*, and the fact that with a density of only 1.03 g/ml this product could be used at concentrations of up to 15 g/litre (90 000 cm^2/litre). However, this preference does not detract from the quality of other microcarriers, most of which have been used with equal success, and many of which offer particular advantages as discussed later.

Culture apparatus

A spinner vessel is not suitable unless the stirring system is modified. Paddles with large surface area are needed. These are commercially available, but can be easily constructed out of a silicone rubber sheet and attached to the magnet with plastic ties. The advantage of constructing one within the laboratory is that the blades can be made to incline 20-30° from the vertical, thus giving much greater lift and mixing than vertical blades. Microcarrier cultures are stirred very

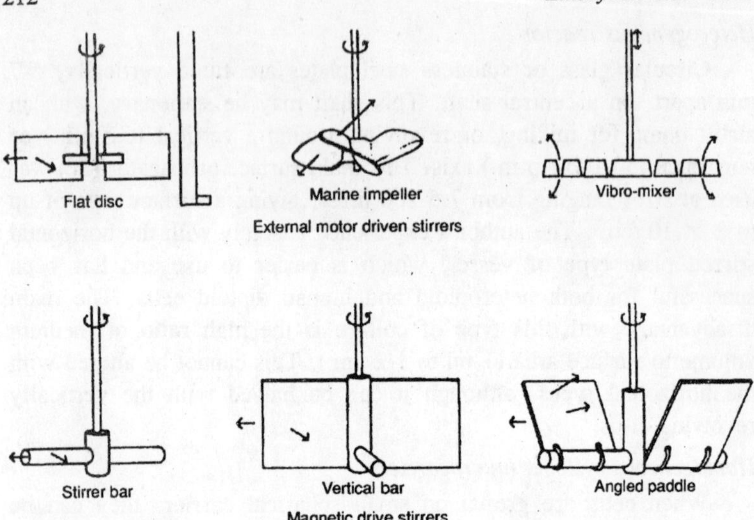

Fig. 9.5. Types of impellers for growing suspension and microcarrier cells.

slowly (maximum 75 r.p.m.) and it is essential to have a good quality magnetic stirrer that is capable of giving a smooth stirring action in the range of 20-100 r.p.m. Never use a stirrer mechanism that has moving surfaces in contact with each other in the medium, otherwise the microcarriers will be crushed. Thus, stirrers which revolve on the bottom of the vessel are unsuitable. As mixing, and thus mass transfer, is so poor in these cultures, the depth of medium should not exceed the diameter by more than a factor of two unless oxygenation systems or regular medium changes are performed. In the basic culture systems, medium changes have to be carried out at frequent intervals. It is worthwhile, therefore, to make suitable connections to the culture vessel to enable this to be done conveniently *in situ*. This will pay dividends in reducing the chances of contaminating the culture.

Initiation of the culture

Many of the factors are critical when initiating a microcarrier culture. Microcarriers are spherical, and cells will always attach to an area of minimum curvature. Therefore a microcarrier surface can never be ideal, however suitable its chemical and physical properties.

Ensure the medium and beads are at a stable pH and temperature, and inoculate the cells (from a logarithmic, not a stationary culture) into a third of the final medium volume. This increases the chances of cells coming into contact with the microcarriers. Microcarrier concentrations of 2-3 g/litre should be used. Higher concentrations

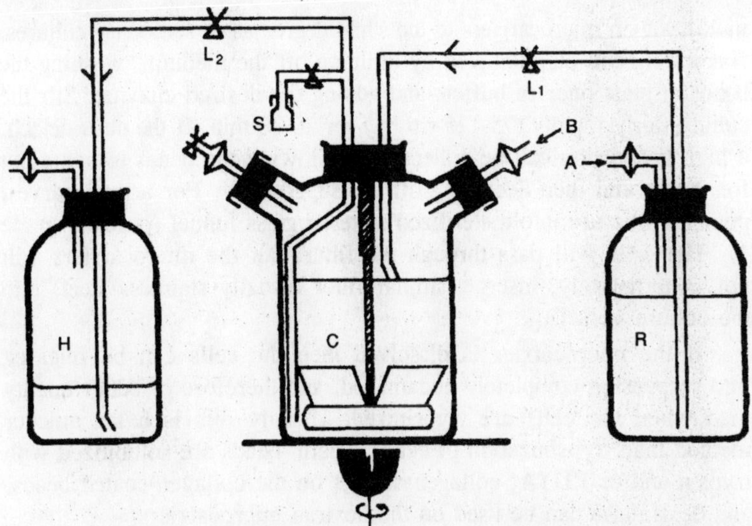

Fig. 9.6. A simple microcarrier culture system allowing easy medium changing. To fill the culture (C) open clamp (L_1) and push the medium from the reservoir (R) using air pressure (at A). To harvest, stop the stirring for 5 min, open L_2, and push the medium from (C) to (H) using air pressure (at B). S is a sampling point.

need environmental control or very frequent medium changes. After the attachment period (3-8 h), slowly top up the culture to the working volume and increase the stirring rate to maintain completely homogeneous mixing. If these conditions are adhered to, and there are no changes in temperature and pH, then all cells that grow on plastic surfaces should readily initiate a microcarrier culture.

Maintenance of the culture

It is very easy to monitor the progress of cell growth in a microcarrier culture. Samples can be easily removed, cell counts (by nuclei counting) and glucose determinations carried out, and the cell morphology examined. As the cells grow, so the beads become heavier and need an increased stirring rate. After three days or so the culture will become acidic and need a medium change. Again, this is an extremely easy routine: turn off the stirrer, allow the beads to settle for 5 min, and decant off as much medium as desired. Top up gently with fresh medium (pre-warmed to 37°C) and restart the stirring.

Harvesting

It is very difficult to harvest many cell types from microcarriers unless the cell density on the bead is very high, and cells do not multilayer on microcarriers to the same degree as in stationary cultures. Harvesting can be attempted by draining off the medium, washing the beads at least once in buffer, and adding the desired enzyme. Stir the culture fairly rapidly (75-125 r.p.m.) for 20-30 min. If the cells detach, a high proportion can be collected by allowing the beads to settle out for 2 min and then decanting off the supernatant. For a total harvest pour the mixture into a sterilized sintered glass funnel (porosity grade 1). The cells will pass through the filter, but the microcarriers will not. Alternatively, insert a similar filter (usually stainless steel) into the culture vessel.

If the microcarrier is dissolved then the cells can be released into suspension completely undamaged, and therefore of better quality than when the cells are trypsinized. Usually this is a tar quicker method than trypsinization of cells. Gelatin beads are solubilized with trypsin and/or EDTA, collagenase acts on the collagen-coated beads, and dextranase can be used on the dextran microcarriers.

Protocol 3. Microcarrier culture

1. Add complete medium to spinner flask (200 ml in a 1 litre flask), gas with 5% CO_2, and allow to equilibrate.
2. Decant PBS from sterilized stock solution of Cytodex-3 and replace with growth medium (1 g to 30-50 ml). Add Cytodex-3 to spinner vessel to give a final concentration of 2 g/litre (1-3 g/litre).
3. Put spinner on magnetic stirrer at 37°C and allow temperature and physiological conditions to equilibrate (minimum 1 h).
4. Add cell inoculum obtained by trypsinization of late log phase cells (pre-stationary phase) at over five cells per bead (optimum to ensure cells on all beads is seven); inoculate at the same density per square centimetre as with other culture types, e.g. $5\text{-}10 \times 10^4/\text{cm}^2$. Cytodex-3 at 2 g/litre inoculated at six cells per bead gives: 8×10^6 microcarriers $= 4.8 \times 10^7$ cells/litre
   $9500\ \text{cm}^2 = 5 \times 10^4$ cells/cm
   $200\ \text{ml} = 2.5 \times 10^5$ cells/ml
5. Place spinner flask on magnetic stirrer and stir at the minimum speed to ensure that all cells and carriers are in suspension (usually 20-30 r.p.m.). It is advantageous if cells and carriers are limited to the lower 60-70% of the culture for this purpose.

   Alternatively either:

   (a) Inoculate in 50% of the final medium volume and add the rest of the medium after 4-8 h.

(b) Or stir intermittently (for 1 mm every 20 min) for the first 4-8 h. However, only use this option for cells with very poor plating efficiency because it causes clumping and uneven distribution of cells per bead.

6. When the cells have attached (90% attachment) the stirring speed can be increased to allow complete homogeneity (40-60 r.p.m.).
7. Monitor progress of culture by taking 1 ml samples at least daily; observe microscopically and carry out cell (nuclei) counts.
8. As the cell density increases there is often a tendency for microcarriers to begin clumping. This can be avoided by increasing the stirring speed to 75-90 r.p.m.
9. After three to four days the culture will become acid. Remove the spinner flask and re-gas the headspace and/or add sodium bicarbonate (5.5% stock solution). With some cells, or at microcarrier densities of 3 g/litre or more, partial medium changes should be carried out. Allow culture to settle (5-10 min), siphon off at least 50% (usually 70%) of the medium, and replace with pre-warmed fresh medium (serum can be reduced or omitted at this stage). Replace spinner flask on stirrer.
10. After four to five days cells reach a maximum cell density (confluency) at the same level as in static cultures, although multilayering is not so prevalent. Thus a cell yield of 1-2 × 10^6/ml (2-4 × 10^5/cm^2) can be expected.
11. Cells can be harvested in the following way:
    (a) Allow culture to settle (10 min).
    (b) Decant off as much medium as possible (> 90%).
    (c) Add warm Ca^2/Mg^2-free PBS (or EDTA in PBS) and mix.
    (d) Allow culture to settle and decant off as much PBS as possible.
    (e) Add 0.2% trypsin (30 ml) and stir at 75-100 r.p.m. for 10-20 min at 37°C.
    (f) Allow the beads to settle out (2 min). Either decant trypsin plus cells or pour mixture through a sterile, coarse-sinter glass filter or a specially designed filter such as the Cellector.
    (g) Centrifuge cells (800 g for 5 min) and resuspend in fresh medium (with serum or trypsin inhibitor, e.g. soybean inhibitor at 0.5 mg/ml).

Note: as a guide to calculating settled volume and medium entrapment by microcarriers, 1 g of Cytodex, for example, has a volume of 15-18 ml.

Scaling-up of microcarrier culture

Scaling-up can be achieved by increasing the microcarrier concentration, or by increasing the culture size. In the first case nutrients and oxygen are very rapidly depleted, and the pH falls to non-physiological levels. Medium changes are not only tedious but provide rapidly changing environmental conditions. Perfusion, either to waste or by a closed-loop, must be used to achieve cultures with high microcarrier concentration. This can be brought about only by an efficient filtration system so that medium without cells and micro-carriers can be withdrawn at a rapid rate. This is constructed of a stainless steel mesh with an absolute pore size in the 60-120 μm range. Attachment to the starrer shaft means that a large surface area filter can be used and the revolving action discourages cell attachment and clogging.

However scaling-up is achieved, oxygen limitation is the chief factor to overcome. This is an especially difficult problem in microcarrier culture because stirring speeds are low. Sparging cannot be used as microcarriers get left above the medium level due to the foaming it causes. The perfusion filter previously described, however,

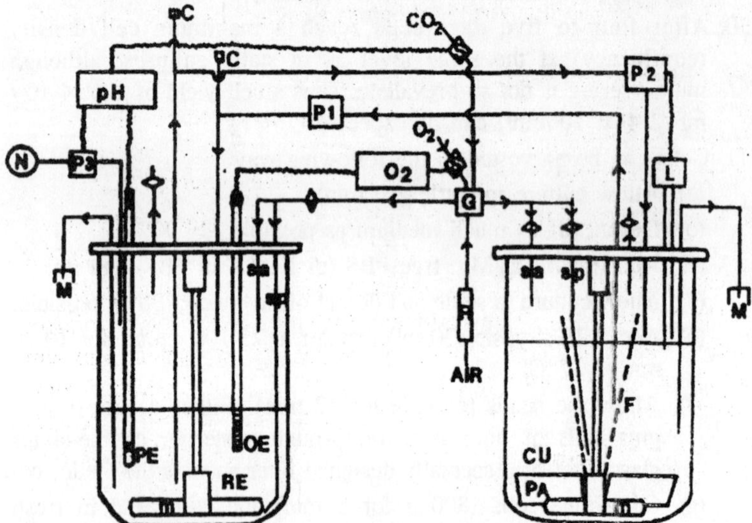

Fig. 9.7. A closed-loop perfusion system with environmental control to allow high cell density microcarrier culture. (CU) culture vessel; (RE) reservoir; (C) connector for medium charge and harvesting; (F) filter; (G) gas blender; (L) level controller; (M) sampling device; (m) magnet; (N) alkali (NaOH) reservoir; (OE) oxygen electrode; (PA) paddle; (PE) pH electrode.

does allow sparging into that part of the culture in which no beads are present. Unfortunately, this means that the most oxygenated medium in the culture is removed by perfusion, but at a sparging rate of 10 cm^3 of oxygen/litre considerable diffusion of oxygenated medium occurs into the main culture. However, modified spin filters are available which have a separate compartment for perfusion and sparging from Braun Biotech. Thus oxygen delivery to a microcarrier culture can utilize the following systems:

(a) Surface aeration.

(b) Increasing the perfusion rate of fully oxygenated medium from the reservoir.

(c) Sparging into the filter compartment.

These three systems are used by the author in the system to run cultures at up to 15 g/litre Cytodex-3 in culture vessels between 2-20 litre working volume.

Summary of microcarrier culture

Microcarrier cultures are used commercially for vaccine and interferon production in fermenters up to 4000 litres. These processes use heteroploid or primary cells. One of the problems of using very large scale cultures is that the required seed inoculum gets progressively larger. Harvesting of cells from large unit scale cultures is not always very successful, although the availability of collagen and gelatin microcarriers eases this problem considerably. In *situ* trypsinization of cells is the usual method of choice, but it is difficult to remove a high percentage of the cells in a viable condition. Filters incorporated into the culture vessel provide the best solution and are most efficient with the 'hard' (glass, plastic) microcarriers which do not block the filter pores like the dextran and gelatin microcarriers. Some microcarriers can be washed and reused but this, of course, is not possible with the gelatin- or collagenase-treated beads.

Suspension Culture

As indicated previously in this chapter, suspension culture is the preferred method for scaling-up cell cultures. Some cells, especially those of haemopoietic derivation, grow best in suspension culture. Others, particularly transformed cells, can be adapted or selected, but some, for example human diploid cell strains (W1–38, MRC-5), will not survive in suspension at all. A further factor that dictates whether suspension systems can be used is that some cellular products are expressed only when the cell exists in a monolayer or if cell-to-cell contact is established (e.g. for the spread of intracellular viruses through a cell population).

Adaptation to Suspension Culture

Cell lines vary in the ease with which they can be persuaded to grow in suspension. For those that have the potential, there are two basic procedures that can be used to generate a suspension cell line from an anchorage-dependent one.

Selection

This method, as demonstrated by the derivation of the LS cell line from L-929 cells by Paul and Struthers. and the HeLa-S3 clone from HeLa cells, depends on the persistence of loosely attached variants within the population. A confluent monolayer culture is lightly tapped or gently swirled, the medium decanted from the culture, and cells in suspension recovered by centrifugation. Cells have to be collected from many cultures to provide a sufficient number to start a new culture at the required inoculum density (at least 2×10^5/ml), This procedure has to be repeated many times over a long period, because many of the cells that are collected are only in the mitotic phase, rather than potential suspension cells. Eventually it is possible with some cell lines to derive a viable cell population that divides and grows in static suspension, just resting on the substrate rather than attaching and spreading out.

Adaptation

This method probably works the same way as the selection procedure, except that a selection pressure is exerted on the culture while it is maintained in suspension mechanically. However, with many cell lines the relatively large number that become anchorage-independent suggests it is more than just selection of a few variant cells. The method given in Protocol 4, used successfully by the author for many cell lines, is based on one published for BHK21 cells.

Protocol 4. Adaptation to suspension culture

1. Prepare the cell suspension by detaching monolayer cells with trypsin (0.1%) and versene (0.01%).
2. Make ready at least two, but preferably three or more, spinner vessels by adding the growth medium (calcium- and magnesium-free modification of any recognized culture medium).
3. Add cells at a concentration of at least 5×10^5/ml and commence stirring (the lowest rate which keeps the cells homogeneously mixed, e.g. 250 r.p.m.).

4. Sample and cany out viable cell counts every 24 h.
5. Every three days remove the medium from the culture, centrifuge (1000 g), and resuspend the cells in fresh medium at a density of at least 2.5×10^5/ml. Depending upon the degree of cell death, cultures will almost certainly have to be amalgamated to keep the cell density at the required level.
6. If, at a medium change, there is significant attachment of cells to the culture vessel, especially at the air-medium interface, then add trypsin (0.01%)/ versene (0.01%) mixture to the empty vessel. Stir at 37°C for 30 min and recover the detached cells by centrifugation. If the cells in suspension are badly clumped, they too can be added to the trypsin/versene solution. This treatment is usually necessary at the first and second medium changes, but very rarely after this point. If cell clumping persists, then add 50 μg/ml trypsin or Dispase (Boehringer) to the growth medium.
7. Successful adaptation is recognized initially by an increase in cell numbers after a medium change, and subsequently by getting consistent cell yields in a given period of time (indicating a constant growth rate).

It is usually necessary to maintain the newly established cell strain in stirred suspension culture, because reversion to anchorage dependence can occur in static cultures. Sometimes cells will adhere to the substrate without becoming completely spread out and continue to grow and divide with a near-spherical morphology.

Static Suspension Culture

Many cell lines will grow as a suspension in a culture system used for monolayer cells (i.e. with no agitation or stirring). Cell lines that are capable of this form of growth include the many lymphoblast lines, hybridomas, and some non-haemopoietic lines, such as the LS cells described in the previous section. However, with the latter type of cells there is always the danger of reversion to a monolayer (e.g. a small proportion of the LS cell line always attaches and is discarded at each subcultivation). Static suspension culture is unsuitable for scale-up, for reasons already stated for monolayer culture.

Small Scale Suspension Culture

For the purpose of this chapter, small scale means under 2 litres. This may seem an entirely arbitrary definition, but it is made on the basis that above this volume additional factors apply. The conventional laboratory suspension culture is the spinner flask, so-called because it

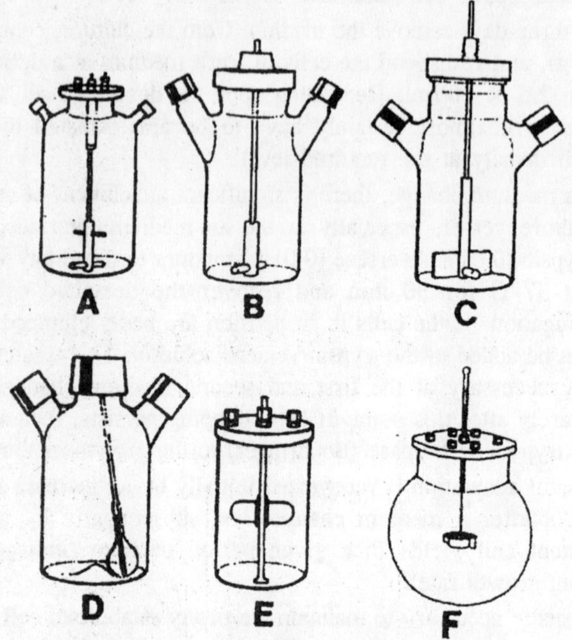

Fig. 9.8. Commercially available spinner cultures.

contains a magnetic bar as the stiirer, and this is driven from below the vessel by a revolving magnet.

The range of options are:

(a) Conventional vessels with spin bar.

(b) Radial (pendulum) stirring system for improved mixing at low speeds (under 100 r.p.m.).

(c) Bellco dual overhead drive system (radial stirring and permits perfnsion).

(d) Techne Br-06 floating impeller system allows working volumes of 500 ml to 3 litres to be used and increased during culture.

Protocol 5. Spinner flask culture

1. Add 200 ml medium to a 1 litre spinner flask, gas with 5% CO_2, and warm to 37°C
2. Add 2-4 × 10^7 cells (1-2 × 10^5/ml) harvested from a growing culture of cells (i.e. not a stationary or dying culture). Note well: this is a higher inoculum than used for stationary cultures.
3. Place spinner vessel on magnetic stirrer at 37°C and set at 100-200 r.p.m. (this is variable depending upon the vessel size,

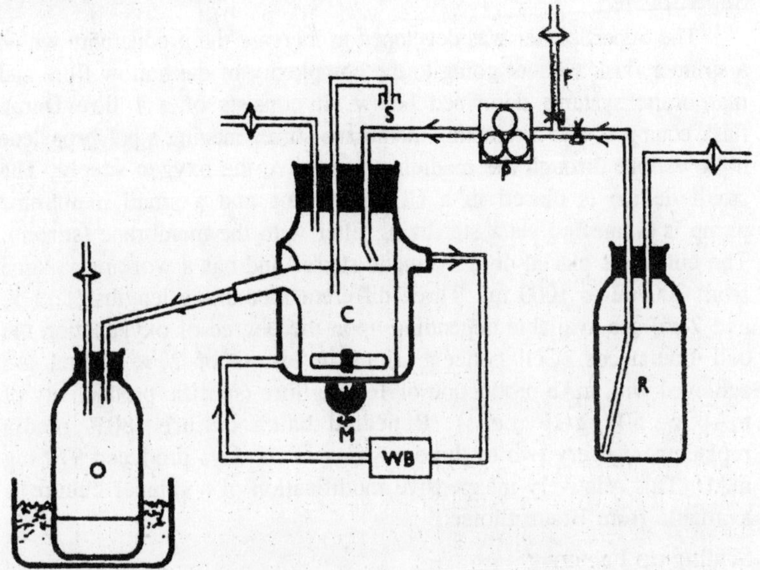

Fig. 9.9. Continuous-flow culture system.

geometry, fluid volume, and cells. Set a speed which visually shows complete homogeneous mixing of the cells throughout the medium—150 r.p.m. is usually a safe choice).

4. Monitor the growth daily by removing a small sample through the side-arm in a Class II cabinet and carrying out a cell count and visual inspection for cell morphology and lack of microbial contamination.
5. Monitor the pH (colour change), especially in closed (non-gassed) systems, and should the culture become acidic (under pH 6.8), add sodium bicarbonate (5.5% stock solution) or re-gas with 5% CO_2. After three days a better option would be to allow the cells to settle out, remove 40-70% of the medium, add fresh pre-warmed medium to the culture, and continue stirring.
6. After four to five days the cells should reach their maximum density (7-10 × 10^5/ml) and can be harvested. The timing will depend upon the kinetics of the cell (whether monoclonal antibody production (mAb) is growth or dying phase associated).

Note: if clumping or attachment to the culture vessel should occur the vessels can be siliconized with Dow Corning 1107 or Repelcotc (dimethyldichlorosilane), or medium with reduced Ca^{2-} and Mg^{2+} can be used.

SuperSpinner

The SuperSpinner was developed to increase the productivity within a spinner flask without going to the complexity of the hollow fibre and membrane systems described below. It consists of a 1 litre Duran flask equipped with a tumbling membrane stirrer moving a polypropylene hollow fibre through the medium to improve the oxygen supply. The small device is placed in a CO_2 incubator and a small membrane pump is connected via a sterilizing filter with the membrane (stirrer). The culture is placed on a magnetic stirrer and has a working volume from 300 ml to 1000 ml. Three different membrane lengths (1, 1.5, and 2 m) are available depending upon the degree of oxygenation the cell line needs. Cell concentrations in excess of 2×10^6/ml are achieved with mAb production of 160 mg/litre (specific productivity of mAb c. 50 μg/10^6 cells). Repeated batch culture (80% media replacement every two to three days) over 32 days produced 970 mg mAb. This relatively inexpensive modification of a spinner culture is available from Braun Biotech.

Scaling-up Factors

In scale-up, both the physical and chemical requirements of cells have to be satisfied. The chemical factors require environmental monitoring and control to keep the cells in the proper physiological environment. These factors, which include oxygen, pH, medium components, and removal of waste products, are described already.

Physical parameters include the configuration of the bioreactor and the power supplied to it. The function of the stirrer impeller is to convert energy (measured as kW/m^3) into hydrodynamic motion in three dimensions (axial, radial, and tangential). The impeller has to circulate the whole liquid volume and to generate turbulence (i.e. it has to pump and to mix) in order to create a homogeneous blend, to keep cells in suspension, to optimize mass transfer rates between the different phases of the system (biological, liquid, and gaseous), and to facilitate heat transfer.

Good mixing becomes increasingly difficult with scaling-up, and the power needed to attain homogeneity can cause problems. The energy generated at the tip of the stirrer blade is a limiting factor as it gives rise to a damaging shear force. Shear forces are created by fluctuating liquid velocities in turbulent areas. The factors which affect this are: impeller shape (this dictates the primary induced flow direction), the ratio of impeller to vessel diameter, and the impeller tip speed (a function of rotation rate and diameter). The greater the turbulence the more efficient the mixing, but a compromise has to be reached so that cells are not damaged. Large impellers running at low speeds give a low shear force and high pumping capacity, whereas smaller impellers need high stirring speeds and have high shear effects.

Magnetic bar stirring gives only radial mixing and no lift or turbulence. The marine impeller is more effective for cells than the flat-blade turbine impellers found in many bacterial systems, as it gives better mixing at low stirring speeds. If the cells are too fragile for stirring, or if sufficient mixing cannot be obtained without causing unacceptable shear rates, then an alternative system may have to be used. Pneumatic energy, for example mixing air bubbles (e.g. airlift fermenter) or hydraulic energy (e.g. medium perfusion), can be scaled-up without proportionally increasing the power. To improve the efficiency of mechanical stirring, the design of the stirrer paddle can be altered (e.g. as described for microcarrier culture), or multiple impellers can be used.

A totally different stirring concept is the Vibro-mixer. This is a non-rotating agitator which produces a stirring effect by a vertical reciprocating motion with a path of 0.1-3 mm at a frequency of 50 Hz. The mixing disc is fixed horizontally to the agitator shaft, and conical-shaped holes in the disc cause a pumping action to occur as the shaft vibrates up and down. The shaft is driven by a motor which operates through an elastic diaphragm; this also provides a seal at the top of the culture. A fermentation system using this principle is available commercially. The advantages of this system are the greatly reduced shear forces, random mixing, reduced foaming, and reduced energy requirement, especially for scaling-up.

The significant effect on vessel design of moving from a magnetic stirrer to a direct drive system is the fact that the drive has to pass through the culture vessel. This means some complexity of design to ensure a perfect aseptic seal while transferring the drive, complete with lubricated bearings, through the bottom or top plate of the vessel. Culture vessels with magnetic coupling are becoming available at increasingly higher volumes, and overcome this problem of aseptic seals.

Scaling-up cannot be proportional; one cannot convert a 1 litre reactor into a 1000 litre reactor simply by increasing all dimensions by the same amount. The reasons for this are mathematical: doubling the diameter increases the volume threefold, and this affects the different physical parameters in different ways. One factor to be taken into consideration is the height/diameter ratio, as this is one of the most important fermenter design parameters. In sparged systems the taller the fermenter in relation to its diameter the better, as the air pressure will be higher at the bottom (increasing the oxygen solubility) and the residence time of bubbles longer. However, in non-sparged systems which are often used for animal cells and rely on surface aeration, the surface area:height ratio is more important and a 2:1 ratio should not be exceeded (preferably 1:1.5).

Mass transfer between the culture phases has been discussed in relation to oxygen. It is this characteristic of scaling-up that demands the extra sophistication in culture design to maintain a physiologically correct environment. This sophistication includes impeller design, oxygen delivery systems, vessel geometry, perfusion loops, or a completely different concept in culture design to the stirred bioreactor.

Stirred Bioreactors

The move from externally driven magnetic spinner vessels to fermenters capable of scale-up from 1 litre to 1000 litres and beyond, has the following consequences at some stage:

(a) Change from glass to stainless steel vessels.

(b) Change from a mobile to a static system: connection to steam for in situ sterilization; requirement for water jacket or internal temperature control; need for a seed vessel, a medium holding vessel, and downstream processing capability.

(c) Greater sophistication in environmental control systems to meet the increasing mass transfer equipment.

In practice, the maximum size for a spinner vessel is 10 litres. Above this size there are difficulties in handling and autoclaving, as well as the difficulty of being able to agitate the culture adequately. Fermenters with motor driven stirrers are available from 500 ml, but these are chiefly for bacterial growth. It is a significant step to move above the 10-20 litre scale as the cost of the equipment is significant and suitable laboratory facilities are required. There is a wide range of vessels available from the various fermenter suppliers which include some of the following modifications:

1. Suitable impeller (e.g. marine)
2. No baffles
3. Curved bottom for better mixing at low speeds
4. Water jacket rather than immersion heater type temperature control (to avoid localized heating at low stirring speeds)

5. Top-driven stirrer so that cells cannot become entangled between moving parts
6. Mirror internal finishes to reduce mechanical damage and cell attachment

As long as adequate mixing, and thus mass transfer of oxygen into the vessel, can be maintained without damaging the cells, there is no maximum to the scale-up potential. Namalva cells have been grown in 8000 litre vessels for the production of interferon. There are many heteroploid cell lines, such as Vero, HeLa, and hybridomas, which would grow in such systems. Whilst regulatory agencies require pharmaceutical products to be manufactured predominantly from normal diploid cells (which will not grow in suspension), the only motive for using this scale of culture was for veterinary products. However, the licensing of products such as tPA, EPO, interferon, and therapeutic monoclonal antibodies from heteroploid lines means that large scale bioreactors are in widespread use. There are many applications for research products from various types of cells, and this is served by the 2-50 litre range of vessels. At present, the greatest incentive for large scale systems is to grow hybridoma cells (producing monoclonal antibody). In tissue culture the antibody yield is 50- to 100-fold lower than when the cells are passaged through the peritoneal cavity of mice, although the purity is greater, particularly if serum-free medium is used. The main need is to supply antibodies to meet the requirements for diagnostic and affinity chromatography purposes, but increasingly as new therapeutic and prophylactic drugs are being developed and licensed much larger quantities are required. Many of these cells are fragile and low yielding in culture, and the specialized techniques and apparatus described below are partly aimed towards this type of cell.

Continuous-flow Culture

At sub-maximal growth rates, the growth of a cell is determined by the concentration of a single growth-limiting nutrient. This is the basis of the chemostat, a fixed volume culture, in which medium is fed in at a constant rate, mixed with cells, and then leaves at the same rate. The culture begins as a batch culture while the inoculum grows to the maximum value that can be supported by the growth-limiting nutrient (assuming the dilution rate is less than the maximum growth rate). As the growth-limiting nutrient decreases in concentration, so the growth rate declines until it equals the dilution rate.

When this occurs, the culture is defined as being in a '*steady state*' as both the cell numbers and nutrient concentrations remain constant. When the culture is in a steady state, the cell growth rate (μ) is equal to the dilution rate (D). The dilution rate is the quotient of the medium flow rate per unit time and the culture volume (V):

$$\mu = D = f/V \text{ day}$$

As the growth rate is dependent on the medium flow rate, the mean generation (doubling) time can be calculated:

$$\mu = \ln 2D$$

An alternative system to the chemostat is the turbidostat in which the cell density is held at a fixed value by altering the medium supply. The cell density (turbidity) is usually measured through a photoelectric cell. When the value is below the fixed point, medium supply is stopped to allow the cells to increase in number. Above the fixed point, medium is supplied to wash out the excess cells. This system really works well only when the cell growth rate is near maximum. However, this is its main advantage over the chemostat, which is least efficient or controllable when operating at the cell's maximum growth rate.

Equipment

Complete chemostat systems can be purchased from dealers in fermentation equipment. However, systems can be easily constructed in the laboratory. The culture vessel needs a side-arm overflow at the required liquid level, which should be approximately half the volume of the vessel. If a suitable 37°C cabinet is not available, then a water-jacketed vessel is needed. Apart from this, all other components are standard laboratory items. Vessel enclosures can be made from silicone (or white rubber) bungs wired onto the culture vessel. A good quality peristaltic pump, such as the Watson-Marlow range, is recommended.

Experimental procedures

Recommended cells are LS, HeLa-S3, or an established lymphoblastic cell line such as L1210 or a hybridoma. Growth-limiting factors can be chosen from the amino acids (e.g. cystine) or glucose.

Protocol 6. Growth of cells in continuous-flow culture

1. Inoculate the culture vessel at 10^5 cells/ml in preferred medium (Eagles MEM with 10% calf serum and the growth-limiting factor).
2. The chosen dilution rate is turned on after 24-48 h of growth. The maximum rate must not exceed the maximum growth of the cell line, which is usually within the range 14-27 h doubling time (although some cells can double their number in 9 h). Thus, the dilution rate will be in the range of O.I/day (ta = 166 h) to 1.2 (td = 14 h).

3. A steady state will become established within 100-200 h, although it may take up to 400 h, especially at the low dilution rates. This will be recognized by the fact that the daily cell counts will not vary by more than the expected counting error.
4. The culture can be maintained almost indefinitely, assuming it is kept sterile and no breakdown in components occurs. Sometimes the culture has to be terminated because of excessive attachment of cells at the interfaces. A duration of 1000 h is considered satisfactory.
5. Once the steady state has been achieved, the flow rate can be altered, and the response of the cell population in establishing new steady state conditions can be studied.
6. As well as carrying out routine cell counts, measurements can also be made to demonstrate the homogeneity of the cells and the medium. For instance, the size and chemical composition of the cells is remarkably consistent. Also, some of the medium components should be measured (e.g. glucose, an amino acid, lactic acid) to demonstrate again the consistency of the culture over a long period of time.

Uses

Continuous-flow culture provides a readily available continuous source of cells. Also, because optimal conditions or any desired physiological environment can be maintained, the culture is very suitable for product generation. For many purposes a two-stage chemostat is required so that optimal conditions can be met for cell growth (first stage) and product generation (second stage).

Airlift Fermenter

The airlift fermenter relies on the bubble column principle both to agitate and to aerate a culture. Instead of mechanically stirring the cells, air bubbles are introduced into the bottom of the culture vessel. An inner (draft) tube is placed inside the vessel and mixing occurs because the air bubbles lift the medium (aerated medium has a lower density than non-aerated medium). The medium and cells which spill out from the top of the draft tube then circulate down the outside of the vessel. The amount of energy (compressed air) needed for the system is very low, shear forces are absent, and this method is thus ideal for fragile animal and plant cells. Also, as oxygen is continuously supplied to the culture, the large number of bubbles results in a high mass transfer rate. Culture units are commercially available in sizes from 2-90 litres. The one disadvantage of the system is that scale-up is more or less linear (the 90 litre vessel requires nearly 4 m headroom). Whether it will be possible to use multiple draft tubes, and thus enable units with greatly increased diameters to be used, remains a developmental challenge. However, 2000 litre reactors are in operation for the production of monoclonal antibodies.

Immobilized Cultures

Immobilized cultures are popular because they allow far higher unit cell densities to be achieved (50-200 $\times$ 10^6 cells/ml) and they also confer stability. and therefore longevity, to cultures. Cells *in vivo* are in a three-dimensional tissue matrix, therefore inmmobilization can mimic this physiological state. Higher cell density is achieved by facilitating perfusion of suspension cells and increasing the unit surface area for attached cells. In addition many immobilization materials protect cells from shear forces created by medium flow dynamics. As cells increase in unit density they become less dependent on the external supply of many growth factors provided by serum, so real cost savings can be achieved.

The emphasis has been on developing systems for suspension cells because commercial production of monoclonal antibody has been such a dominant factor. Basically two approaches have been used: immurement (confining cells within a medium-permeable barrier), and entrapment (enmeshing cells within an open matrix through which medium can flow unhindered).

Immobilization techniques that can be scaled-up, such as the porous carriers described already, will become a dominant production technology once the beads and process parameters have been optimized. They are currently excellent laboratory systems for the manufacture of cell products and, with gelatin carriers, production of cells.

Immurement Cultures

Hollow fibres

These have been discussed previously, and are very effective for suspension cells at scales up to about 1 litre (1-2 $\times$ 10^8 cells/ml). Simple systems can be set up in the laboratory by purchasing individual hollow fibre cartridges (e.g. Amicon, Microgon). All that is needed is a medium reservoir and a pump to circulate the medium through the intracapillary section of the cartridge, and a harvest line from the extracapillary compartment in which the cells and product reside. To get better performance there are a number of '*turn-key*' systems that can be purchased, from relatively simple units, e.g. Kinetek, Cellco, Cell-Pharm, Amicou, Asahi Medical Co, to extremely complex and sophisticated units capable of producing up to 40 g of monoclonal antibody per month.

The Acusyst system made the breakthrough by using pressure differentials to simulate in vivo arterial flow. The unit has a dual medium circuit, one passing through the lumen, the other through the extracapillary (cell) space. By cyclically alternating the pressure between the two circuits, medium is made to pass either into, or out of, the lumen. This allows a flushing of medium through the cell compartment and overcomes the gradient problem, and also gives the possibility of concentrating the product for harvesting. There are a range of Acusyst units, the Acusyst MiniMax and Maximiser being the laboratory models using respectively 2-3 and 3-8 litres of medium per day, and capable of running continuously for 50-150 days (four to five months being a typical run length).

There is an initial growth phase of 10-25 days depending upon inoculum size, medium composition, and cell line, and when a density of 1-2 $\times$ 10^{10} cells is reached (1.1 m^2 cartridge, e.g. Acusyst Maximiser) production starts and daily harvesting of mAb begins. Cells cannot be sampled during the culture so reliance is placed upon a number of process measurements made on a regular basis—on-line pH and oxygen; off-line glucose, lactate, and if necessary ammonia and LDH (for loss of viability) at least weekly. The cultures are purchased as a turn-key unit with all accessories and full operating instructions so it is inappropriate to produce a protocol.

The Tecnomouse is another hollow fibre bioreactor containing up to five flat culture cassettes containing hollow fibres surrounded by a silicone membrane that gives uniform oxygenation and. nutrient supply of the culture and ensures homogeneity within the culture. The system comprises a control unit, a gas and medium supply unit, and the five culture cassettes.

The culture is initiated with cells from 3 $\times$ 225 cm^2 flasks (5 $\times$ 10^7) and grows to 5 $\times$ 10^8 cells. The continuous medium supply is controllable and programmable with typical flow rates (perfusion or recirculation) of 30 ml/h increasing to 70 ml/h in 5 ml steps. After six days harvesting can be started and repeated every two to three days for 30-70 days. Each cassette gives 7 ml culture with 2,5-5.5 mg/ml mAb (equivalent to an average of 10 mg/day). Thus about 10 litres will yield 400 mg mAb/month and 1.5 g/month using the five cassettes.

Membrane culture systems (miniPERM)

There have been a number of membrane-based culture units developed, many based on dialysis tubing, and even available as large fermenters (Bioengineering AG Membrane Laboratory Fermenter with a Cuphron dialysis membrane of 10000 Dalton molecular weight cut-off forming an inner chamber). One of the more successful and currently available systems is described here—the miniPERM Bioreactor.

Protocol 7. MiniPERM Bioreactor

1. Add 35 ml of cell suspension at 1-2 × 10^6 cells/ml in a syringe to the production module through the Luer lock.
2. Add 300-400 ml culture medium to the nutrient module.
3. Place the bioreactor on the bottle turning device in the gassed incubator and set the speed to 10 r.p.m.
4. Sample daily by removing aliquots of culture with a syringe via the Luer lock and carry out cell counts (and optionally glucose measurements).
5. At a cell density of 6 × 10^6/ml replace the medium in the nutrient module every four days, at higher densities every day. Serum-free medium can be used at this stage.
6. Typically a cell density of 15-30 × 10s cells/ml is maintained after day 6 with a daily yield of 0.8 mgmAb/ml (range for various hybridomas 0.5-12 mg/day).
7. Keep the culture going until viability is lost (after 40 days).

The miniPERM Bioreactor consists .of two components, the production module (40 ml) containing the cells and the nutrient module (600 ml of medium). The modules are separated by a semi-permeable dialysis membrane (MWCO 12.5 kDa) which retains the cells and mAb in the production module but allows metabolic waste products to diffuse out to the nutrient module. There is a permeable silicone rubber membrane for oxygenation and gas exchange in the production module. The whole unit rotates (up to 40 rp.m.) within a CO_2 incubator. It can be purchased as a complete disposable ready to use unit or the nutrient module (polycarbonate) can be autoclaved and reused at least ten times.

Encapsulation

The entrapment of cells in semi-solid matrices, or spheres, has many applications, but the basic function is to stabilize the cell and thus protect it from suboptimal conditions. Cells can be immobilized by adsorption, covalent bonding, cross-linking, or entrapment in a polymeric matrix. Materials that can be used are gelatin, polylysine, alginate, and agarose; the choice largely depends upon the problem being addressed. The matrix allows free diffusion of nutrients and generated product between the enclosed microenvironment and the external medium, Alginate is a poly sac charide and is cross-linked with Ca^{2+} ions. The rate of cross-linking is dependent on the concentration of Ca^{2+} (e.g. 40 min with 10 mM $CaCl_2$). A recommended technique is to suspend the cells in isotonic NaCl buffered with Tris (1 mM) and 4% sodium alginate, and to add this mixture dropwise into a stirred solution of isotonic NaCl, 1 mM Tris, 10 mM Ca^{2+} at pH 7.4. The resulting spheres are 2-3 mm in diameter. The entrapped cells can be harvested by dissolving the polymer in 0. 1 M EDTA or 35 mM sodium citrate. Disadvantages of alginate are that calcium must be present and phosphate absent and that large molecules, such as monoclonal antibodies, cannot diffuse out. For these reasons, agarose in a suspension of paraffin oil provides a more suitable alternative. 5% agarose in PBS free of Ca^{2+} and Mg^{2+} is melted at 70°C, cooled to 40°C, and mixed with cells suspended in their normal growth medium. This mixture is added to an equal volume of paraffin oil and emulsified with a Vibro-mixer. The emulsion is cooled in an ice-bath, growth medium added, and, after centrifugation, the oil is removed. The spheres (80-200 um) are washed in medium, centrifuged and, after removing the remaining oil, transferred to the culture vessel.

A custom-made unit can be purchased. This 3 litre fluidized bed culture (bubble column) actually encapsulates the cells in hydrogel beads within the culture vessel and comes with a range of modular units to control all process parameters.

Entrapment Cultures

Opticell

This system has already been described. The special ceramic cartridges for suspension cells (S Core) entrap the cells within the porous ceramic walls of the unit. They are available in sizes from 0,42 m^2 to 210 m^2 (multiple cartridges), which will support 5×10^{10} cells with a feed/harvest rate of 500 litres/day and give a yield of about 50 g of monoclonal antibody per day.

Fibres

A simple laboratory method is to enmesh cells in cellulose fibres (DEAE, TLC, QAE. TEAE; all from Sigma). The fibres are autoclaved at 30 mg/ml in PBS, washed twice in sterile PBS, and added to the medium at a final concentration of 3 mg/ml in a spinner stirred bioreactor. This method has even been found suitable for human diploid cells.

Porous Carriers

Microcarriers and glass spheres are restricted to attached cells and, because a sphere has a low surface area/volume ratio, restricted in their cell density potential. A change from a solid to porous sphere of open, interconnecting pores, increases their potential enormously. There are various types of porous (micro)carrier commercially available. A characteristic of these porous carriers is their equal suitability for suspension tells (by entrapment) and anchorage-dependent cells (huge surface area).

Table 9.3. Porous carriers—advantages compared to solid carriers

(a) Unit cell density 20- to 50-fold higher.
(b) Support both attached and suspension cells.
(c) Immobilization in 3D configuration easily achieved.
(d) Short diffusion paths into a sphere.
(e) Suitable for stirred, flttidized, or fixed bed reactors.
(f) Good scale-up potential by comparison with analogous systems (e.g. microcarrier at 4000 litre).
(g) Cells protected from shear.
(h) Capable of long-term continuous culture.

The problem with many immobilization materials is that diffusion paths become too long, preventing scale-up, A sphere is ideal in that cells and nutrients have only to penetrate 30% of the diameter to occupy 70% of the total volume. This facilitates scale-up as each sphere, whether in a stirred, fluidized, or fixed bed culture, can be considered an individual mini-bioreactor.

Fixed bed (porosphere)

The apparatus and experimental procedures described above for solid spheres is equally suited to porous Siran spheres of 4-6 mm diameter. The only differences are that the bed should be packed with oven-dried spheres and the void volume of cells plus medium inoculated (2×10^6/ml) directly to the bed (dry beads permit better penetration of cells). A larger medium volume is required, and faster perfusion rates (5-20 linear cm/min) should be used (cells are protected from medium shear within the matrix), e.g. 1 litre packed bed needs at least a 15 litre reservoir. After the initial 72 h period, 10 litres of fresh medium is added daily.

Protocol 8. Fixed porous bead (Siran) bed reactor

This protocol is based on a 1 litre fixed bed (water jacket temperature control) of 5 mm Siran beads (Schott) perfused from a 15 litre (11 litre working volume) Applikon BV stirred tank with pH and oxygen control as a medium reservoir.

1. Preliminary preparation:
   (a) Beads. Boil in distilled water for 2 h, rinse, and oven dry (4 h at 100°C). Rinse again and heat dry, then heat sterilize (180°C for 2 h) in a sealed beaker. Add aseptically to autoclaved culture vessel.
   (b) Reservoir vessel. Calibrate probes, assemble, and autoclave.
2. Suspend inoculum (5×10^9 cells per litre bed volume) in 800 ml pre-warmed medium (it is recommended that the medium is circulated for several hours through the bed to both equilibrate the system, check the probes, and condition the beads before inoculation).
3. Add cell suspension to the inoculation vessel, connect to the bottom connector of the glass bead vessel, and slowly add the inoculum by air pressure (by hand pump) being careful not to introduce air bubbles.
4. Drain and refill the bed twice.
5. Start perfusing the medium at 40 ml/min or a linear flow velocity of 2 cm/min.
6. Monitor the reservoir daily for glucose concentration and when the level falls below 2 mg/ml change the medium for a fresh supply (the harvest). Alternatively the system can be run in a continuous mode where the medium feed rate is adjusted to maintain a glucose concentration of 2 mg/ml.
7. Run the culture until the cells lose viability or the bed becomes blocked with cell debris, usually after 50-60 days. This can be delayed further if occasionally the pel-fusion rate is significantly increased for a short time, or the culture is set up for alternate up and down-flow.
8. At the end of the culture, should a cell count be needed, remove a measured aliquot of beads (e.g. 20 cc). add 100 ml of 0.1% citric acid and 0.2% Triton X-100, place on an orbital shaker (ISO rev/min) at 37°C for 2 h. Collect supernatant, measure the volume, wash beads with PBS, add crystal violet to the supernatant (0.1% final concentration), and count on a haemocytometer.

Typical results in the author's laboratory (38-40) with Protocol 8 are 2.75×10^{10} viable cells/litre giving an average yield of 166 mg/litre/day (compared to stirred reactor and airlift cultures of the same hybridoma of 25.5 and 18.5 mg/litre respectively). This method is a low investment introduction to high productivity production of mAb which is simple to use and reliable with low maintenance, at least for the first 50 days of culture.

Fluidized beds

Porous microcarrier technology, currently the most successful scale-up method for high density perfused cultures, was pioneered by the Verax Corporation. Turn-key units were available from 16 ml to 24 litres fluidized beds. The smallest system in the range, Verax System One, is a bench-top continuous perfusion fluidized bioreactor suitable for process assessment and development, and also for laboratory scale production of mAbs. Cells are immobilized in porous collagen microspheres, weighted to give a specific gravity of 1.6, which allows high recycle flow rates (typically 75 cm/min) to give efficient fluidization. The microspheres have a sponge-like structure with a pore size of 20-40 um and internal pore volume of 85% allowing immobilization of cells to densities of $1\text{-}4 \times 10^8$/ml. They are fluidized in the form of a slurry.

The Verax system comprises a bioreactor (fluidization tube), a control system (for pH, oxygen, medium flow rates), gas and heat exchanger and medium supply, and harvest vessels. The system is run continuously for long periods (typically over 100 days). In the author's laboratory it produced 15×10^{10} cells/litre and 540 mg mAb/litre/day (compared to 166 in the fixed bed described above, 25.5 in a stirred reactor, and 18.5 mg in an airlift fermenter). Protocols for its operation come with the equipment and versions have been published. In summary, it is probably the most productive system available, giving the cells a very high specific production rate, but does require some skill to operate to its maximum potential.

An alternative commercial system is the Cytopilot which is a fluidized system using polyethylene carriers (Cytoline) and supports 12×10^7 cells/ml carrier. It is available as the Cytopilot-Mini (400 ml bed) for laboratory scale operation as well as sizes up to 25 litres. The unit has a magnetic stirrer that drives the medium up through a distribution plate into the upper chamber in which the microcarriers are lifted by hydrodynamic pressure. The degree of fluidization is controlled by the stirrer speed and a clear boundary layer is kept at the top of the culture so that clear medium can flow through the internal central circulation tube (loop) back to the stirrer. The culture is oxygenated by means of a mini-sparger delivering microbubbles and a medium feed rate of up to 25 bed matrix volumes per day giving high productivity. The unit can also be used as a packed bed bioreactor if the circulation system is reversed.

There is a range of porous microcarriers available for fluidized systems. Alternatively, some of the carriers are designed for stirred cultures.

Stirred cultures

The Cultispher-G (gelatin), Cellsnow (cellulose), and ImmobaSil (silicone rubber) microcarriers are the most suited to stirred bioreactors, and can be used in an identical manner to solid microcarriers, i.e. 2 g/litre in shake flasks, spinner flasks, or stirred fermenters. The silicone rubber of the ImmobaSil microcarriers facilitates oxygen diffusion and this offers a great advantage over other formulations. About a 40-fold higher concentration of attached cells, and even greater densities of suspension cells, can be achieved over solid microcarriers. The cells can be released from the microcarriers by collagenase treatment.

10

FATE MAPPING OF STEM CELLS

To discuss stem cell "*fate*," a working definition of a stem cell must be utilized. For the purposes of this discussion, a practical definition is a cell with the capacity for prolonged or unlimited self-renewal, combined with the capacity to produce at least one type of highly differentiated progeny. In the context of that definition, the term "*stem cell fate mapping*" implies that a stem cell has a specific fate that can be discovered in experimental systems directed toward fate mapping, and once discovered, the same fate can be extrapolated to other stem cells of that type. Short of the ultimate common fate of cell death, however, stem cells may have widely variable fates depending on the parameters and conditions imposed by the mapping system. This is one of the current challenges to our understanding of stem cell biology, and a primary source of controversy regarding the identity and potential of specific stem cells. There is increasing evidence for remarkable plasticity in some stem cells even when derived from adult tissues. To cite one of many recent examples, it appears that hematopoietic stem cells give rise to muscle under circumstances of muscle injury or disease. A few years ago, this would have been heretical based on data from previously utilized fate mapping systems. It is clear that as new fate mapping systems are utilized, the currently accepted map for many stem cell candidate populations may need to be changed.

Nevertheless, it seems intuitive that in biological systems, specific stem cells do have defined and predictable fates. This may range from true pluripotentiality for embryonic stem cells to a relatively limited fate, under normal biologic circumstances, for multipotential stem cells in tissues of nonembryonic origin. Perhaps for the purpose of this discussion, the definition of fate mapping should be restricted to determining a "normal" fate for a defined stem cell, i.e., determination of the fate that a stem cell would be expected to have in its normal environment under the regulatory influences of homeostasis. These fates would include the participation of the cell in normal development and the response of the cell to biologic perturbations such as tissue injury, senescence of the organism, and disease. It follows that a complete mapping system of the normal fate of a stem cell would require recapitulation of all of the variables to which that cell might be exposed, during the lifetime of the organism, a daunting task for all but the simplest biological systems.

From this discussion, it is apparent that the reductionism inherent in in vitro systems is fundamentally flawed for fate mapping applications and could misrepresent the normal fate of a stem cell. For example, if long-term bone marrow cultures were used for fate mapping of *hematopoietic stem cells* (HSC), a skewed, predominantly myelopoietic perspective of HSC fate would be assumed. Rather than reductionism, the conceptual framework of network theory is much more relevant to fate mapping applications. A stem cell is a single unit in a complex biologic network of other stem cells. The maintenance of the stem cell compartment and provision of differentiated progeny depend on a myriad of cell-autonomous regulators modulated by external regulatory signals. Stem cell quiescence or induction of symmetric or asymmetric cell division is the result of the summation of these regulatory inputs. It is obvious that the complexity of a biologic network can only be reproduced by an in vivo system. The problem is that, at the present time, such systems are beyond our analytical capacity. In the future, analytical methods applicable to networks will evolve and allow prediction of stem cell behavior in the context of normal or abnormal perturbations in the network, i.e., true fate mapping. Until network theory is applicable to analysis of biological systems, fate mapping will be intrinsically analogous to assessment of a few frames of a feature-length film. Interpretation of the map will depend on which frames are selected, i.e., the mapping system utilized.

Within the confines of our current capabilities, how can one determine the fate, or multitude of fates, of an individual stem cell? One of two general strategies has been most frequently applied: (1) in situ labeling and subsequent fate mapping of a resident population of cells or (2) transplantation of a labeled cell population with subsequent mapping of its fate. In both strategies, labeling may be based on specific cell markers, or extrinsically applied labels such as retroviral transduction with a marker gene or labeling with thymidine or *bromodeoxyuridine* (BrdU). Each approach has its relative strengths and weaknesses, and the results must be taken in appropriate context. A major limitation of both strategies is that the readout is dependent on the ability to label a single or homogeneous population of stem cells. The in situ approach has been useful in the rare circumstances where stem cells can be identified precisely by their morphology or location, for instance in the *Drosophila* gonad or peripheral nervous system where stem cells and their non-stem progeny have a well-defined orientation relative to surrounding cells. Although stem cell markers may be useful in some instances to label stem cells in situ, a hallmark of most stem cells is the lack of expression of a specific stem cell marker.

The use of marker genes or BrdU is limited by the low efficiency of labeling quiescent stem cell populations and nonspecificity of cell type labeled, and by down-regulation with terminal differentiation (marker genes) or dilution by multiple cell divisions (BrdU). Similarly, strategies to map the fate of a stem cell after transplantation ideally would transplant either a single definitively isolated stem cell or a homogeneous population of stem cells. Since, in general, current methodology lacks the sensitivity for extreme low-frequency analysis, cell populations rather than single cells must generally be used as the stem cell input. The ability to extrapolate fate mapping results to a single cell within a population is dependent on the homogeneity of the starting cell population. In reality, cell populations isolated using current approaches are rarely, if ever, homogeneous. It therefore becomes critically important to recognize the limitations of current mapping systems in the context of the input population and to interpret the results as a population readout, rather than a single cell readout.

A second limitation is the receptive environment. Even if a single cell or completely homogeneous cell population is transplanted, one must ask whether the receptive environment perturbs the normal stem cell fate. The concept of external control of stem cell fate mediated by microenvironmental signals present in a stem cell "*niche*" is well established. Therefore, normal control of stem cell fate would require engraftment in the appropriate niche. The primary advantage of an in situ strategy is that the cell is labeled in a normal environment without the need for cell processing or transplantation, either of which might affect cell fate. However, the in situ approach is not applicable to analysis of human stem cells, for obvious reasons. Therefore, surrogate biological systems need to be developed for assessment of normal human cell potential. In the remainder of the chapter, we discuss one such system as applied to fate mapping of the human hematopoietic and mesenchymal stem cell.

Human-sheep System for Fate Mapping of Human Stem Cells

The use of surrogate animal systems for the study of human cell biology has had variable success. The primary problems with this approach are the immune response against xenogeneic antigen, and issues of species specificity of microenvironmental factors that affect donor cell function. The immune response can be avoided by the use of sufficiently immunodeficient animal models, most notably the nude mouse, SCID mouse, or NOD/SCID mouse. Unfortunately, the mouse is relatively short-lived, making long-term assessment, even under the ideal circumstances of long-term human cell engraftment, impossible. In addition, the mouse models as an assay of human hematopoiesis remain problematic with respect to the long-term maintenance of multilineage, balanced hematopoiesis. Even when human micro-environments have been co-transplanted as in the SCID-hu systems, the models have not proven to be a reliable reflection of normal human hematopoiesis.

An alternative immunodeficient environment is that of the early gestational fetus. Since the classic observations by Billingham and Medawar of "*acquired*" immunologic tolerance, the phenomenon of fetal tolerance has been recognized. Evidence is now overwhelming that the fetal thymic microenvironment plays a primary role in determination of self-recognition and repertoire of response to foreign antigen. Pre-T-cells undergo positive and negative selection during a series of maturational steps in the fetal thymus that are controlled by thymic stromal cells. The end result is deletion of T-cell clones with high affinity for self-antigen in association with self-MHC, and preservation of a T-cell repertoire against foreign antigen. Therefore, theoretically at least, introduction of foreign antigen prior to thymic processing should result in presentation of donor antigen in the thymus with clonal deletion of alloreactive T-cells. Although less well defined than the mouse system, the immunology of the fetal sheep has been investigated. The fetal sheep is immunologically tolerant of allogeneic skin grafts or of allogeneic or xenogeneic hematopoietic cells, prior to 75 days gestation, avoiding the immunologic barriers present in postnatal models. In addition to immunologic tolerance, there may be other advantages for the fetal sheep model as a fate mapping system. No irradiation or other conditioning regimen is used, so there is no perturbation of the normal receptive environment. The transplantation of cells during development may offer a maximal opportunity for distribution of stem cells into normal niches, as it recapitulates the developmental process of hematogenous distribution and migration of hematopoiesis into developing hematopoietic environments. The sheep lives for many years, allowing true long-term assessment of transplanted cell populations. Finally, because human and sheep DNA and proteins are widely disparate with respect to sequence homology, human-specific markers can be utilized for the unequivocal detection and characterization of human cells by a variety of methodologies.

Fate Mapping of Human Hematopoietic Stem Cells in the Fetal Sheep Model

The HSC has been rigorously defined as a multipotential cell capable of both self-replication and differentiation into all of the hematopoietic lineages. Implicit in this definition is the capacity for long-term, multilineage repopulation of primary and secondary recipients following transplantation. The isolation of HSCs by physical, functional, and phenotypic characteristics has made tremendous progress over the past few decades with characterization of human HSC paralleling, but lagging somewhat behind, that in the mouse for practical reasons. A primary obstacle to validation of human HSC candidate populations has been the absence of biologically relevant in vivo assay systems to rigorously assess HSC content. In part to address this need, we have investigated the fetal sheep as an engraftment model for human HSCs.

We initially assessed the ability of human fetal liver-derived HSC to engraft early gestational fetal sheep and demonstrated long-term multipotential human chimerism in the bone marrow and peripheral blood. We subsequently separated human HSC from long-term chimeric animal bone marrow and demonstrated multilineage engraftment of second-generation recipients for over 1 year, proving unequivocally that human HSC engraft in the sheep bone marrow. In subsequent studies, for the sake of efficiency, the assay has been abbreviated so that human cells are harvested just before term (~2 months after transplantation) and retransplanted into second-generation fetal sheep. In this abbreviated assay, multilineage engraftment of the second-generation animal that persists for 2 months or more after transplantation confirms the presence of human HSC in the donor population. The assay has now been utilized to assess a number of human cell sources and adult bone-marrow-derived HSC candidate populations and has demonstrated excellent sensitivity as an HSC assay. Human HSC demonstrate balanced multilineage engraftment with lymphoid, myeloid, and erythroid expression at the bone marrow progenitor level. Peripheral blood analysis using lineage markers confirms circulating multilineage human cells in the first few years after birth. In contrast to mouse models, the human–sheep model of xenogeneic hematopoietic chimerism after in utero transplantation has allowed assessment of the durability of chimerism with analysis of some animals ongoing at nearly 10 years after transplantation. The results from experiments directed toward assaying the presence of HSC have obvious implications for the model as a fate mapping system for other stem cells.

An interesting aspect of the model relevant to all xenogeneic surrogate systems is the species specificity of the receptive microenvironment. On in vitro assessment, human cells are minimally responsive to sheep cytokines (with the exception of erythropoietin) and vice versa. The administration of human recombinant cytokines IL-3/GM-CSF or SCF to long-term chimeric sheep results in dramatic increases in human hematopoiesis. These findings suggest that the sheep microenvironmental milieu is capable of long-term support of human HSC viability, but incapable of normal support of definitive hematopoiesis, at least at a level that is competitive with host hematopoiesis. It also raises the interesting possibility that the microenvironment can be manipulated, for instance by cotransplantation of normal or genetically manipulated human stromal elements, to create normal or perturbed "human" receptive environments. Such strategies may further improve the utility of the model for human HSC fate mapping and allow unique insight into microenvironmental control of fate decisions.

Mesenchymal Stem Cells

The existence of a population of multipotent *mesenchymal stem cells* (MSCs) that reside in adult tissues and are capable of participation in repair of tissue injury related to disease, aging, or trauma was championed by Caplan as the concept of "*mesangenesis*". Supporting evidence that such a cell exists has been drawn primarily from studies on cells isolated from bone marrow by plastic adherence. Freidenstein et al. originally described such a population isolated by placing whole bone marrow in plastic culture dishes and pouring off the nonadherent (hematopoietic) cells after 4 hours. This left a heterogeneous population of cells, the most tightly adherent of which were spindle-shaped and formed foci of 2–4 cells. These cells remained dormant for 2–4 days and then began to multiply rapidly, ultimately, after several passages, giving rise to confluent layers that were more uniformly spindle cell in appearance.

Most importantly, these cells had the ability to differentiate into colonies that resembled small deposits of bone or cartilage. Subsequent studies by other investigators with cells isolated by protocols similar to Freidenstein's confirmed that this population of cells could be induced to differentiate in vitro into a variety of mesenchymal tissues, including fat, bone, cartilage, bone marrow stroma, and myotubules. The conditions required for induction of differentiation varied somewhat between species. Nevertheless, mesenchymal progenitors could be isolated from mouse, rat, rabbit, and human bone marrow that readily gave rise to adipocytes, chondrocytes, and osteoblasts under specific conditions. Most studies that have been performed with mesenchymal progenitors have been performed with these relatively heterogeneous cell populations. Although efforts have been made to isolate a more homogeneous MSC population, in most cases it remains to be proven whether these more highly purified populations maintain the full multipotentiality of those isolated by simple plastic adherence.

It has been assumed that these heterogeneous mesenchymal progenitor populations contain a subpopulation of true mesenchymal stem cells, and the term stem cell. has been loosely applied in many of these studies. In reality, the mesenchymal stem cell has not yet been isolated or definitively characterized. For practical purposes, however, a definition for MSC that is presently applied is adult bone-marrow-derived populations enriched for progenitors that can be induced by specific in vitro conditions to commit to differentiated mesenchymal cell phenotypes and/or tissues. Obviously, this definition does not have the rigor required for the definition of many other populations of stem cells discussed in this text. On the other hand, it took years to define the hematopoietic stem cell with the rigor currently applied, and the previous definitions were steps toward our current understanding. Nevertheless, from the perspective of fate mapping of the mesenchymal stem cell, this definition leaves much to be desired. From the perspective of the input cell, the populations are crudely heterogeneous. From the perspective of the mapping system, the in vitro potential may or may not represent a "normal" in vivo fate for MSCs.

There is some information available regarding the behavior of MSC populations in vivo. Much of the information lies in the realm of tissue engineering and is more applicable to the question of what MSCs will do when placed directly into an in vivo site, with or without a biosynthetic construct, than to what the normal fate of an MSC might be. Experiments utilizing porous diffusion chambers or ceramic cubes have documented the capacity of MSCs to form fibrous tissue, cartilage, or bone in vivo. For instance, autologous canine MSCs isolated from bone marrow, grown in culture, and loaded onto porous ceramic cylinders resulted in healing of induced bone defects that would otherwise result in non-union. Another interesting study documents astroglial differentiation of murine MSC after direct injection into the lateral ventricles of neonatal mice. Although such evidence implies that MSCs may participate in such processes in vivo, it in no way proves that they, in fact, normally do so. In a separate system, the osteochondrogenic behavior of MSCs in vivo on subcutaneous ceramic implants was found to be independent of their in vitro osteochondrogenic behavior, once again demonstrating that the environment in which the MSCs are assayed modulates their ultimate fate.

There are a number of reports documenting donor-derived stromal elements after bone marrow transplantation, indirectly supporting the presence of a non-hematopoietic stem cell in whole bone marrow that gives rise to stromal supporting elements. Similarly, a report documenting donor-derived skeletal muscle in a muscle injury model after bone marrow transplantation supports a bone marrow resident stem cell with mesenchymal differentiative capacity. Further indirect evidence supporting the presence of osteoprogenitors in bone marrow was recently reported by Horwitz et al. (1999) demonstrating clinical evidence of improvement in three patients with osteogenesis imperfecta after standard bone marrow transplantation with documentation of 1.5–2% donor-derived osteoblasts in two of the patients. All of these studies utilized whole bone marrow as a donor source rather than a defined stem cell population and thus can only be used as indirect support for the existence of a multipotent MSC that has the normal fate of stromal cell, myocyte, or osteocyte differentiation. More recently, Gussoni et al. have reported muscle reconstitution in irradiated *mdx* mice after transplantation of highly enriched hematopoietic stem cells. Their findings suggest that the MSC may be an intermediate population and that the hematopoietic stem cell is more pluripotent than previously appreciated. Once again, however, the use of irradiation in a mouse model of ongoing muscle damage may induce transdifferentiation that is not a normal fate of an HSC.

Of greater relevance to this discussion are studies that have followed the fate of MSC-containing populations after systemic transplantation. There have been two studies in mice in which cultured mouse-adherent cell populations have been transplanted systemically and documented to persist following transplantation. In the first, cells from transgenic mice expressing a human minigene for collagen I were used as mesenchymal progenitor donors, and the fate of the cells was followed after transplantation into irradiated mice. Donor cells were detected in bone marrow, spleen, bone, cartilage, and lung up to 5 months later by PCR for the human minigene, and a PCR in situ assay on lung indicated that the donor cells diffusely populated the parenchyma. Reverse transcription-PCR assays indicated that the marker collagen I gene was expressed in a tissue-specific manner.

A second study transplanted either cultured adherent cells or whole bone marrow into irradiated mice with a phenotype of fragile bones resembling osteogenesis imperfecta caused by expression of the human minigene for type I collagen. With either source of cells, a similar distribution of engraftment was documented as observed in the previous study and, in addition, fluorescense in situ hybridization assays for the Y chromosome indicated that, after 2.5 months, donor male cells accounted for 4–19% of the fibroblasts or fibroblast-like cells obtained in primary cultures of the lung, calvaria, cartilage, long bone, tail, and skin. Although these studies suggest the presence of a mesenchymal stem cell, no immunohistochemical assessment was performed to assess what the engrafted donor cells actually were. The use of a mouse population of MSC is inherently problematic, since it is exceedingly difficult to eradicate hematopoietic elements from the culture. In the absence of immunohistochemical characterization, the presence of contaminating hematopoietic elements and fibroblasts in the donor cell preparation could account for the observed distribution of donor cells. In addition, the requirement for radiation in the model could potentially alter the receptive microenvironment, influencing the observed cell fate.

Fate Mapping of Human MSC

From the preceding discussion, it is clear that there are significant gaps in our understanding of MSCs and MSC fate. Basic questions of what constitutes an MSC and how it can be defined, where it resides, and what biologic developmental and homeostatic mechanisms it participates in remain to be defined. Similarly, the clinically important questions of the transplantability of MSCs, their ability to home and engraft to appropriate sites, their ultimate capacity to differentiate and function after transplantation, their immunogenicity, and the regulatory factors that control MSC proliferation and differentiation are relatively unknown. To approach these questions one needs, at a minimum, to have a defined input population and an appropriate assay system for fate analysis.

In the studies discussed above, a major problem has been the lack of homogeneity and degree of characterization with respect to multipotentiality of the input populations used. Although a true MSC remains to be defined, much less isolated, it is clear that a greater degree of characterization and homogeneity can be currently achieved in some species than in others. Murine MSCs, for instance, have been exceptionally difficult to isolate and expand in culture, without contaminating hematopoietic elements, whereas human MSC are relatively easy to isolate and expand. As described in the preceding chapter, human MSCs have been relatively well characterized by their ability to proliferate in culture with homogeneous morphology, by the uniform presence of a consistent set of surface marker proteins, and by their consistent differentiation into multiple mesenchymal lineages under controlled in vitro conditions. Analysis of colonies derived from individual cells from cultured human MSCs confirms the presence of a subpopulation of cells with at least tripotential differentiative capacity (bone, cartilage, and fat). Thus, from a fate mapping perspective, the use of a relatively defined, although undoubtedly heterogeneous, input population can be achieved with available human MSC.

However, the use of human MSCs involves obvious practical limitations with respect to the fate mapping system. Ideally, one would want to place MSCs into an unperturbed system that is immunologically inert and in which MSC would engraft in their normal locations and subsequently respond to normal growth and regulatory signals. One would also need to have donor discriminatory markers to identify the MSC and their differentiated progeny. We have developed a system that satisfies some of these requirements.

We reasoned that the fetal sheep model might also offer advantages for the engraftment of human MSC. We have therefore modified the model for the analysis of engraftment and ultimate cell fate of human MSCs. Human MSCs were transplanted into fetal sheep at either 65 days gestation (term = 145 days) or 85 days gestation. These time

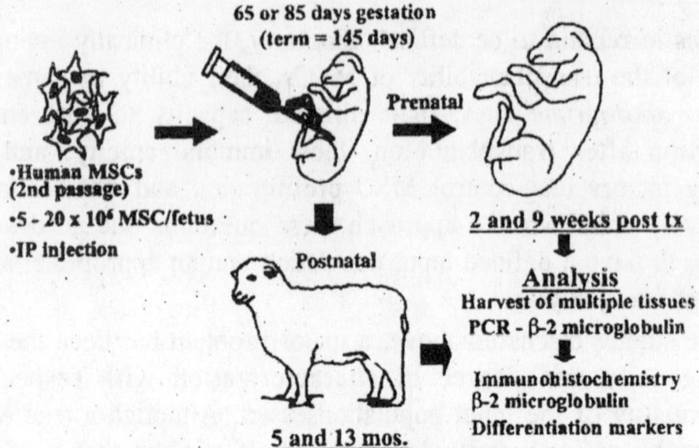

Fig. 10.1. Schematic of the human-sheep model modified for analysis of human mesenchymal stem cells.

points were chosen to assess the effect of two developmental events on MSC distribution and engraftment. First, hematopoiesis is derived entirely from the liver in the fetal sheep at 65 days with only minimal development of the bone marrow. By 85 days, the bone marrow is densely populated with hematopoietic cells. Following in utero hematopoietic cell transplantation, it has been observed that donor cells preferentially home to the fetal liver up until the time of bone marrow formation, after which they home almost exclusively to the bone marrow. Therefore, in the case of hematopoietic engraftment, the receptive environments strongly influence the distribution of engraftment. The second event is the development of immunologic competence for rejection of skin grafts or hematopoietic grafts. Fetal sheep develop the capacity to reject allogeneic skin grafts and demonstrate allogeneic or xenogeneic hematopoietic engraftment failure, which appears to be immunologically mediated, after 75 days gestation. In view of evidence that MSCs have the capacity to ablate allogeneic and xenogeneic response in vitro, we wondered whether they might be immunomodulatory in vivo.

Distribution of Engrafted Human MSCs in the Sheep Model

To assess distribution of engraftment, and to screen for human cell engraftment in tissues prior to immunohistochemical assessment, we utilized PCR for human-specific β_2-microglobulin DNA sequences. Multiple tissues were harvested at intervals of 2 weeks, 2 months, 5 months, or 13 months after in utero transplantation. Human cells were widely distributed in most tissues assessed by PCR at 2 weeks after transplantation, including liver, spleen, bone marrow, thymus, lung, brain, muscle, heart, and blood. After 2 months, human DNA was still detected in all tissues examined from fetuses transplanted at 65 days gestation, including cartilage, with the exception of brain. In fetuses transplanted at 85 days gestation, human DNA was detected after 2 months in the spleen, bone marrow, thymus, heart, and blood. Five months after in utero transplantation (3 months after birth), human DNA was detected in the bone marrow, thymus, spleen, lung, cartilage, and blood of fetuses transplanted at 65 days and in the heart, brain, skeletal muscle, and blood of fetuses transplanted at 85 days. In two animals transplanted at 85 days gestation 13 months earlier, human cells could still be detected by PCR in liver, bone marrow, heart, cartilage, and blood. The pattern of engraftment differed between animals, but in total, 28 of 29 injected animals demonstrated engraftment of human cells in one or more tissues.

From these results, it can be surmised that MSCs, despite their very large size, can be transplanted and are capable of engraftment in multiple tissues, even when transplanted into the fetal peritoneal cavity. This requires migration across endothelial barriers, integration into host tissue microenvironments, and survival with available growth and regulatory signals. Our findings of a variable pattern of long-term MSC engraftment, following detection of MSCs at 2 weeks in nearly all tissues studied, supports a model of nonselective hematogenous distribution, with subsequent selective long-term survival in specific tissues. This may be a function of the ability of specific microenvironments to support the engraftment and differentiation of MSC, or alternatively, the loss of engraftment from some tissues may be due to heterogeneity of the transplanted population with respect to differentiation potential or replicative capacity and longevity. A third possibility is that the xenogeneic microenvironment can support the viability and differentiation of human MSCs, but not their self-replication.

Does the distribution of MSCs observed in this model replicate the distribution of MSCs under normal circumstances? The answer will be unknown until the distribution of MSCs in human tissues is understood. If, in fact, MSCs are distributed to multiple tissues early in gestation to engraft and remain resident for repair of tissue injury, then in utero transplantation may recapitulate that ontogenic event. The fact that MSCs were found in most organs, including the bone marrow, is supportive of such a model. The results of our DNA PCR were confirmed by immuno-histochemistry using antihuman β_2-

microglobulin. Negative controls consisted of tissues of transplanted sheep that were PCR negative, or the same tissues from non-transplanted, age-matched controls. In all but a few tissues, immunohistochemistry confirmed the presence of human cells, and all of the negative control tissues were negative by immunohistochemistry. Many human MSCs were seen in pre- and postnatal hematopoietic and lymphopoietic tissues, including the fetal liver, bone marrow, spleen, and thymus. Multiple human MSCs could often be appreciated in a single high-power field in these tissues. Human cells were also identified in non-lymphohematopoietic sites, including the heart, skeletal muscle, cartilage, and lung. Five and thirteen months after transplantation, human cells continued to be present in multiple tissues, including the bone marrow, thymus, cartilage, heart, skeletal muscle, and brain.

Site-specific Differentiation of MSCs in the Sheep Model

The obvious next question was, What had the human MSCs become? Differentiation of human MSCs in various tissues was assessed by one of three techniques: (1) characteristic morphology with antihuman β_2-microglobulin staining; (2) immunohistochemical double staining for anti-human β_2-microglobulin and a second nonhuman-specific differentiation marker; or (3) when available, positive staining with human-specific differentiation markers proven not to cross-react with sheep cells. Using these techniques, site-specific differentiation was confirmed for human cardiomyocytes, chondrocytes, adipocytes, bone marrow stromal cells, thymic stromal cells, and skeletal myocytes. Although PCR results suggested that human cells were engrafted in the CNS, double-staining immunohistochemistry using antihuman β_2-microglobulin and anti-GFAP (glial fibrillary acid protein) showed that the human cells were not differentiated glial cells, but rather were perivascular cells surrounding blood vessels in the gyral sulci.

The results show that MSCs are capable of site-specific, multipotential differentiation and tissue integration following transplantation. Human MSCs have been shown in vitro to differentiate into adipocytic, chondrocytic, or osteocytic lineages. Less well characterized MSC populations from other species have been induced in vitro toward myocytic differentiation. Our study confirms in vivo chondrocytic differentiation and for the first time clearly demonstrates cardiomyocytic and myocytic differentiation of a defined human MSC population. MSCs derived from bone marrow from multiple species have been demonstrated to support hematopoiesis with equal or greater efficacy than stromal layers formed in long-term Dexter cultures. Our study upholds the role of MSCs in stromal support of hematopoiesis, both in the fetal liver and postnatal bone marrow. We found multiple large human cells intimately associated with clusters of hematopoiesis in the fetal liver at 2 and 9 weeks after transplantation. In addition, large cells that stained positively for human-specific CD23 were identified in the bone marrow at 9 and 22 weeks after transplantation. CD23 has been identified as a low-affinity IgE receptor as well as a functional CD21 ligand present on a variety of hematopoietic cells and bone marrow stromal cells. Our interpretation of CD23-positive cells in this study as "*stromal*" is based on the large size of the cells and the absence of human hematopoietic cells in either the donor cell population or the recipient bone marrow.

A relatively surprising finding was the presence of large thymic cells that stained positive for human-specific β_2-microglobulin and CD74. CD74 is a cell-surface MHC class II-associated invariant chain molecule that is expressed on B cells, Langerhans cells, dendritic cells, activated T cells, and thymic epithelium. The morphology of $CD74^+$ cells in this study appears similar to the ovine thymic epithelial cells in the surrounding thymus. The precursor of thymic dendritic cells is thought to be the HSC, whereas the origin of the thymic epithelial cell is controversial. Our data support a mesenchymal origin for the thymic epithelial cell as a "*stromal*" supporting cell in the thymus.

The persistence of human cells observed in this xenogeneic model, even when transplanted after the development of immunocompetence in the sheep fetus, is intriguing. Potential mechanisms for tolerance include failure of immune recognition, local immune suppression, or thymic deletional tolerance. Human MSCs are known to express class I HLA antigen but do not express class II, which may limit immune recognition. Although thymic stromal cells are known to participate in thymocyte positive and negative selection, and host thymic antigen-presenting cells are capable of facilitating clonal deletion of donor-reactive lymphocytes after in utero HSC transplantation, neither mechanism would account for tolerance after the appearance of mature lymphocytes in the peripheral circulation. In vitro, MSCs added to mixed lymphocyte cultures, however, have been shown to non-specifically ablate alloreactivity by an as-yet-unknown mechanism. We speculate that the persistence of MSCs in this model results from a combination of minimal immunogenicity and local immune suppression.

These results support the in vivo multipotentiality of human MSCs when transplanted into a developing receptive environment. The engraftment observed in multiple tissues with site-specific differentiation supports the concept that the fate of MSCs is determined by the environment in which they engraft rather than by an intrinsically programmed fate. Is this the full repertoire of MSC fate? Probably not. We observed a number of cells in a variety of tissues that we are currently unable to characterize. The most interesting of these are perivascular cells that persist as long as 13 months after transplantation in multiple tissues. These maintain fibroblast-like morphology and may represent the undifferentiated tissue resident MSC that has been postulated. We also have not observed CNS engraftment or glial differentiation of the MSCs used in this study. This could be a function of restriction of engraftment by the blood–brain barrier, or the cell population we are utilizing may not have neural potential.

Our results also support, but do not prove, the hypothesis that MSCs are distributed hematogenously during ontogeny to receptive tissues where they reside and ultimately differentiate. An intriguing observation relevant to this hypothesis is that circulating "*stromal*" elements can be isolated from human fetal blood up to but not beyond 15 weeks gestation. These cells on early characterization have many properties analogous to MSCs and may prove to be the early cells that "*seed*" tissues, including hematopoietic stromal environments with stromal and mesenchymal progenitors.

The current results provide little insight into the normal regulation of MSC fate in vivo. What regulatory signals cause MSCs to proliferate and differentiate? Do they participate in normal tissue repair? Can they "*rejuvenate*" to any degree organs or tissues as they age? Do they respond to tissue injury in a beneficial manner? Our hope in the future is to utilize the sheep system to explore stimuli that may induce MSC response and to investigate other populations of MSC, such as those derived from the early gestational human fetus, in our model to determine whether developmental age of MSCs affects their fate. An improved understanding of the regulation of MSCs in vivo will help determine strategies to manipulate MSCs for therapeutic benefit. It is clear that in order to utilize MSCs for tissue engineering, cellular, and gene therapy applications, we will need to understand how to deliver a large number of MSCs to specific sites and induce appropriate differentiation. This will require further insight into the normal and disease-induced regulation of MSC proliferation and differentiation, the transplant immunology of MSCs, as well as the development of strategies for diffuse or site-specific delivery for specific applications.

The fate mapping of MSCs as well as other stem cell types is currently rudimentary and is inherently limited by our current methodologic and analytical capabilities. Future advances in stem cell understanding will require better characterization of stem cell populations, further definition of the signaling pathways that regulate stem cell behavior and the interaction between intrinsic and extrinsic regulatory influences, and advances in our ability to analyze complex biologic systems. The development of in vivo surrogate systems that can accurately recapitulate the normal biologic events in the life of a stem cell will be essential for unraveling the intricacies of stem cell fate. The human–sheep model may be particularly useful for such studies, as it represents a relatively unperturbed developmental model that has thus far demonstrated long-term engraftment and multipotential differentiation of two human stem cell types.

11

EMBRYONIC STEM CELLS AND TRANSGENESIS

The use of mouse *embryonic stem* (ES) cells to alter the mouse genome has become a routine method to study gene function alongside classical transgenesis. Gene targeting by homologous recombination is the most widely used method of ES cell-mediated genome alterations. It allows the introduction of specific mutations of a gene of interest into the mouse germ line. ES cells can be used to mediate random insertional transgenesis as well, which is perhaps an alternative to the pronuclear DNA injection method. However, the combination of recently developed different site-specific recombinase systems combined with ES cell technology makes it possible to create more sophisticated conditional transgenic and conditional gene knock-out mouse models.

The site-specific recombinases catalyze the recombination between two consensus DNA sequences. If these sites are properly designed into a transgene or a gene-targeting vector, the site-specific recombination event can trigger the expression of a transgene. It can also create a loss-of-function allele of the targeted gene in a specific cell lineage and/or in a specific time of development, conditional to the expression of the recombinase. Cre recombinase of the bacteriophage P1 is the most widely used in mouse genetics. This prokaryotic enzyme was first shown to work in the mouse by Lakso et al.. Cre protein recombines DNA between two 34-bp long *loxP* recognition sites. If these sites are placed in the same DNA strand, in the same orientation, the recombination results in the excision of the intervening sequence, leaving a single *loxP* site behind.

The second most popular is Flp recombinase from yeast. Concerning its mechanism of action, it is similar to *Cre/loxP*, but so far appears to be less efficient. However, recently developed enhanced Flp (Flpe) might change this situation. The 34-bp consensus recombination site of Flp is called FRT.

The third potential enzyme is still in the earliest stage of its development for use in mammalian genome alterations, but certainly holds promise that it could become an additional recombinase system. This is an integrase from Streptomyces phage ϕC31, which was recently demonstrated to function in human cells. It carries out an efficient site-specific unidirectional recombination between attP and attB sites. It presently seems as if this integrase can be used to delete sequences flanked by attP and attB sites. It is also possible that this could become the system of choice for recombinase-mediated site-specific insertion into the genome.

After mediating an insertion into the genome, both Cre and Flp, create two functional *loxP* or FRT sites, respectively, that flank the inserted sequence. Obviously, these sites could be the target of a second recombination, which will excise the inserted sequence. In contrast, ϕC31-mediated recombination between the attP and attB sites does not recreate either site, and therefore, no further recombination that could remove the inserted sequence will occur.

In the ideal but not at all unrealistic picture for the near future, if two recombinase systems are working in addition to the *Cre/loxP* system, there will be more freedom for the Cre system to be used as a postintegration switch. There is already a large collection of transgenic mouse lines expressing Cre recombinase at a high level and specificity, and there are more to come. The database presenting a collection of Cre transgenic lines created through the effort of many researchers can be found at. Such transgenic lines expressing Cre recombinase driven by lineage–tissue-specific promoters can be derived by classical transgenesis as well as by the relatively new approach of gene-targeted knock-in of the Cre recombinase into an endogenous gene for a sometimes more reliable expression.

To be able to control the time of the switch, independently of the endogenous regulatory elements and lineage specificity, recombinase systems are combined with inducible gene expression systems. The recombinase gene can be placed under the control of an inducible promoter (ubiquitous or lineage-specific) or constructed as an inducible fusion protein.

Tamoxifen and RU-486 inducible systems use the nuclear localization capability of estrogen or a progesterone receptor ligand-binding domain in the presence of the ligand. The Cre recombinase is fused to a mutant ligand-binding domain, which has lost its ability to bind endogenous estrogen or progesterone, but still binds tamoxifen (an estrogen antagonist) or RU-486 (a synthetic steroid), respectively. In the presence of the synthetic ligand, the Cre fusion protein translocates into the nucleus and executes its function. So far, only partial tamoxifen-inducible Cre-mediated excision has been obtained.

The tetracycline-inducible gene expression system uses the DNA-binding domain of the bacterial tet-repressor protein and a strong transcriptional activator domain (VP16 from the herpes virus), which are fused together. Such a heterologous protein can bind to the tetracycline operator element and activate transcription depending on the presence of tetracycline. The combination of *Cre/loxP* and the doxycycline (Dox) system has proven promising. However, problems with mosaic expression, toxic effects of the trasnsactivator, and a high background recombination level have been encountered.

The fine tuning of that system has required quite a bit of time. Recently, however, ubiquitous Cre recombination was achieved upon per oral administration of antibiotic using a Dox inducible single-construct Cre transgene. The authors placed both the Cre recombinase and the *reverse tetracycline-dependent transactivator* (rtTA) under the control of the same bidirectional Dox responsive promoter. In this arrangement, the transcription is auto-inducible depending on the availability of Dox and minimal amounts of rtTA. Both rtTA toxicity and background Cre recombination were found minimal in the absence of Dox.

The efficiency and specificity of any recombinase system needs to be tested at the cellular level using a transgenic reporter mouse line (i.e., a transgenic line expressing a reporter gene in response to Cre-mediated recombination). For that, a single-copy transgene should have a ubiquitous promoter, followed by a *loxP*-flanked transcriptional STOP region, and then the coding region of a reporter gene. The reporter initiates expression under the control of the promoter only after Cre excision removes the STOP region. Such a conditional *lacZ* reporter construct was introduced into the ROSA26 gene-trap integration site. The other recently developed system (Z/AP) uses two reporters: cells express *lacZ* before Cre excision and heat-resistant human alkaline phosphatase after excision. This was achieved by random insertion of the transgene and single-copy integration by ES cell-mediated transgenesis.

ES cell-mediated as opposed to classical transgenesis has an advantage of a higher frequency of single-copy integration. It is an important issue, since multiple-copy integration can create more than two *loxP* sites and a potentially unpredictable outcome of the Cre excision and chromosome instability.

Conditional transgenesis is based on a similar strategy to that behind the Cre reporter lines: the promoter and the coding region for the gene of interest are separated by a *loxP*-flanked STOP region that does not allow any transcription or translation of the gene of interest. The gene is expressed when this region is removed by Cre-mediated excision. A ubiquitously expressed conditional transgenic line with single-copy integration is produced using ES cells. The transgene is activated when crossed with various Cre transgenic lines of different lineage specificity. Cre-mediated recombination, and therefore transgene activation, can also be inducible as discussed above.

Here, we describe our example applied in the design of the Z/AP reporter, which can be followed for any conditional transgenesis. The βgeo-*lacZ*/neo^R fusion coding sequence is placed upstream of a triple repeat of the simian virus 40 (SV40) polyadenylation signal (3xpA). The βgeo/3xpA is inserted between two *loxP* sites that are placed in a pCAGG vector containing the *cytomegalovirus* (CMV) enhancer/chicken β-actin promoter in front of it. This vector is referred as pCALL. The coding sequence of the transgene can be inserted downstream of *loxP*-flanked βgeo/3xpA to produce an expression vector of interest.

There are several variations of the construct for random integration into ES cells to generate a conditional transgenic mouse line. Alterations of the Z/AP construct can be customized for specific needs. Different positive selectable markers, such as hygromycin B and puromycin, allowing the survival of cells integrating the marker into the genome, can be used instead of neomycin. Among the possible reporter systems, bacterial β-galactosidase (β-gal) and human alkaline phosphatase allow the detection at a single cell level in histological sections. *Green fluorescent protein* (GFP) cloned from jellyfish, *Aequorea victoria* as well as its variants, yellow and cyan fluorescent proteins, permit the detection of gene expression and protein localization in living cells.

Perhaps the same principles are used for targeted insertion of conditional transgenes into an endogenous gene, but it requires significant target vector building and screening for targeted events. One of the strategies for the creation of a conditional gene knockout is to place two *loxP* sites around a functionally essential part of the gene of interest, using gene targeting in ES cells. Such minimal modification should leave the gene functional until lineage specific and/or inducible Cre recombinase is applied. This strategy requires a sophisticated gene targeting design, involving the proper insertion of *loxP* sites around the functional part of the gene to create the null allele. The selectable marker needs to be flanked with target sites for its later removal, preferably by a different recombinase system (Flp/FRT) to avoid multiple *loxP* sites.

Here, we give a brief description of the steps involved in the creation of conditional transgenic animals.

MATERIALS

ES Cell Tissue Culture

Equipment

1. Tissue culture facility, preferably used only for ES cells, including a laminar flow cabinet, humidified incubator (37°C, 5% CO_2), inverted phase-contrast microscope with 4×,10×, 20–25× objectives, stereomicroscope (dissecting) with transmitted light base, table-top centrifuge, water bath (optional), –70°C freezer, liquid nitrogen tanks. Microscope equipped with fluorescence, appropriate filters, and camera is necessary if fluorescent proteins are to be used as reporters.
2. Sterile disposable cell culture plasticware (100-, 60-, 35-mm dishes, 6-well, 24-well, 4-well, flat and V-bottom 96-well plates, centrifuge tubes, cryovials). Various sterile disposable or reusable detergent-free glass pipets.
3. Electroporation apparatus, and capacitance extender. 4-cm electrode gap electroporation cuvettes.
4. Multichannel pipettor with volume adjustable up to 200 μL, sterile disposable reagent reservoirs for multichannel pipettor, and sterile pipet tips.
5. Multichannel aspirator system (optional).
6. Isopropanol freezing container (optional) and/or styrofoam box with lid.

ES cells and feeders

Several ES cell lines allowing efficient germline transmission have been developed. To maintain pluripotency, ES cells are cultured on feeder cells: primary *embryonic fibroblasts* (Emfi) or STO fibroblast cell line; or on gelatinized plates in the presence of *leukemia inhibitory factor* (LIF). Growth media and culture conditions should be used as suggested for each ES cell line. Here we describe the protocols established and currently used for R1 ES cells. Mitotically inactivated Emfiare used as feeders only for a long-term culture of R1 ES cells (typically before and after cryopreservation). Otherwise, culture of R1 ES cells for electroporation and during the selection is done on gelatinized plates. Emfi cells can be made from any strain of mice including transgenic mice that express bacterial neomycin or hygromycin genes depending on the choice of selectable markers used for altering the ES cell genome. Neo^R and $Hygro^R$ mice are available from Jackson Laboratories. Detailed protocols for preparation of Emfi stocks and mitomycin C-treated feeders for ES cell culture are presented in many publications.

ES cell culture media and reagents

It is recommended to use only tissue culture grade or, preferably, ES cell-qualified reagents whenever available. The water quality is a critical factor for optimal culture. It should be from a regularly maintained Milli-Q (Millipore) filtration system, preferably pretreated by deionization. Commercially available ultrapure water can be used as an alternative. If large quantities of media and solutions are necessary, it can be prepared from powder and filter-sterilized. Otherwise, commercially available ES cell-qualified solutions are recommended.

1. *Complete ES Medium (ES-DMEM):*
   (a) *Dulbecco's modified Eagle's medium* (DMEM) with 4.5 g/L D-glucose (high glucose), buffered with 2.2 g/L sodium bicarbonate. Store in the dark at 4°C.

   Prior to use, DMEM should be supplemented with the components listed below. If the complete media is stored for longer than 2 wk, it should be supplemented with additional 2 m*M* L-glutamine from 100X stock, as L-glutamine is unstable. GlutaMAX media from Gibco contains L-glutamine in a stabilized form of the dipeptide and can be used without extra supplements of L-glutamine.

   (b) 15% ES cell-qualified *fetal bovine serum* (FBS). Store main stock in the dark at –20°C. Heat-inactivate at 56°C for 30 min (optional).
   (c) 0.1 m*M* Nonessential amino acids stored at 4°C.
   (d) 1 m*M* Sodium pyruvate stored at 4°C.

(e) 0.1 m*M* β-mercaptoethanol stored as aliquots at –20°C.

(f) 2 m*M* L-Glutamine stored as aliquots at –20°C.

(g) 50 U/mL Penicillin and 50 μg/mL streptomycin stored as aliquots at –20°C. Alternatively, 100X Pen–Strep–L-Glu combo available from Gibco.

(h) LIF, stored as aliquots at –20°C.

2. *Feeder medium:* DMEM, 1X Pen–Strep, 10% FBS.
3. *ES cell freezing medium:* freezing medium should be prepared fresh prior to use and kept on ice. We commonly use 22% FBS as a final concentration. It is possible to increase the concentration of FBS in a freezing media up to 40% (e.g., 2X ES cell-culture freezing medium from Specialty Media) for better recovery of small amount of cells in 96-well plates. 2X: 60% ES-DMEM, 20% FBS, 20% dimethyl sulfoxide (DMSO). 1X: 80% ES-DMEM, 10% FBS, 10% DMSO.
4. *0.1% Gelatin:* 0.1% solution in water, autoclaved, and stored at 4°C. Alternatively ES cell-qualified 0.1% gelatin.
5. *Trypsin-EDTA:* 0.05% trypsin, 0.53 m*M* EDTA. 0.25% Trypsin, EDTA (1 m*M*). Store at 4°C (main stock at –20°C).
6. *Phosphate-buffered saline (PBS)* without Ca^{2+} and Mg^{2+}.
7. *Selection reagents:* Geneticin (G418), puromycin, hygromycin B. Working concentration for selection agents must be determined by killing curves. We routinely use 150–200 μg/mL G418 for R1 cells.
8. *Reagents for the lipofection:* LipofectAMINE; Opti-MEM I reduced serum medium.

Isolation of DNA from 96-Well ES Cell Colonies

1. Lysis buffer: 10 m*M* Tris-HCl, pH 7.5, 10 m*M* EDTA, 10 m*M* NaCl, 0.5% sarcosyl, 1 mg/mL proteinase K added before use.
2. NaCl-ethanol mixture: 150 μL of 5 *M* NaCl per 10 mL of cold 100% ethanol (prepared fresh).
3. Restriction digestion mixture (per well): 1X appropriate restriction buffer, 1 m*M* spermidine, 100 μg/mL *bovine serum albumin* (BSA), 50–100 μg/mL RNase A, 10–20 U of enzyme. Use 35–40 μL per sample.

β-Gal Staining of ES Cells and Postimplantation Stage Embryos

1. PBS without Ca^{2+} and Mg^{2+}.
2. Stock solutions: 10% Nonidet P-40 (NP40); 1% Na deoxycholate; 0.5 *M* EGTA; 1 *M* $MgCl_2$; 0.5 *M* K_3 $[Fe(CN)_6]$, 0.5 *M* K_4

[Fe(CN)$_6$] (kept light protected at room temperature); 25–50 mg/mL 5-bromo-4-chloro-3-indolyl-β-O-galactopyranoside (X-gal) in DMSO or *dimethyl formamide* (DMF) kept light-protected at –20°C.

3. Fix solution: 0.2% glutaraldehyde in PBS (for cells); 0.2% glutaraldehyde in PBS, containing 5 m*M* EGTA and 2 m*M* $MgCl_2$ (for embryos). Glutaraldehyde is added fresh prior to use. 2% Glutaraldehyde, 2 or 4% *paraformaldehyde* (PFA) are also used for fixation of embryos older than E12.5.
4. Wash solution: PBS for cells; PBS, containing 0.01% Na-deoxicholate, 0.02% NP40, 5 m*M* EGTA, 2 m*M* $MgCl_2$ for embryos.
5. Stain solution: 1 mg/mL X-gal, 5 m*M* K_3 [Fe(CN)$_6$] , 5 m*M* K_4 [Fe(CN)$_6$] in wash solution (prepared fresh prior to use). Stain solution can be used several times if filtered after use and stored at –20°C.

Ready to use reagents for β-gal expression are available from Specialty Media.

Preimplantation Embryo Culture

Equipment

1. Stereomicroscope(s) (dissecting) with both transmitted and reflected lights or fiber optics light source with gooseneck. The use of two microscopes is convenient for 2-cell stage embryo fusion and embryo transfer into pseudopregnant females, however, one microscope is also enough. We find that the frosted glass, instead of the more common, transparent glass in the base of the microscope, gives better view of the zonae pellucidae of preimplantation stage embryos necessary for zona removal.
2. Humidified incubator (37°C, 5% CO_2).
3. Sterile Petri dishes of different size (60, 100 mm), organ culture dishes. We find the plastic of 35-mm Easy Grip Falcon tissue culture plates to be best suited for making depressions for aggregations.
4. Sterile 1-mL syringes, 26-gauge 1/2 inch long needles (26G1/2), 30G1/2 needles. To make a flushing needle, the sharp tip of 30G1/2 needle is first cut off and then polished on a sharpening stone or sand paper. The flushing needle is flushed with 70% ethanol before and after use.
5. Bunsen or alcohol burner.

6. Pasteur pipets that are drawn by hand over a flame and broken to produce pipets for embryo manipulation, with a diameter of a capillary slightly larger than an embryo. It is important to flame-polish the tip of the pipet as the embryos without zonae are easily damaged. Embryo manipulating pipets are connected through a latex tubing to an aspirator mouthpiece. Such a mouthcontrolled pipet is used for all embryo manipulations and embryo transfer. Alternatively, a finger-controlled pipet (small piece of tubing closed at one end or small bulb connected to the drawn capillary) is used for embryo manipulations.
7. Surgical instruments (e.g., *Fine Scientific Tools* [FST]): sharp fine-pointed scissors, fine forceps, straight or curved blunt forceps with serrated tips, forceps with 1 × 2 teeth, serrefine; wound clips and Autoclip applier.
8. Cell-fusion instrument CF-150B, aggregation needle DN-09.

Mouse stock

C57 Bl/6 mice are the most common strain used as donors of host embryos for chimera production by blastocyst injection of different ES cell lines. In our facility, random-bred ICR (CD-1) mice have been successfully used for many years as both donors of host embryos and recipients of manipulated embryos. The mouse colony necessary for the creation of chimeras should contain a stock of females, stud males, and vasectomized males. The details of maintenance of such a colony as well as all procedures involved in the production of superovulated and pseudopregnant animals are described in multiple publications.

Preimplantation embryo culture media and reagents

Since embryos are cultured for 24 or even 48 h in aggregation experiments, the quality of culture conditions is more critical than in blastocyst injection. Embryo culture media is commercially available. However, as culture media can not be used for longer than 2, or a maximum of 3 wk, it is often necessary to prepare media from scratch or from stocks, as described previously. We keep all our concentrated stocks at –70°C for a few months. BSA is kept desiccated at 4°C.

1. M2 is a HEPES-buffered media that is used during embryo collection and other manipulations in room atmosphere (fusion, zonae removal). Filtered aliquots are stored at 4°C and brought up to room temperature prior to use. Embryos should not be kept in M2 for prolonged periods of time and are rinsed well with a few drops of equilibrated KSOM media before being placed into the incubator.

2. KSOM is a bicarbonate-buffered media used for embryo culture and was developed through a simplex optimization procedure by Lawitts and Biggers. It may also be supplemented with both essential and nonessential amino acids, which were shown to improve the development in vitro. Filtered aliquots are stored at 4°C and equilibrated by placing the tube, a 1-mL syringe, or the culture dish containing KSOM in the incubator well in advance before use.
3. Embryo-tested light mineral oil.
4. Acidic tyrode solution for removing zonae. Main stock aliquots are stored at –20°C. An aliquot is thawed and kept at 4°C when needed, and it is brought to room temperature prior to use.
5. 0.3 mol/L Mannitol in ultrapure water containing 0.3% BSA for embryo fusion. Aliquots are stored at –20°C. A freshly thawed aliquot is used for fusion, and the unused portion is discarded.

METHODS

ES Cell Culture

Passage of ES cells

1. Prepare the necessary number of gelatinized plates by coating them with 0.1% gelatin: rinse the plate with 0.1% gelatin solution covering the surface (5 mL/100-mm plate), leave for a few minutes, aspirate, and allow to dry for a few minutes, add ES-DMEM, and place in the incubator. Alternatively, replace the media on the appropriate number of prepared feeder plates to ES-DMEM.
2. Aspirate the growth medium, rinse twice with PBS, add trypsin (1.5–2 mL/100-mm, 1–1.5 mL/60-mm, 0.5 mL/35-mm dish, 0.25–0.3 mL/well for 4- or 24-well plates), place in the incubator for 5 min.
3. Swirl the plate to detach clumps from the bottom of the plate, pipet the cells gently to break the clumps (optional). Add an equal volume of ES-DMEM to neutralize trypsin, pipet up and down several times, transfer the suspension into a 12-mL tube. Pellet the cells by low-speed centrifugation (270*g*) for 5 min at room temperature.
4. Aspirate the supernatant, add 1 drop of PBS or ES-DMEM to the pellet, flick the tube to resuspend the cells before adding ES-DMEM (optional).
5. Add 5–7 mL of ES-DMEM to the tube, pipet gently to mix well, and split the contents at a 1:5 or 1:7 ratio into the new plates

containing a sufficient volume of medium (5 mL/60-mm, 10 mL/100-mm plate). About 1 × 10^6 cells onto 60-mm plate, 2 × 10^6 cells onto 100-mm plate.

6. Change the medium the next day and split every second day as described.

Freezing and thawing of ES cells in cryovials

Usually ES cells are frozen at about 5 × 10^6 cells/mL of 1X freezing media (approx 4 vials from 100-mm dish).

1. Change growth media 2 to 3 h before freezing the cells (optional).
2. Freshly prepare 2X or 1X freezing media.
3. Harvest the cells in a 12-mL tube containing ES-DMEM. Pellet the cells at 270*g* for 5 min at room temperature.
4. Remove the supernatant, resuspend the cells gently in half of the final volume required using ES-DMEM, gradually add an equal volume of 2X freezing medium while shaking the tube, and mix by pipetting up and down several times. Alternatively, gently resuspend the pellet in 1X freezing media.
5. Quickly aliquot 1 mL of the cell suspension into labeled cryovials and immediately place them in a precooled styrofoam box that will allow them to cool down gradually. Alternatively, isopropanol containers purchased from a number of manufacturers can be used.
6. Immediately place the container in a –70°C freezer for 1 to 2 d, then transfer cryovials into a liquid nitrogen tank for long-term storage.

A vial, frozen in such a way, can be thawed onto a 60-mm plate. Freezing and thawing are usually counted as one passage.

1. Thaw the vial by quickly warming it at 37°C.
2. When ice crystals almost disappear, aseptically transfer the cell suspension into a 12-mL tube using the pipet filled with ES-DMEM to slowly dilute DMSO.
3. Pellet the cells at 270*g* for 5 min and aspirate the supernatant.
4. Resuspend the pellet in fresh ES-DMEM, plate on a gelatinized or feeder plate, gently swirl the plate bidirectionally to evenly distribute the cells, and place in the incubator.
5. The next day, remove floating dead cells and change the media. If the correct procedure was used, cells should be ready for passage in 2 to 3 d.

Introduction of DNA into ES Cells

Electroporation is the most common way of in vitro introduction for both stable integration and transient expression in ES cells. Recently, we have successfully used lipofection for transient transfection of Cre recombinase into single-copy integrants. It is important to test the excision of the STOP region in vitro before in vivo experiments. In our example, ES cells become neo-sensitive and lose *lacZ* activity after Cre excision.

Electroporation of ES cells

Cells are routinely passaged 2 d prior to electroporation. Usually one 10-cm plate at approx 80% confluency (15–20 × 10^6 cells/mL) will provide enough cells for 1 to 2 electroporations. We regularly electroporate 20–40 μg of DNA into cells at the density of 7 × 10^6 cells/mL. Vector DNA is linearized by restriction enzyme digestion, extracted twice with phenol:chloroform (optional), ethanol-precipitated, washed twice in 0.5–1 mL of 70% ethanol, and resuspended in sterile 0.1 × Tris-EDTA (TE), PBS, or water at a concentration of 1 μg/μL. For transient expression of Cre recombinase, circular plasmid is used for electroporation. Electroporated cells are then plated very sparsely (approx 1000 cells/100-mm dish).

1. Change medium on ES cells at least 2 h prior to electroporation.
2. Switch on the electroporation apparatus and set up conditions in advance. We routinely use 250 *V*, 500 μF for the Bio-Rad GenePulser and capacitance extender.
3. Harvest cells as described above. It is critical to get a single-cell suspension. Pool the cells from all dishes into one tube.
4. Resuspend the pellet in a minimal volume of ice-cold PBS (about 1 mL/100-mm plate). Determine the cell density with a hemocytometer and dilute with a volume of PBS that will give the required density of a cell suspension. Keep the suspension on ice. Recently, we found that ES cell electroporation buffer available from Specialty Media, used instead of PBS in this step, gives significantly better recovery of electroporated ES cells.
5. Gently mix 0.8 mL of the ES cell suspension and 20–40 μg of DNA in a precooled electroporation cuvette.
6. Electroporate the cells, then place the cuvettes on ice for 20 min (optional).
7. Prepare the appropriate number of gelatinized plates. The number of plates will depend on the cells' survival, specific vector,

selection approach, and desired density of colonies. We routinely plate cells from one cuvette onto 1.5–2 × 100-mm plates.

8. Transfer electroporated cells from the cuvette into the tube with the appropriate volume of ES-DMEM (15–20 mL/cuvette). Cells from several cuvettes can be pooled into one tube and gently mixed by pipetting into a uniform suspension. Transfer cell suspension onto gelatinized plates, swirl bidirectionally to evenly distribute cells across the surface. Incubate overnight and change the medium the next day.
9. Start drug selection 24–48 h after electroporation. Change the selection media every day for the first few days, then every other day. Continue the selection until colonies become apparent and ready to pick i.e., visible to the naked eye, it usually takes 6–10 d.

Lipofection for transient expression of transgenes

1. Plate ES cells to a 35-mm or 6-well gelatinized dish (2 × 10^5 cells/dish/well) using ES-DMEM.
2. Incubate overnight, the cells should reach 30–40% confluency.
3. Prepare the following solutions in 5-mL polystyrene tubes. It is possible to use serum-free or reduced serum ES-DMEM instead of OPTI-MEM. Antibacterial agents should not be present in either media during the transfection.

   Solution A: For each well of a 6-well plate or 35-mm dish, mix 1 μg of circular plasmid containing an expression vector DNA with 300 μL of OPTI-MEM reduced serum medium.

   Solution B: For each well of a 6-well plate or 35-mm dish, mix 8 μL of Lipofectamine reagent (LF) with 300 μL of OPTI-MEM.
4. Combine necessary volumes of solutions A and B, mix gently, and incubate at room temperature for 20 min.
5. Place the solution (500 μL/well/dish) onto ES cells rinsed with OPTI-MEM. Incubate for 5 h at 37°C.
6. Replace the media with drug-free ES-DMEM or proceed with selection.

Growing Drug-Resistant ES Cell Clones

Picking ES cell clones into 96-well plates

Well-separated colonies of similar size with a defined perimeter and a compact center with undistinguishable individual cells should be picked. Large colonies and those with large distinguishable cells are differentiating and should be avoided if possible. Colonies can either be picked with the naked eye or by using a dissecting microscope with transmitted light. They can also be circled on the bottom of the plate with a marker for easier visualization.

1. Using the multichannel pipettor, prepare the appropriate number of gelatinized flat-bottom 96-well plates. Aliquot 30–50 μL of trypsin into the wells of a V-bottom 96-well plate.
2. Aspirate growth media from the plate with ES colonies, rinse it twice with PBS, add 6–8 mL of PBS to completely cover the dish
3. Using a drawn Pasteur pipet or a Gilson P20 or P200 set at 15 μL, carefully dislodge the colony from the dish and pull it into the pipet tip with as little volume of PBS as possible (usually 3–5 μL).
4. Transfer each individual colony in a minimal volume of PBS into one of the wells of a V-bottom 96-well plate containing trypsin.
5. Using a new tip for each colony, proceed with the rest. This process should not take longer than 30–60 min, 48 or 96 colonies should be picked at a time, depending on the picking speed.
6. Place a 96-well plate in a 37°C incubator for 10 min.
7. Working row by row with a multichannel pipettor, add 50–70 μL of selective ES-DMEM to each well with trypsin (to 100 μL/well). Gently pipet up and down several times to disaggregate the cells. Transfer the suspension to the equivalent row of gelatinized plates. Alternatively, add ES-DMEM to all the wells to neutralize the trypsin first, and then proceed with pipetting and transfer.
8. Using the multichannel pipettor, wash each well of the V-bottom plate with another 100 μL of medium, and transfer the volume to the equivalent row in a flat-bottom 96-well gelatinized plate (to total vol of 200 μL/well). Place the plate into the 37°C incubator.
9. Change the media daily until the cells are ready for passage (80% confluency).

Passage of ES cells in 96-well plates

Optimally, 3 or 4 d after picking colonies into the 96-well plates, the cells reach the density required for passage. Replica plates are used for the preparation of DNA for screening, X-gal staining, and for creating frozen stocks. It is generally recommended to have two confluent replica plates for DNA preparation and one or two replica plates to keep as a frozen stock. It is also important to identify strong overall expresser clones using the reporter system of choice (X-gal staining in our example). The wells equivalent to such identified strong expresser clones are used for later isolation of DNA for Southern blotting to test single site–copy integration. Confirmed single-copy integrants are thawed and expanded for further in vitro and in vivo analysis.

1. Prepare the required number of gelatinized flat 96-well plates. Add 150 μL of ES-DMEM per well and place in a 37°C incubator until necessary.
2. Aspirate the medium from the plate to be split and wash twice with 200 μL of PBS using the multichannel pipettor.
3. Add 50 μL of trypsin per well. Incubate at 37°C for 10 min. The cells should detach with gentle tapping on the plate.
4. Add 50 μL of medium into each of the wells to stop trypsinization. Pipet up and down at least five times to mix well. Working row by row, transfer the cell suspension into 2 or 3 new gelatinized plates containing ES-DMEM. Place in the 37°C incubator and change the media the next day. Alternatively, one half of such a cell suspension can be frozen right away in one new V-bottom or flat 96-well plate containing 2X freezing media. At least one more passage is required to create two replica plates for DNA preparation and one for X-gal staining in addition to one or two frozen plates.

Freezing and thawing of ES cells in 96-well plates

1. Freshly prepare 2X cell freezing media and keep it on ice.
2. Trypsinize the cells of an 80% confluent 96-well plate as described above. The final volume of the cell suspension is 100 μL/well.
3. Working quickly on ice, aliquot 100 μL of 2X freezing media into each well. Pipet the cells up and down several times to get a homogeneous suspension. Alternatively, transfer the cell suspension into the new V-bottom 96-well plate, containing cold 2X freezing media.
4. Add 50 μL of cold sterile mineral oil to each well on top of the cell suspension.
5. Wrap the plates in parafilm and foil (the latter is optional). Place in a precooled styrofoam box, and store in a –70°C freezer, preferably not longer than 2 mo, until ready for thawing and expansion.

The first frozen plates are considered as master plates. The unfrozen replica plates are used for further characterization of the clones. After strong expresser and single-copy integrant clones have been identified, the frozen stock can be thawed and expanded for further analysis as described below.

1. Prepare the necessary number of 4- or 24-well feeder plates containing ES-DMEM.
2. Remove the plate containing the identified clones from the freezer. Unwrap the plate and place in the incubator to thaw.
3. When ice crystals almost disappear, wipe the outside of the plate with 70% ethanol.
4. Transfer the content of the well into the wells of prepared feeders plate.
5. Rinse the original wells of the 96-well plate with more ES-DMEM and transfer to the same wells. Change the media after overnight culture and daily.
6. Passage the cells when they reach 70–80% confluency to a larger plate (i.e., 35 mm). If cells do not reach confluency in a few days, but form few colonies in a well, they can be trypsinized, broken into smaller cell clumps, and plated on the same plate.
7. Passage every other day, freeze the cells in vials as described above. Use them for in vitro tests (i.e., transient expression of Cre recombinase by electroporation or lipofection).

β-Gal Staining of ES Cells and Postimplantation Stage Embryos

lacZ gene expression is detected by the enzymatic activity of the gene product β-gal in both embryos and cells. Generally, the staining becomes visible after a 15-min or an overnight incubation, depending on the level of expression.

1. Wash cells or embryos twice in PBS.
2. Add freshly prepared fixative solution to completely cover the cells or embryos, incubate at room temperature for 5–7 min (for cells), 5 min, and 15–20 min.
3. Wash three times with PBS (cells) or wash buffer (embryos).
4. Replace the wash buffer with X-gal stain to completely cover the cells or embryos and incubate at 37°C light protected with optional shaking.
5. Replace the X-gal stain with PBS (cells) or wash buffer (embryos) and store at 4°C. For longer storage, embryos can be refixed in a fresh 4% solution of PFA or formaldehyde.

DNA Isolation from ES Clones in 96-Well Plates

This procedure was established by Ramirez-Solis et al. and allows cell lysis, DNA precipitation, and the restriction digestion in the original

96-well plate in which ES cells were growing. Usually two replica plates are used for DNA isolation, then one is processed for Southern blot analysis, leaving the second plate as a back up. The cells should be lysed, and genomic DNA isolated from them when the majority of clones are confluent. The yellow color of the media within 24 h of its change indicates cell confluency.

1. Aspirate media from each well, wash twice with PBS.
2. Add 50 μL of lysis buffer to each well.
3. Incubate the plates overnight at 55°C in a humid atmosphere (wrap the plates sealed with the parafilm with wet paper towels and place in a plastic container or sealed plastic bag).
4. The next day, add 100 μL of cold NaCl-ethanol mixture to each well.
5. Leave the plate undisturbed at room temperature for 30–60 min (or more if necessary) until the precipitated DNA attached to the dish is visible against a dark background.
6. Gently invert the plate on the paper towel to drain the liquid.
7. Rinse 3 times with 150–200 μL of 70% ethanol per well, inverting the plate each time. After this step, DNA can be stored in 70% ethanol at -20°C.
8. Invert the plate after the final wash and allow to air-dry for 10–15 min (it is important for all the ethanol to dry out).
9. Add 34–40 μL of restriction digest mixture per well, mix, seal the plate, and incubate overnight at 37°C in a humid atmosphere.
10. Proceed with Southern blot analysis.

Introduction of ES Cells into Mice

As our ES cell line of choice is R1, we use aggregation of ES cells with cleavage stage host ICR embryos to create the chimeras, although blastocyst injections may be used as well. Concerning the conditional transgenic lines, a faster way to test the in vivo expression of transgenes introduced into several candidate ES cell clones is by analyzing chimeras produced by aggregation with tetraploid or diploid embryos. Clones that give the expected results are then chosen for germline transmission. After successful germline transmission, conditional transgenic animals can be crossed with switch transgenic lines, where Cre recombinase is driven by lineage-specific promoters.

Preparation of ES cells for aggregation

It is important to maintain optimal culture conditions for all ES cell cultures, but particularly for ES cell clones to be introduced into mice. At least one passage on a gelatinized plate is required before aggregation. Sparser than usual, passage 1 or 2 d before aggregation produces the colonies of 8–15 cells that are lifted by gentle trypsinization.

1. Three or four days prior to aggregation, thaw a vial of ES cells or passage as described above. Change the media the next day.
2. One or two days prior to aggregation, trypsinize ES cells as described above and ensure a single-cell suspension. Twenty-four hours growth is enough for most clones, but 48 h are necessary for slower growing clones.
3. Resuspend the pellet in ES-DMEM. Leave the tube undisturbed for a few minutes to allow for large clumps and feeders to settle.
4. Seed the cell suspension from the top portion onto a few gelatinized plates using different dilutions (e.g., 1:10 to 1:50).
5. On the day of aggregation, after preparation of the embryos, wash the cells first with PBS, then with trypsin (optional).
6. Add minimal amount of trypsin to just cover cells (0.5 mL/60-mm plate) and place in the incubator for 1 to 2 min or leave at room temperature until colonies start to detach from the plate.
7. Watch under the microscope for colonies to detach from the plate, gently swirl the plate, or tap at the microscope stage. Do not over-trypsinize, as cells will become sticky and hard to manipulate. Add the required volume of ES-DMEM to the plate. Do not pipet. ES cells are now ready for aggregation during the next hour or two. Keep the plate at room temperature, as cells will start attaching to the plate in the incubator.

Recovery of 8- and 2-cell stage embryos

It is preferable to use only noncompact 8-cell stage embryos for aggregation. However, we routinely use all embryos with intact blastomeres, from the 8-cell to morula stage, collected at 2.5 d postcoitum (dpc). Two-cell stage embryos, collected at 1.5 dpc, are used for production of tetraploid embryos. The recovery procedure for 2-cell stage embryos is essentially the same.

1. Prepare the culture plates using KSOM media (organ culture or microdrops overlayed with mineral oil). Place the tube or the syringe with KSOM into the incubator.
2. Dissect the oviducts from 2.5 dpc pregnant females, leaving the upper part of the uterus attached, and place in the drop of M2.
3. Transfer one oviduct into a small drop of M2.

4. Using the dissecting microscope, insert the flushing needle attached to a 1-mL syringe filled with M2 into the infundibulum. The use of fine forceps helps to place the needle in the right position. Flush M2 media through the oviduct and observe its swelling.
5. Proceed with the remaining oviducts, keeping the time of manipulations in M2 minimal.
6. Collect embryos into a fresh M2 drop using embryo-manipulating mouth- or finger-controlled drawn Pasteur pipet. Wash them through several M2 drops to get rid of debris.
7. Wash embryos through few drops of equilibrated KSOM and place them in the culture dish.

Aggregation of ES cells with cleavage stage embryos

Preparation of the aggregation plate

1. Using a 1-mL syringe filled with KSOM media, place microdrops (approx 3 mm in diameter) on a 35-mm dish. We usually place 2 rows of 4 to 5 drops in the middle of the plate and 2 more rows of 3 drops on each side. Cover with mineral oil.
2. Wipe the aggregation needle with 70% ethanol. Press the needle into the plastic, make a slight circular movement. We usually make 6 depressions per KSOM drop. Three or 6 drops on the sides are left without depressions. They are used for the final selection of ES cell clumps. The plate holds 40–60 aggregates.
3. The plate is placed into the incubator to allow good equilibration of the media.

Zona removal

In order to allow the attachment of ES cells to the blastomeres of the embryos as an obvious precondition of aggregation, the zonae pellucidae of the embryos need to be removed. We use acidic Tyrode solution to dissolve this glycoprotein membrane. Since the acid diluted with buffered solution will not work as efficiently and the acid transferred into culture media will damage the embryos, it is important to transfer a minimal amount of solutions between drops and use multiple washes. The number of embryos manipulated at a time depends on the speed of manipulations.

1. Place M2 and acid Tyrode's drops into a 100-mm Petri dish. The temperature of the acid tyrode should not exceed room temperature.
2. Pick 20–50 embryos with a minimal volume of media and wash them through one acid drop. Transfer embryos to a fresh drop of acid. Move them by pipetting and observe zonae dissolution.

3. As soon as the zonae dissolve, immediately transfer the embryos with a minimal volume of acid to a M2 drop. Rinse through several M2 drops. Do not allow the embryos to touch each other. Proceed with the remaining embryos.
4. Wash the embryos through several drops of equilibrated KSOM and place them into the aggregation plates.

Assembly of aggregates

We aggregate an ES cell clump of 8–15 cells with one diploid embryo. Two tetraploid embryos are also usually aggregated with such a clump of ES cells. It is possible to use one tetraploid embryo for aggregation, but it might be necessary to culture aggregates for one more night to ensure blastocyst formation. Aggregates are assembled in either of two ways: (1) the embryo is first placed into the depression, and the clump of ES cells is placed next to it, or (2) the clumps of ES cells are first distributed into depressions, and then the embryos are placed next to the clumps. Both ways work equally well.

1. After zona removal, embryos are placed into the aggregation plates inside the depressions or beside them, making sure they are not touching each other. The plates are kept in the incubator until the ES cells are ready.
2. Prepare ES cells for aggregation as described above.
3. Under the dissecting microscope, choose a number of clumps of required size and transfer them into the microdrops, not containing depressions, for final selection.
4. Select a few clumps of 8–15 cells and carefully transfer them individually into the depressions, next to the embryos. Alternatively, distribute clumps into all depressions in a plate, then place the embryos next to each clump.
5. Assemble all the aggregates in this manner, check the plate to ensure that all ES cell clumps are in contact with embryos, and incubate overnight. The next day, the majority of aggregates will be blastocysts or late morulae. They can be transferred into 2.5 dpc pseudopregnant females as described previously.

Generation of tetraploid embryos

The electrofusion of blastomeres of 2-cell stage embryos for the generation of tetraploid embryos was first developed by Kubiak and Tarkowski. Most tetraploid embryos die shortly after implantation, but when aggregated with diploid embryos, they can contribute to the primitive endoderm and trophoectoderm lineages and are excluded from the primitive ectoderm lineage. When the tetraploid embryos are aggregated with ES cells, the resulting fetuses are ES cell derived, as ES cells do not contribute to trophectoderm and primitive endoderm. Thus, ES cell and tetraploid components complement each other. Therefore, the method of ES cell ↔ tetraploid embryo aggregation can be used, for example, as a rapid test for the developmental potential as well as for the transgene expression of an ES cell clone. The application of ES cell ↔ tertaploid embryo chimeras is, however, much broader.

1. Prepare the culture plate (microdrops of KSOM overlayed with mineral oil). Thaw a frozen aliquot of mannitol.
2. Collect 2-cell stage embryos as described above.
3. Turn on the cell-fusion instrument and set up the parameters. The fusion of blastomeres of 2-cell stage embryos occurs when a DC electric pulse is applied perpendicular to the plane of the blastomeres' contact. We apply one or two pulses of 30 V and 40 μs for the fusion in nonelectrolyte solution (mannitol) using the CF-150B cell-fusion instrument from BLS Ltd. The adjustable 1 MHz AC field (1 to 2 V) allow the orientation of 2-cell stage embryos in the electrode chamber (distance between electrodes is 250 μm). The actual parameters might vary and need to be determined in a pilot experiment. The goal is to reach 90% fusion in 30–60 min without embryo lysis.
4. Place an electrode chamber connected to the pulse generator into the 100-mm Petri dish. Use the same dish for embryo washes; alternatively, use another plate on a second microscope.
5. Place two large drops of M2 and a drop of mannitol solution in the dish. Place the mannitol drop over the electrode chamber.
6. Pick 25–30 embryos and rinse them well with mannitol by quickly pipetting up and down. Place the embryos between the electrodes, spacing them from each other. Manually move the embryos that are not oriented.
7. When all the embryos are properly oriented, push the trigger pulse.
8. Immediately transfer the embryos into an M2 drop. Wash the embryos through a few drops of equilibrated KSOM. Place them into the culture plate in the incubator.
9. Proceed with the rest of the embryos. The mannitol drop over the electrode chamber should not be used for longer than 15 min and should be replaced with a fresh one after that time. The number

of embryos handled in 15 min depends on the speed of manipulations.

10. It is very important to select perfectly fused tetraploid embryos 30–60 min after application of the pulse and transfer them into a fresh drop. Since embryos are recovered at the late 2-cell stage, the second mitotic division is expected soon after fusion. If not checked in time, fused and cleaved tetraploid embryos could be confused with non-fused diploid 2-cell stage embryos. Under optimal conditions, around 90–95% of embryos should fuse.
11. After overnight incubation, 3- to 4-cell stage embryos are used for sandwich aggregation as described above.

Notes

1. The quality of FBS is the most important factor for successful ES cell culture. ES cell qualified pretested serum is available from some suppliers. It is still recommended to test different lots of serum from different companies before ordering large quantities. It is also possible to use Knockout Serum Replacement (SR) from Gibco in combination with regular DMEM or Knockout D-MEM for ES cell culture. However, in this case, ES cells need to be grown on feeders all the time.
2. It is possible to purify LIF from bacteria transfected with recombinant LIF constructs. Each new batch of LIF should be tested for the final concentration necessary to maintain pluripotency of ES cells. For R1 ES cells, we use final concentration of 1000 U/mL.
3. The quality of water is possibly even more critical for embryo culture than for ES cell media. Disposable plasticware is highly recommended, or if clean glassware is used, it should never be exposed to detergent or organic solvents. All chemicals should be of highest grade, or embryo-tested (available from Sigma) and used only for media preparation. Embryos are cultured in organ culture dishes or in microdrops of KSOM media covered with embryo-tested light mineral oil. A majority of E0.5 dpc embryos reaching blastocyst stage demonstrates the optimal culture conditions. The minimal time between sacrificing the embryo donors and putting the embryos in the culture dish and the proper removal of debris after flushing also contributes to successful embryo culture.
4. It is important to carefully follow the ES cell culture protocol to maintain their pluripotency and ability to contribute to the germline.

Typically, ES cells are kept at relatively high density and should be passaged when they reach a subconfluent state of 70–80%, i.e., tightly packed colonies almost touch each other. Media should be changed every day. For regular maintenance ES cells should never be seeded too sparsely (when 4 to 5 d is required to reach subconfluency), nor should they grow past 90% confluency before passage. Both conditions induce cell differentiation. Usually ES cells are split at 1:5 to 1:7 dilution depending on their growth rate, ideally every other day (about 1×10^6 cells are seeded onto 60-mm plate, 2×10^6 cells onto 100-mm plate). Cells should be trypsinized to a single-cell suspension as large clumps might differentiate. The passage number of ES cells must be kept as low as possible with a stock of frozen vials kept in liquid nitrogen. It is recommended to create a frozen pool of low passage number of ES cells that serve as the only regular source of ES cells for many years of experiments.

5. For example, for 2 wells: × 2.5, i.e., 2.5 μg DNA in 600 μL OPTI-MEM$^+$ 20 μL LF in 600 μL of OPTI-MEM.
6. Often cells in different wells might not grow at a synchronous rate. Some wells might not be subconfluent but have one or two colonies. Such wells can be trypsinized individually to allow more even growth after plating back the single-cell suspension. It is best to choose a time for the passage when the majority of the wells have reached 80% confluency. Ninety-six-well plates are split into 2 or 3 replica plates, which can then be passaged one more time if necessary.
7. For E12.5 and older embryos as well as tissues, the samples can be sectioned sagitally using a razor blade after 30 min of incubation at room temperature in freshly prepared prefix solution (2% PFA in PBS) and then fixed for an extra 30–60 min on ice.
8. If a plate has too many ES cell colonies that are also larger than necessary, they can be gently resuspended in trypsin. Aliquots of cell suspension can be transferred into a new plate with ES-DMEM after pipetting up and down 1–3 times to achieve clumps of the right size.
9. The goal is to create a small depression with a smooth surface, deep enough to hold the aggregate safely even when moving the plate to the incubator.
10. We usually aggregate 150–200 embryos (transferred into 8–10 recipients) per ES cell clone for germ line transmission. If chimeric

embryos are dissected at mid-gestation to assess in vivo expression, two ES cell clones can be done in one experiment. As ES cells are derived from a pigmented mouse strain (in the case of R1 from an F1 hybrid of 129Sv-cp × 129 SvJ cross), and host embryos are derived from albino mice, the degree of chimerism can be estimated by eye pigmentation. Pigmented cells in the eyes are ES cellderived and indicate overall ES cell contribution. By analyzing such chimeric embryos, a few candidate ES cell clones can be chosen for best contribution potential and gene expression for further aggregation for germ line transmission.

11. The AC field can be set up in advance; alternatively, slowly increase the AC field, so the embryos will orient properly.

12

EMBRYONIC ANTIGENIC CELLS

DETECTION AND VISUALIZATION OF PROTEIN INTERACTIONS WITH PROTEIN FRAGMENT COMPLEMENTATION ASSAYS

A first step in defining the function of a novel gene is to determine its interactions with other gene products in an appropriate context; that is, because proteins make specific interactions with other proteins as part of functional assemblies, an appropriate way to examine the function of the product of a novel gene is to determine its physical relationships with the products of other genes. This is the basis of the highly successful yeast two-hybrid system. The central problem with two-hybrid screening is that detection of protein-protein interactions occurs in a fixed context, the nucleus of *S. cerevisiae*, and the results of a screening must be validated as biologically relevant using other assays in appropriate cell, tissue or organism models. While this would be true for any screening strategy, it would be advantageous if one could combine library screening with tests for biological relevance into a single strategy, thus tentatively validating a detected protein as biologically relevant and eliminating false-positive interactions immediately. It was with these challenges in mind that our laboratory developed the *protein-fragment complementation assay* (PCA) strategy. In this strategy, the gene for an enzyme is rationally dissected into two pieces. Fusion proteins are constructed with two proteins that are thought to bind to each other, fused to either of the two probe fragments. Folding of the probe protein from its fragments is catalyzed by the binding of the test proteins to each other and is detected as reconstitution of enzyme activity. We have already demonstrated that the PCA strategy has the following capabilities: (1) Allows for the detection of protein–protein interactions in vivo and in vitro in any cell type; (2) allows for the detection of protein–protein interactions in appropriate subcellular compartments or organelles; (3) allows for the detection of induced versus constitutive protein–protein interactions that occur in response to developmental, nutritional, environmental, or hormone-induced signals; (4) allows for the detection of the kinetic and equilibrium aspects of protein assembly in these cells; and (5) allows for screening of cDNA libraries for protein–protein interactions in any cell type.

In addition to the specific capabilities of PCA described above, are special features of this approach that make it appropriate for screening of molecular interactions, including: (1) PCAs are not a single assay but a series of assays; an assay can be chosen because it works in a specific cell type appropriate for studying interactions of some class of proteins; (2) PCAs are inexpensive, requiring no specialized reagents beyond those that are necessary for a particular assay and off the shelf materials and technology; (3) PCAs can be automated and high-throughput screening could be done; (4) PCAs are designed at the level of the atomic structure of the enzymes used; because of this, there is additional flexibility in designing the probe fragments to control the sensitivity and stringencies of the assays; and (5) PCAs can be based on enzymes for which the detection of protein–protein interactions can be determined differently, including by dominant selection or production of a fluorescent or colored product.

The selection of enzymes and design of PCAs have been discussed in detail, and here, we will review only the most basic ideas. Polypeptides have evolved to code for all of the chemical information necessary to spontaneously fold into a stable, unique 3-dimensional structure. It logically follows that the folding reaction can be driven by the interaction of two peptides that together contain the entire sequence, and in the correct order, that a single peptide will fold. This was demonstrated in the classic experiments of Richards and Taniuchi and Anfinsen. In practice, this is not easily done. Protein fragments will tend to aggregate rather than fold together into the three dimensional structure of the complete protein. However, if one uses soluble binding proteins to the fragments, which by interacting with each other increase the effective concentration of the fragments, correct folding could be favored over any other nonproductive process. If the protein that folds from its constitutive fragments is an enzyme, whose activity could be detected in vivo, then the reconstitution of its activity can be used as a measure of the interaction of the fused binding proteins. Further, this binary all or none folding event provides for a very specific measure of protein interactions dependent on not mere proximity, but the absolute requirement that the peptides must be organized precisely in space to allow for folding of the enzyme from the polypeptide chain. We select to dissect proteins into fragments that are not capable of spontaneously folding along with their complementary fragment, into a functional and complete protein. These facts distinguish the PCA strategy from complementation of naturally occurring and weakly associating subunits of enzymes, in which some spontaneous assembly occurs.

All applications of PCA, thus far, have been performed in model cell lines as described below. However, it is not difficult to imagine how the basic approach could be combined with functional genomics approaches using *embryonic stem* (ES) cells. The most obvious applications would be in gene-trap strategies, in which a reporter PCA fragment is inserted randomly or directionally into specific genes in the same way that an individual reporter gene such as lacZ or *green fluorescent protein* (GFP) might be incorporated. The PCA reporter fragments would be inserted into the genome of ES cells, and the cells induced to differentiate in vitro. The goal, then, would be to identify differentiated cells in which two marked gene products interact and then to identify the interacting partners. We have already developed 5 PCAs based on dominant-selection, colorimetric, or fluorescent outputs. All of these assays have potential applications to gene trapping in ES cells. Here, we will discuss the most well developed PCA, based on the enzymes *murine dihydrofolate reductase* (mDHFR) and *tiethylene-melamine* (TEM) β-lactamase.

The DHFR PCA can be used in a variety of applications to perform simple survival-selection as readout, but simultaneously, as a fluorescent assay allowing quantitative detection of protein interactions as well as the determination of the cellular location of protein interactions can be performed. The β-lactamase assay can be used as a very sensitive in vivo or in vitro quantitative detector of protein interactions as, unlike DHFR, one measures the continuous conversion of substrate to colored or fluorescent product. However, it should be noted that the generation of a product by an enzyme does not guarantee that signal-to-background would be superior to that of fixed fluorophore reporters like GFP and fluorescein-conjugated methotrexate (fMTX) bound to DHFR. Observable signal-to-background depends, for example, on the quantum yield of the fluorophore, retention of fluorophore by a cell, the optical properties of the cells used, and the extent to which fluorophores are retained in individual cellular compartments. For instance, in spite of no enzymatic amplification, the DHFR fluorescence assay only requires between 1000 to 3000 molecules of reconstituted DHFR to clearly distinguish a positive response from background.

Reconstitution of DHFR activity can be monitored in vivo by cell survival in DHFR-negative cells (CHO-DUKX-B11, for example) grown in the absence of nucleotides. The principle of the DHFR PCA survival assay is that cells, simultaneously expressing complementary fragments of DHFR fused to interacting proteins or peptides, will survive in media depleted of nucleotides. Alternative, recessive selection can be achieved in DHFR-positive cells by using DHFR PCA fragments containing one or more of several mutations that render the refolded DHFR resistant to the anti-folate drug methotrexate. The cells are then grown in the absence of nucleotides with selection for methotrexate resistance.

The survival DHFR PCA is an extraordinarily sensitive assay. In mammalian cells, survival is dependent only on the number of molecules of DHFR reassembled, and we have determined that this number is approx 25–100 molecules of DHFR per cell. The second approach is a fluorescence assay based on the detection of fMTX binding to reconstituted DHFR. The basis of the DHFR PCA fluorescence assay is that complementary fragments of DHFR, when expressed and reassembled in cells, will bind with high affinity (K_d = 540 p*M*) to fMTX in a 11 complex. fMTX is retained in cells by this complex, while the unbound fMTX is actively and rapidly transported out of the cells. In addition, binding of fMTX to DHFR results in a 4.5-fold increase in quantum yield. Bound fMTX, and by inference reconstituted DHFR, can then be monitored by fluorescence microscopy, *fluorescence -activated cell sorting* (FACS), or spectroscopy.

It is important to note that, though fMTX binds to DHFR with high affinity, it does not induce DHFR folding from the fragments in the PCA. This is because the folding of DHFR from its fragments is obligatory; if binding of the oligomerization domains does not induce folding, no binding sites for fMTX are created. Therefore, the number of complexes observed, as measured by number of fMTX molecules retained in the cell, is a direct measure of the equilibrium number of complexes of oligomerization domain complexes formed, independent of binding of fMTX. The other obvious application of the DHFR PCA fluorescence assay is in determining the location in the cell of interactions, as illustrated in a number of cell types.

β-Lactamase is strictly a bacterial enzyme and has been genetically deleted from many standard *Escherichia coli* strains. It is not present at all in eukaryotes. Thus, the β-lactamase PCA can be used universally in eukaryotic cells and many prokaryotes, without any intrinsic background. Also, assays are based on catalytic turnover of substrates with rapid accumulation of product. This enzymatic amplification should allow for relatively weak molecular interactions to be observed. The assay can be performed simultaneously or serially in a number of modes, such as the in vitro colorimetric assay, the in vivo fluorescence assay, or the survival assay in bacteria. Assays can be performed independent of the measurement platform and can easily be adapted to high-throughput formats requiring only one pipetting step.

Materials

DHFR PCA survival assay

1. 12-Well plates, tissue culture treated; 6-well plates, tissue culture treated.
2. Minimum essential medium: α-medium without ribonucleosides and deoxyribonucleosides (α-MEM).
3. Dialyzed fetal bovine serum (FBS).
4. Adenosine; desoxyadenosine; thymidine.
5. Lipofectamine Plus reagent.
6. Trypsin-EDTA.
7. Cloning cylinders.

DHFR PCA fluorescence assay

1. 12-Well plates, tissue culture treated.
2. Dulbecco's modified Eagle medium (DMEM); α-MEM.
3. Cosmic calf serum; dialyzed FBS.
4. Lipofectamine Plus reagent.
5. fMTX.
6. Dulbecco's phosphate-buffered saline (PBS).
7. Geltol aqueous mounting medium.
8. Trypsin-EDTA.
9. Microcover glasses, 18-mm circles, no 2.
10. Microscope slides, glass, 25 × 75 × 1.0 mm.
11. 96-Well black microtiter plates.
12. Bio-Rad protein assay.

β-Lactamase PCA colorimetric assay

1. 12-Well plates, tissue culture treated.
2. DMEM.
3. Cosmic calf serum.
4. Fugene 6 transfection reagent.
5. Trypsin-EDTA.
6. Dulbecco's PBS.
7. Nitrocefin.
8. 96-Well plate.

β-Lactamase PCA Fluorometric Assay

1. 12-Well plates, tissue culture treated.
2. DMEM.
3. Cosmic calf serum.
4. Fugene 6 transfection reagent.
5. Trypsin-EDTA.
6. Dulbecco's PBS.
7. CCF2-AM.
8. 96-Well white microtiter plates.
9. Normal saline: 140 m*M*, NaCl, 5 m*M* KCl, 2 m*M* $CaCl_2$, 10 m*M* Hepes, 6 m*M* Sucrose, 10 m*M* Glucose, pH 7.35.
10. Physiological saline solution: 10 m*M* HEPES, 6 m*M* sucrose, 10 m*M* glucose, 140 m*M* NaCl, 5 m*M* KCl, 2 m*M* $MgCl_2$, 2 m*M* $CaCl_2$, pH 7.35.
11. 15-mm Glass coverslip.

Method

DHFR PCA survival assay

1. Split CHO DUKX-B11 (DHFR-negative; could also be done in other cells lines) cells 24 h before transfection at 1×10^5 in 12-well plates in α-MEM medium enriched with 10% dialyzed FBS and supplemented with 10 μg/mL of adenosine, desoxyadenosine, and thymidine.
2. Co-transfect cells with the PCA fusion partners using Lipofectamine Plus reagent according to the manufacturer's instructions.
3. Forty-eight hours after the beginning of the transfection, split cells at approx 5×10^4 in 6-well plates in selective medium consisting of α-MEM enriched with dialyzed FBS, but without addition of nucleotides.

4. Change medium every 3 d. The appearance of distinct colonies usually occurs after 4 to 10 d of incubation in selective medium. Çolonies are observed only for clones that simultaneously express both interacting proteins fused to one or the other complementary DHFR fragments. Only interacting proteins will be able to achieve normal cell division and colony formation.

   For further analysis of the interacting protein pair:

5. Isolate 3 to 5 colonies per interacting partners by trypsinization (trypsin-EDTA) using cloning cylinders and grow them separately.
6. Select the best expressing clone by immunoblot (Western blot) or using the DHFR PCA fluorescence assay. Amplification of the expressed gene using methotrexate resistance can be done afterwards if desired, to obtain clones with increased expression.
7. Carry out functional analysis of the clone stably expressing your interacting proteins pair fused to the complementary DHFR fragments by using the DHFR PCA fluorescence assay.

DHFR PCA fluorescence assay

Fluorescence microscopy

1. Split COS cells (this assay can be used with any other cell line) 24 h before transfection at 1×10^5 on 18-mm circles glass coverslips in 12-well plates in DMEM medium enriched with 10% Cosmic calf serum.
2. Transiently co-transfect cells with the PCA fusion partners using Lipofectamine Plus reagent according to the manufacturer's instructions.
3. The next day, change medium and add fMTX to the cells at a final concentration of 10 μM.

   For stable cell lines:

   For CHO DUKX-B11 cells (or other cell line) stably expressing PCA fusion partners, seed cells to approximately 2×10^5 on 18-mm glass coverslips in 12-well plates in α-MEM medium enriched with 10% dialyzed FBS. The next day, fMTX is added to the cells at a final concentration of 10 μM.

4. After an incubation with fMTX of 22 h at 37°C, remove the medium, and wash the cells with PBS 1X, and re-incubate for 15–20 min at 37°C in the culture medium to allow for efflux of unbound fMTX. Remove the medium and wash the cells four times with cold PBS 1X on ice, and finally mount the coverslips on microscope glass slides with an aqueous mounting medium.

5. Fluorescence microscopy is performed on live cells. These experiments must be performed within 30 min of the wash procedure. If the negative control (untransfected cells treated with fMTX) is too fluorescent, the wash procedure must be modified.

Flow cytometry analysis

Preparation of cells for FACS analysis is the same as described for fluorescence microscopy, except that following the PBS 1X wash (just two times in this case), cells are gently trypsinized (trypsin-EDTA), suspended in 500 μL of cold PBS 1X, and kept on ice prior to flow cytometric analysis within 30 min. Data are collected on a FACS analyzer with stimulation with an argon laser tuned to 488 nm with emission recorded through a 525 nm band width filter.

Fluorometric analysis

Preparation of cells for fluorometric analysis is the same as described for fluorescence microscopy, except that following the PBS 1X wash (just 2 times in this case), cells are gently trypsinized (trypsin-EDTA). Plates are put on ice, and 100 μL of cold PBS is added to the cells. The total cell suspensions are transferred to 96-well black microtiter plates, and keep on ice prior to fluorometric analysis. The assay can be performed on any microtiter plate reader; we use a Perkin-Elmer HTS 7000 Series Bio Assay Reader in the fluorescence mode. The excitation and emission wavelengths for the fMTX are 497 nm and 516 nm, respectively. Afterward, the data are normalized to total protein concentration in cell lysates (Bio-Rad protein assay).

In vitro β -lactamase PCA colorimetric assay

1. Split COS or HEK 293 T cells (this assay can be use with any other cell line) 24 h before transfection at 1×10^5 in 12-well plates in DMEM medium enriched with 10% Cosmic calf serum.
2. Transiently co-transfect cells with the PCA fusion partners using Fugene 6 Transfection reagent according to the manufacturer's instructions.
3. Forty-eight hours after transfection, cells are washed three times with cold PBS, resuspended in 300 μL of cold PBS, and kept on ice. Cells are then centrifuged at 4°C for 30 s, the supernatant discarded, and cells resuspended in 100 μL of cold phosphate buffer 100 m*M*, pH 7.4 (β-lactamase reaction buffer).
4. Freezing in dry ice-ethanol for 10 min and thawing in a water bath at 37°C for 10 min then lyse cells with three cycles of freeze and thaw. Cell membrane and debris are removed by

centrifugation at 4°C for 5 min (10,000*g*). The supernatant whole cell lysate is then collected and stored at –20°C until assays are performed.

5. Assays are performed in 96-well microtiter plates. For testing β-lactamase activity, 100 μL of phosphate buffer 100 m*M*, pH 7.4, is allocated into each well. To this, add 78 μL of H_2O and 2 μL of 10 m*M* Nitrocefin (final concentration of 100 μ*M*). Finally, add 20 μL of unfrozen cell lysate (final buffer concentration of 60 μ*M*).
6. The assays can be performed on any microtiter plate reader; we use a Perkin-Elmer HTS 7000 Series Bio Assay Reader in the absorption mode with a 492-nm measurement filter.

In vivo enzymatic assay and fluorescent microscopy with CCF2/AM

1. Split COS or HEK293 T cells 24 h before transfection at 1×10^5 in 12-well plates in DMEM enriched with 10% Cosmic calf serum.
2. Transiently co-transfect cells with the PCA fusion partners using Fugene 6 Transfection Reagent according to the manufacturer's instructions.
3. Twenty-four hours after transfection, cells are split again to assure 50% confluency the following day (1.5×10^5). They are split either onto 12-well plates for suspension enzymatic assay or onto 15-mm glass coverslips for fluorescent microscopy.
3. Forty-eight hours after transfection, cells are washed three times with PBS to remove all traces of serum.
4. Cells are then loaded with the following: 1 μ*M* of CCF2/AM diluted into a physiologic saline solution for 1 h.

   For in vivo enzymatic assay:
5. Cells are then washed twice with the physiologic saline. The cells are then resuspended into the same solution, and 1×10^6 cells are aliquoted into a 96-well fluorescence white plate and are read for blue fluorescence with a Perkin Elmer HTS 7000 Series Bio Assay Reader in the fluorescence Top reading mode with a 409-nm excitation filter and a 465-emission filter.

   For fluorescence microscopy:
6. Cells are washed twice with the physiologic saline as in step 5, prior to examination under the microscope.

We used two substrates to study the β-lactamase PCA. The first one is the cephalosporin called *nitrocefin*. This substrate is used in the in vitro colorimetric assay. β-lactamase has a *kcat*/km of 1.7 ×

10^4 m$M^{-1*}s^{-1}$. Substrate conversion can be easily observed by eye; the substrate is yellow in solution, while the product is a distinct ruby red color. The rate of hydrolysis can be monitored quantitatively with any spectrophotometer by measuring the appearance of red at 492 nm. Signal-to-background, depending on the mode of measurement, can be greater than 30 to 1.

We have also developed an in vivo fluorometric assay using the substrate CCF2/AM. While not as good a substrate as nitrocefin (kcat/km of 1260 m$M^{-1*}s^{-1}$) CCF2/AM has unique features that make it a useful reagent for in vivo PCA. First, CCF2/AM contains butyryl, acetyl, and acetoxymethyl esters, allowing diffusion across the plasma membrane, in which cytoplasmic esterases catalyze the hydrolysis of its ester functionality releasing the polyanionic (4 anions) β-lactamase substrate CCF2. Because of the negative charge of CCF2, the substrate becomes trapped in the cell. In the intact substrate, fluorescence resonance energy transfer (FRET) can occur between a coumarin donor and fluorescein acceptor pair covalently linked to the cephalosporin core. The coumarin donor can be excited at 409 nm with emission at 447 nm, which is within the excitation envelope of the fluorescence acceptor (maximum around 485 nm), leading to remission of green fluorescence at 535 nm. When β-lactamase catalyzes hydrolysis of the substrate, the fluorescein moiety is eliminated as a free thiol. Excitation of the coumarin donor at 409 nm then emits blue fluorescence at 447 nm, whereas the acceptor (fluorescein) is quenched by the free thiol.

Notes

1. Alternatively, recessive selection can be achieved in eukaryotic cells by using DHFR fragments containing one or more of several mutations (for example F31S mutation) that reduce the affinity of refolded DHFR to the anti-folate drug methotrexate and growing cells in the absence of nucleotides with selection for methotrexate resistance. This would obviously be necessary in working with mouse ES cells as, with all eukaryotes, DHFR activity is present.
2. The best orientations of the fusions for the DHFR PCA are: protein A-DHFR(1,2) plus protein B-DHFR(3) or DHFR(1,2)-protein A plus protein B-DHFR(3), where proteins A and B are the proteins to test out for interaction. We typically insert a 10 amino acid flexible polypeptide linker consisting of $(Gly.Gly.Gly.Gly.Ser)_2$ between the protein of interest and the DHFR fragment (for both fusions). DHFR(1,2) corresponds to amino acids 1–105, and DHFR(3) corresponds to amino acids 106–186 of murine DHFR.

The DHFR(1,2) fragment that we use also contains a phenylalanine to serine mutation at position 31 (F31S), rendering the reconstituted DHFR resistant to MTX treatment.

3. It is crucial that cell density is kept to a minimum and that cells are well separated when split, to avoid cells "*harvesting*" nutrients from adjacent cells on dense plates or colonies that might appear to be forming from clumps of cells that were not sufficiently separated during the splitting procedure.
4. The choice of dialyzed FBS manufacturer is crucial. Cells need very little nucleotide in the medium to propagate, and this will result in false positives. The Hyclone dialyzed FBS has proven a particularly reliable source.
5. The fluorescence DHFR PCA assay is universal and, in theory, can be used in any cell type or organism. This assay has already been shown to work in several mammalian cell lines as well as in plant cells and insect cells.
6. A stock solution of 1 m*M* fMTX should be prepared as follows: dissolve 1 mg of fMTX in 1 mL of *dimethyl formamide* (DMF). To facilitate the dissolution, incubate 15 min at 37°C and vortex mix every 5 min. Protect the tube from light. Keep at –20°C.
7. Complementary fragments of DHFR fused to interacting protein partners, when expressed and reassembled in cells, will bind with high affinity (K_d = 540 p*M*) to fMTX in a 1:1 complex. fMTX is retained in cells by this complex, while the unbound fMTX is actively and rapidly transported out of the cells.
8. All of the work reported to date has been performed in live cells. While cells can be fixed, there is a significant reduction in observable fluorescence.
9. Particular attention must be given to optimizing the fMTX load and "*wash*" procedures. Important variables include the time of loading, temperatures at which each wash step is performed, the number and length of wash steps and the time between washing and visualization. Too little washing will mean that background cannot be distinguished from a positive result. One should scrutinize the relevant parameters in the same sense as one would for say, a Western blot. Results may also vary with the way the cells are plated and the types of cells used. Generally, as in other fluorescent microscopy procedures, the shape of cells and the localization of the fluorophore will result in better or worse results. For stable cell lines, the intensity of fluorescence will also depend on the

levels of expression of the fusion proteins. The loading times and concentrations of fMTX (22 h and 10 μM) used may result in a nonspecific and punctate fluorescence that is observed with any filter set. We do not know the source of this background, but it should not be mistaken for the real fluorescence signal produced by the PCA, which should be observed strictly with a filter that is optimal for observation of fluorescein. We have observed that loading fMTX for between 2 and 5 h at lower (5 μM) concentrations prevents this nonspecific signal, although fewer cells are labeled. Loading times and concentrations must be optimized for specific cell types.

10. The best orientations of the fusions for the β-lactamase PCA are: protein A-BLF[1] plus protein B-BLF(2) or BLF(1)-protein A plus protein B-BLF[2], where proteins A and B are the proteins to test out for interaction. We typically insert a 15 amino acid flexible polypeptide linker consisting of $(Gly.Gly.Gly.Gly.Ser)_3$ between the protein of interest and the β-lactamase fragment (for both fusions). BLF(1) corresponds to amino acids 26–196 (Ambler numbering), and BLF(2) corresponds to amino acids 198–290 of TEM-1 β-lactamase.
11. The maximum loading efficiency of CCF2-AM is observed at 50% confluence.
12. Serum may contain esterases that can destroy the substrate.
13. We perform fluorescence microscopy on live HEK 293 or COS cells with an inverse Nikon Eclipse TE-200 (objective plan fluor 40X dry, numerically open at 0.75) Images were taken with a digital *change-coupled device* (CCD) cooled (–50°C) camera, model Orca-II (Hamamatsu Photonics (exposure for 1 s, binning of 2 × 2 and digitalization 14 bits at 1.25 MHz). Source of light is a Xenon lamp Model DG4. Emission filters are changed by an emission filter switcher. Images are visualized with ISee software on an O2 Silicon Graphics computer. The following selected filters are used: filter set no. 31016; excitation filter: 405 nm (passing band of 20 nm); dichroic mirror: 425 nm DCLP; emission filter no. 1: 460 nm (passing band of 50 nm); emission filter no. 2: 515 nm (passing band of 20 nm).

Isolation of Antigen-Specific Intracellular Antibody

The control of gene function by the modulation of mRNA translation, stability, or protein activity has important implications for the treatment of disease, as well as in research programs designed to study the biological role of proteins. For instance, preventing protein function in specific cell types in development can produce a phenotypic knock-out due to the effective protein loss. Antibodies are ideally suited for this purpose, as they have evolved to bind macromolecules with high affinity and can neutralize their function as a result.

Ablation of protein function in vivo with antibodies has been achieved by microinjecting whole antigen-specific immunoglobulin molecules into the cell cytoplasm. However, complete antibody comprises four chains (two heavy and two light chains) held by interchain disulfide bonds. This is not suitable for expression from DNA vectors, as the individual chains would not assemble in the cytoplasm. Therefore, antibody fragments have been employed in the single chain Fv (scFv) format, which are single polypeptide chains comprising a *heavy chain* variable region (VH) and a *light chain* variable region (VL) held by a short linker sequence. ScFv folds into an antibody combining site (which binds antigen) but no effector function is present. Many uses of scFv expressed inside cells (herein called *intracellular antibody* or ICAbs) have been described in cell systems and in whole organisms. For example, in vivo expression of a scFv directed against the coat protein of artichoke mottle crinkle virus has conferred resistance to viral infection in plant cells.

The potential for isolation of scFv, or single VH (or perhaps VL) domains, which specifically bind antigen, should make it possible to derive a spectra of scFv for any in vivo antigenic target. Consequently, expressing scFv inside a given cell to ablate the target antigen is possible. The initial source of cDNA for scFv expression can be from phage display libraries or from immunized mouse spleen. The antigen-specific scFv could subsequently be expressed from a vector (e.g., retrovirus) and any effects on the cell could be assessed. If the scFv are used in *embryonic stem* (ES) cells with tissue-specific transgene-targeted genes, the effect of scFv-protein binding during development could be determined. A further potentially major area of use will be in functional genomics. The human and mouse genome sequences are almost complete, and the full set of gene sequences will soon be known. Many of these derived mRNAs will have no specified function other than those guessed from sequence homologies. Using intracellular scFv to abolish specific protein function will be a powerful tool to obtain first generation data on the role of specific proteins in cells and, later, could be used for phenotypic ablation in mice by transgenic expression of antigen-specific scFv. Finally, intracellular scFv have great potential in human disease treatment by serving as antigen-specific reagents against disease-related proteins inside cells.

Despite this potential of ICAbs, there are few antibodies that can function effectively in an intracellular environment. When scFv are expressed in the cell cytoplasm, folding and stability problems often occur, which results in a nonfunctional low level of expression and a short half-life. These problems are caused by the reducing environment of cell cytoplasm, which hinders the formation of the intrachain disulfide bonds in the immunoglobulin VH and VL domains. Although some scFv can tolerate the absence of the disulfide bond, there is no general rule that predicts such characteristics. Two main approaches can be adopted to circumvent these problems. One would be to derive scFv scaffold(s), which effectively fold in vivo and are expressed well, and then to use these as backbones onto which antigen-specific *complementarity-determining regions* (CDRs) could be introduced. A second more simple and generic approach is to use selection methods that allow isolation of those scFv, which will bind in vivo from a repertoire of different scFv. Taking the latter approach, we have recently developed an in vivo antibody–antigen interaction screening method to identify functional intracellular binders.

Our selection method, *intracellular antibody capture* (IAC) technology, takes advantage of the ability of protein interactions to be detected in cellular environments, such as shown for the yeast two-hybrid system. It was based on the fact that some antibody scFv fragments can fold adequately in vivo to bind antigen in a VH-VL dependent way (i.e., via the antibody combining site) and, thus, using a library of diverse antibody specificities should facilitate their identification, if sufficient scFv could be screened. In addition, ideally, one would like to isolate a small repertoire of intracellular scFv, which would bind to different epitopes on the target antigen. Our screening system comprised of yeast cell expression of a "bait" antigen fused to the LexA DNA binding domain and a library of "prey" scFv fused to the VP16 transcription activation domain. Interaction between the antigen bait and any specific antibody scFv fragment in the yeast intracellular environment results in the formation of a protein complex, in which the DNA binding domain and the activation domain are in close proximity. This results in the activation of yeast chromosomal reporter genes, such as *HIS3* and *LacZ*, facilitating the identification and thus isolation of the yeast carrying the DNA vectors encoding the scFv, which in turn can be isolated to yield the DNA sequence of the antigen-specific scFv. The main limitation of this approach is the number of scFv-VP16 fusion preys that can be screened in yeast antibody–antigen interaction system (conveniently up to 2–5 $\times$ 10^6). This figure is well below the size of scFv repertoires. Thus, it is necessary to limit the numbers of scFv to be screened in vivo in yeast. We currently use one round of in vitro phage scFv library screening (panning) using, for instance, bacterially produced antigen coated on a surface, prior to the in vivo yeast antibody–antigen interaction screening. With this combined approach, we have successfully isolated intracellular scFv specific to the *breakpoint cluster region* (BCR) antigen.

In this section, we describe a protocol to isolate functional intracellular scFv binders to specific antigens, *Intracellular Antibody Capture* (IAC). In summary, the overall strategy involves:

1. (A) In vitro screening of a phage scFv display library (which should have a diversity $>10^9$) with the target antigen (e.g., made by bacterial or bacullovirus expression) to produce a sublibrary enriched for antigen-specific scFv.
2. (B) Subcloning the first round scFv DNA from the phage vector into the yeast vector to make a library of the enriched scFv fused with VP16 transactivation domain.
3. Selection of intracellular scFv binders by performing the antibody–antigen interaction screening in yeast using the scFv-VP16 prey library together with the antigen–DNA binding domain bait.
4. When intracellular scFv binders have been isolated, these are validated by binding of the isolated scFv to the antigen by Western blotting (in vitro). Other functional or in vivo assays may then be tested.

Materials

In vitro phage antibody library screening with antigen

1. Nunc Maxisorp Immunotubes.
2. Phosphate-buffered saline (PBS): 0.1 *M* NaCl, 17 m*M* Na_2HPO_4, 8 m*M* $NaH_2PO_4 \bullet 2\ H_2O$, adjust pH to 7.0–7.2 with HCl.
3. 100 m*M* Triethylamine solution (freshly prepared).
4. Dried milk powder.
5. 1 *M* Tris-HCl, pH 7.4, at room temperature.
6. TG1 *Escherichia coli* bacteria.
7. TYE agar plate (per liter): 15 g agar, 140 m*M* NaCl, 10 g Bacto-tryptone, 5 g Bacto-yeast extract.

8. 2X TY medium (per liter): 16 g Bacto-tryptone, 10 g Bacto-yeast extract, 85 m*M* NaCl, pH adjusted to 7.4 with HCl.

Preparation of the scFv sublibrary, enriched for the antigen-specific scFv, from the first round phage infected TG1

1. Glycerol.
2. QIAGEN Plasmid Midi Kit.
3. QIAEX II Gel Extraction Kit.
4. *Not*I and *Sfi*I restriction enzymes, NEB buffer 2.
5. Elution buffer: 10 m*M* Tris-HCl, pH 8.5, at room temperature.

Construction of yeast scFv-VP16 in vivo expression library for antibody–antigen yeast interaction selection

1. Yeast pVP16 vector.
2. T4 DNA ligase and ligase buffer.
3. DH5α *E. coli*.

Hanahan method of bacterial transformation

1. TFB solution: 10 m*M* 2-(N-morpholino)ethanesulfonic acid (adjusted to pH 6.2 with potassium hydroxide), 100 m*M* KCl, 45 m*M* manganese chloride, 10 m*M* $CaCl_2$, 3 m*M* hexamine cobalt trichloride.
2. Dimethylformamide (DMF).
3. 2.25 *M* Dithiothreitol (DTT) in 40 m*M* potassium acetate, pH 6.0.

Yeast antibody–antigen interaction screening of scFv-VP16 library with the antigen "bait"

1. pBTM116 yeast expression vector.
2. L40 yeast: *Mata* his3 Δ200 trp1-901 leu2-3, 112 ade2 LYS:: $(lexAop)_4$-HIS3 URA3:: $(lex\text{-}Aop)_8$-LacZ GAL4.
3. YPD medium (per liter): 10 g yeast extract, 20 g Bacto-peptone, 2% glucose.
4. YC medium (per liter): 1.2 g yeast nitrogen base without amino acid and ammonium sulfate, 5 g ammonium sulfate, 10 g succinic acid, 6 g NaOH, 0.75 g amino acid mixture (1 g each of adenine sulfate, arginine, cysteine, threonine, and 0.5 g each of aspartic acid, isoleucine, methionine, phenylalanine, proline, serine, and tyrosine), 2% glucose, 0.1 g of tryptophan (W), 0.1 g of leucine (L), 0.05 g of histidine (H), 0.1 g of uracil (U), and 0.1 g of lysine (K). YC–WLHUK is YC medium lacking tryptophan (W), leucine (L), histidine (H), uracil (U), and lysine (K). YC-WL is

YC medium lacking tryptophan (W), leucine (L). YC-L is YC medium lacking leucine (L).

5. 10X TE buffer: 0.1 *M* Tris-HCl, 10 m*M* EDTA, pH 7.5, at room temperature.
6. 10X LiAc: 1 *M* lithium acetate, adjusted to pH 7.5 with acetic acid.
7. PEG-LiAc: 40% polyethylene glycol (PEG) 4000, 1× TE buffer, and 1× LiAc.
8. 100% Dimethyl sulfoxide (DMSO).
9. 20 mg/mL Salmon sperm DNA: sheared with fine bore needles (gauge 25) and boiled for 5 min before use. Denatured salmon sperm DNA can be stored at –20°C.

Confirmation of antibody–antigen interaction in yeast clones by β-galactosidase filter assay

1. Z buffer (per liter): 16.1 g $Na_2HPO_4 \bullet 7\ H_2O$, 5.5 g $NaH_2PO_4 \bullet H_2O$, 0.75 g KCl, 0.246 g $MgSO_4 \bullet 7\ H_2O$, adjusted to pH 7.0 with HCl. Can store at room temperature for at least 1 yr.
2. X-gal stock solution: 20 mg of 5-bromo-4-chloro-3-indolyl-β-D-galactopyranoside (X-gal) in DMF. Store at –20°C and can keep for 1 yr.
3. Z buffer/X-gal: 100 mL Z buffer, 0.27 mL β-mercaptoethanol, 1.67 mL X-gal stock solution. Prepare fresh.
4. Liquid nitrogen.

Plasmid DNA extraction from individual selected yeast colonies

1. Yeast lysis solution: 2% Triton X-100, 15% sodium dodecyl sulfate (SDS), 100 m*M* NaCl, 10 m*M* Tris-HCl, pH 8.0, and 1 m*M* EDTA.
2. Phenol.
3. Chloroform.
4. Acid washed glass beads (425–600 μm).
5. 3 *M* sodium acetate, pH 5.2 (adjusted with glacial acetic acid).
6. 70% and 100% Ethanol.
7. Dry ice.
8. QIAprep Spin Miniprep Kit.

Characterization of the antigen-specific intracellular scFv

1. Sequencing primers
   Forward primer: 5'-TTG TTT CTT TTT CTG CAC AAT-3'
   Back primer: 5'-CAA CAT GTC CAG ATC GAA-3'

2. AmpliTaq DNA polymerase, 10X polymerase chain reaction (PCR) buffer, 25 m*M* $MgCl_2$ and 20 m*M* dNTP.
3. *Bst*N I restriction enzyme.
4. NuSieve 3:1 agarose.
5. Thermal cycler.

Binding of scFv to antigen by immunodetection assay (Western blot)

1. pHEN2 vector.
2. HB2151 *E. coli.*
3. 1X TES buffer: 0.2 *M* Tris-HCl, 0.5 m*M* EDTA, 0.5 *M* sucrose.
4. 9E10 Anti-myc monoclonal antibody and horseradish peroxidase (HRP)-conjugated anti-mouse antibody.

Methods

In vitro phage antibody library screening with antigen

The method for in vitro phage screening below is adapted from Dr. Heather Griffin.

1. Coat an immunotube with 4 ml of the antigen (concentration 50–100 μg/mL in PBS) and leave it standing at 4°C overnight. Also, inoculate a single colony of TG1 into 5 mL of 2X TY and grow at 37°C overnight with shaking.
2. Pour off the antigen the next day and wash the immunotube three times with PBS.
3. Fill the tube with 5 mL of PBS containing 2% (w/v) dried milk powder (M-PBS) for blocking to avoid nonspecific binding of phage particles to the surface of the immunotube. Incubate at 37°C for 2 h and wash the tube three times with PBS.
4. Dilute the phage scFv aliquot containing 10^{i2} to 10^{13} phage in 4 mL of 2% M-PBS and add it to the immunotube. Seal the tube with Parafilm, rotate it on a turntable wheel at room temperature for 30 min, and stand for a further 90 min at room temperature.
5. Pour off the unbound phage and wash the immunotube ten times with PBS containing 0.1% Tween-20 and ten times with PBS (all at room temperature).
6. Add 1 mL of freshly made 100 m*M* triethylamine to elute the bound phage. Seal the tube with Parafilm and rotate on a turntable wheel for 10 min at room temperature.
7. Pipet the eluted phage into a tube containing 0.5 mL of 1 *M* Tris-HCl, pH 7.4 for neutralization. Store it at 4°C.

8. Dilute 0.5 mL of the overnight TG1 bacterial culture into 50 mL of fresh 2X TY and grow at 37°C with shaking until OD_{600} is between 0.5–0.6. This usually takes 1.5–2 h.
9. Add the eluted phage to 8.5 mL of the exponentially growing culture of TG1. Incubate at 37°C for 30 min without shaking, during which phage infection of the TG1 bacteria occurs.
10. Titer bacteria by taking 100 μL of the infected TG1 bacteria and make 10-fold serial dilutions. Plate these dilutions on TYE agar plates containing 100 μg/mL ampicillin and 1% glucose. Grow overnight at 30°C and count the number of colonies for each dilution.
11. Pellet the remaining infected TG1 culture by centrifuging at 2000*g* for 10 min at 4°C. Resuspend cells in 1 mL of 2X TY and plate on a large square plate (23 × 23 cm) with TYE agar containing 100 μg/mL ampicillin and 1% glucose. Grow overnight at 30°C.
12. Add 5 mL of 2X TY with 100 μg/mL ampicillin and 15% glycerol to the large square plate containing the infected TG1 colonies and scrape off the colonies with a glass spreader.

Preparation of the scFv sublibrary, enriched for antigen-specific scFv, from the first round phage infected TG1

The in vitro selection of scFv displayed on phage, reduces the initial complexity of the phage to an enriched population, some of which bind antigen (if a phage library of 5 × 10^9 is used, about 10^5 phage will be recovered). It is then necessary to isolate the phagemid clones encoding these scFv and transfer these to the yeast vector (using *Sfi*I/*Not*I restriction sites) for the in vivo selection of intracellular scFv, which bind antigen.

1. Take 100–200 μL from the 5 mL bacteria recovered by scraping and dilute into 10 mL of 2X TY containing 100 μg/mL ampicillin. This is plated onto ten large square plates of TYE agar containing 100 μg/mL ampicillin and 1% glucose. Grow at 30°C overnight.
2. Store the remaining bacteria at –70°C as glycerol stock in aliquots. This is the sublibrary stock enriched for phage expressing antigen-specific scFv, this will include those scFv, which do or do not bind the target antigen in vivo.
3. Add 5 mL of 2X TY to each plate prepared in step 1 and scrape off all the bacteria colonies. Phagemid DNA can be extracted from the recovered bacteria using QIAGEN Plasmid Midi Kit (Qiagen) according to the manufacturer's instructions. Quantitate

the amount of phagemid DNA obtained by measuring the UV spectrum between 220 and 300 nm. The peak at 258 nm is used to calculate the amount of DNA. DNA concentration (ng/μL) = $OD_{258} \times 50$.

4. Digest 6 μg of phagemid DNA with 20 U of *Not*I enzyme in a 50-μL reaction in a 1.5-mL Eppendorf tube with 1X NEB buffer 2. Incubate at 37°C for 3 h.
5. Add 40 U of *Sfi*I enzyme to the *Not*I-digested DNA and mix. Cover with a drop of mineral oil and incubate overnight at 50°C.
6. Run the digested sample in a 1.5% agarose gel at 50 V for about 45 min and cut out the gel band containing the scFv DNA fragment mixture (800–900 bp). Extract the DNA using QIAEX II Gel Extraction Kit according to manufacturer's instructions. Elute DNA with 20 μL of elution buffer. This is the purified scFv (*Sfi*1/*Not*1) DNA fragment, which is used to construct the yeast scFv-VP16 expression library.

Construction of yeast scFv-VP16 in vivo expression library for antibody–antigen yeast interaction screening

When the extraction of phagemid DNA encoding the sublibrary of scFv has been accomplished, it is subcloned into the yeast VP16 antibody–antigen interaction assay vector. The size of the yeast scFv-VP16 library necessary will depend on the number of phage that was obtained in the first round in vitro screening. Ideally, the yeast library should represent three to ten times the number of phage to include as many potential binders in the yeast library as possible.

1. Digest 1.5 μg of yeast pVP16 vector DNA with 20 U of *Not*I enzyme in a 50-μL reaction with 1X NEB buffer 2. Incubate at 37°C for 3 h.
2. Add 40 U of *Sfi*I enzyme to the *Not*I-digested vector DNA and mix. Cover with a drop of mineral oil and incubate overnight at 50°C.
3. Run the digested sample in a 1% agarose gel at 50 V for about 45 min. Cut out the gel band containing the yeast pVP16 DNA of 7.5 kb. Extract the DNA using the QIAEX II Gel Extraction Kit according to manufacturer's instructions. Elute DNA with 28 μL of elution buffer. This is the purified yeast pVP16 vector (*Sfi*I/*Not*I).
4. Set up a 10-μL ligation reaction containing 1X T4 DNA ligase buffer, 1 μL of purified yeast pVP16 vector (Sfi/Not) DNA, 0.5

μL of purified scFv (Sfi/Not) DNA, and 400 U of T4 DNA ligase. Also set up another 10-μL ligation reaction without scFv DNA to check for the presence of incompletely cut vector background. Incubate overnight at 15°C.

5. Transform 5 μL of ligation reaction into 100 μL DH5α bacteria by the method of Hanahan. Plate the transformed bacteria onto large square plate (23 × 23 cm) of TYE agar containing 100 μg/mL ampicillin. Incubate overnight at 37°C and count the number of colonies obtained, typically around 10,000.
6. Set up more ligations and perform transformation to achieve the required number of colonies. Generally, 40–50 ligations are required (i.e., 40–50 × 10^4 colonies).
7. After overnight growth on the large plates, scrape off all the bacteria by adding 5 mL of 2X TY with 100 μg/ml ampicillin and 15% glycerol, and by using a sterile glass spreader. Dilute 100–200 mL of the recovered bacteria into 10 mL of 2X TY and plate on 10 large square plates of TYE agar containing 100 μg/mL ampicillin. Incubate overnight at 37°C. Store the remaining bacteria at –70°C in aliquots. This is the antigen-specific scFv sublibrary.
8. Recover the bacteria by scraping into 5 mL 2X TY, and extract the plasmid DNA from it using QIAGEN Plasmid Midi Kit. Quantify the DNA by the UV spectrum from 220 to 300 nm. This DNA is the yeast scFv-VP16 library DNA to be used for the yeast antibody–antigen two-hybrid screening. The presence of the scFv insert can be tested by digesting 0.5 μg of DNA with *Sfi*I and *Not*I. This should yield a vector band of approx 4 kb and an insert band (a mixture of scFv fragments) of approx 800 kb.

Hanahan method of bacterial transformation

The bacterial transfection method described by Hanahan allows for a higher efficiency than the standard $CaCl_2$ approaches. Other methods such as electroporation may also be used to achieve high efficiency cloning.

1. Inoculate a single colony of DH5α into 5 mL 2X TY and grow overnight at 37°C.
2. Dilute 0.5 mL of overnight culture into 50 mL of 2X TY and grow for about 2 h with shaking until an OD_{600} of 0.4–0.6 is obtained. Pellet the bacteria by centrifuging at 1000*g* for 8 min at 4°C, and resuspend in 20 mL of TFB solution. Incubate on ice for 20 min.

3. Pellet the bacteria again by centrifuging at 1000*g* for 8 min at 4°C, and gently resuspend in 4 mL of TFB solution. Add 140 μL of DMF and mix gently. Incubate on ice for 5 min.
4. Add 140 μL of 2.25 *M* DTT in 40 m*M* potassium acetate, pH 6.0, and mix gently. Incubate on ice for 10 min.
5. Add 140 μL of DMF and mix gently. The bacteria are now ready for transformation. Only use on the day of preparation.
6. Mix 5 μL of ligation reaction with 200 μL of competent bacteria. Incubate on ice for 1 h.
7. Heat shock for 2 min at 42°C and incubate on ice for 1 min.
8. Add 200 μL of 2X TY and incubate for 20 min at 37°C.
9. Plate all bacteria onto a TYE agar plate containing 100 μg/mL ampicillin. Incubate overnight at 37°C.

Yeast antibody–antigen interaction screening of the scFv-VP16 library with the antigen bait

The yeast antibody–antigen in vivo interaction screen followed the strategy of Fields and Song in having a bait, in this case antigen linked to a DNA binding element, such as LexA DNA binding domain (for instance in the fusion vector, pBTM116). To detect those antibody fragments (scFv), which can bind antigen in vivo, the scFv is cloned as a fusion with the VP16 transcriptional activation domain. Yeast containing the plasmid that expresses scFv specific to the antigen and with which it interacts, can grow on medium without histidine, due to transcriptional activation of the *HIS3* gene, and express β-galactosidase, which can be assayed histochemically.

1. Inoculate about 5 colonies of L40 yeast into 150 mL of YPD medium, and vortex mix to ensure that there are no clumps of yeast. Incubate overnight with shaking at 30°C.
2. Add 50–100 mL of this overnight yeast culture into 1 L of prewarmed YPD medium (in a 2-L Erhlenmeyer flask) to produce an OD_{600} between 0.2 to 0.3. Incubate at 30°C for about 3 h with shaking until OD_{600} reaches 0.4 to 0.6.
3. Pellet the yeast by centrifuging at 1000*g* for 5 min at 21°C, and resuspend in 500 mL of sterile distilled water.
4. Centrifuge the yeast cells at 1000*g* for 5 min at 21°C, and resuspend in 8 mL of freshly made sterile 1X TE-LiAc. This is the competent yeast ready for transformation. Keep at 4°C until use (preferably not more than 1 h).

5. Mix 1 mg of bait DNA (LexA-DBD-antigen), 500 μg of yeast scFv-VP16 library DNA, and 20 mg of salmon sperm carrier DNA. Add the mixture of DNA to 8 mL of competent yeast and mix well.
6. Add 60 mL of freshly prepared sterile PEG-LiAc solution and mix vigorously. Incubate at 30°C for 30 min with shaking.
7. Add 7 mL of DMSO and mix gently by inversion. Do not vortex mix.
8. Heat shock at 42°C for 15 min and swirl to mix every 2 to 3 min.
9. Transfer to ice for 2 min and pellet the cells by centrifuging at 1000*g* for 5 min. Resuspend in 500 mL of YC medium (-WLHUK) and incubate in a 1-L flask for 3 h at 30°C with shaking.
10. Pellet the cells by centrifuging at 2000*g* for 10 min and resuspend in 10 mL of 1X TE. Plate 10 μL onto a 100-mm dish of YC(-WL) agar and incubate at 30°C for 3 to 4 d. Count the number of yeast colonies and multiply this number to 1000 to give the number of yeast screened.
11. Plate the remainder of the transformed yeast onto 10 large square plates of YC(-WLHUK) agar. Incubate at 30°C for 4 to 5 d. Pick those yeast colonies that have grown to about 1 to 2 mm in diameter and separately streak onto a new YC(-WLHUK) plate.

Confirmation of antibody–antigen interaction in yeast clones by β-galactosidase filter assay

A β-galactosidase filter assay is performed to confirm the interaction between the antigen and the scFv intracellularly in those colonies that were able to grow in the absence of histidine.

1. Streak the histidine-independent yeast colonies onto a sterile Whatman no. 1 filter paper.
2. The yeast are lysed by immersing in liquid nitrogen. Use a pair of forceps, hold the filter paper and submerge it into a pool of liquid nitrogen for 15 s. Allow to thaw at room temperature for 1 min.
3. Place the filter, colony side up, onto another filter paper presoaked with Z buffer/X-gal solution. Incubate the filters at 30°C and check periodically for the appearance of blue colonies, usually within 1 h. Yeast colonies that are both histidine independent and β-galactosidase positive are further analyzed.

Plasmid DNA extraction from individual selected yeast colonies

After the identification of yeast clones, which can grow in the absence of histidine and activate β-galactosidase, the yeast scFv-VP16 expression plasmid is extracted and retested in fresh transfections. This eliminates false positives due to spontaneous mutation in yeast, giving rise to histidine independence and β-galactosidase activation, independent of the interaction between the antigen bait and the antibody prey.

1. Inoculate the positive yeast colony into 2 mL of YC(-L) medium and incubate overnight at 30°C with shaking in a 20-mL tube.
2. Pellet the cells by spinning in a benchtop microfuge at 11,600*g* for 1 min at room temperature, and resuspend in 200 μL of yeast lysis solution.
3. Add 200 μL phenolchloroform (1:1 v:v) and 0.3 g of acid-washed glass beads. Vortex mix hard for 2 min, and spin at 11,600*g* for 5 min.
4. Carefully pipet out the supernatant (plasmid-containing) without disturbing the interphase. Add 20 μL of 3 *M* sodium acetate, pH 5.2, and 500 μL of 100% ethanol to the supernatant to precipitate the plasmid DNA.
5. Incubate on dry ice for 1 h, and spin again at 11,600*g* for 5 min.
6. Wash the DNA pellet once with 500 μL 70% ethanol, respin, and dry at room temperature for 10–15 min. Do not over-dry using a vacuum.
7. Resuspend the pellet with 20 μL of sterile water. Transform the 20 μL into DH5α cells using the Hanahan method and plate on TYE with 100 μg/mL ampicillin. Incubate the plate overnight at 37°C.
8. Pick several bacterial colonies and inoculate individually into 2 mL of 2X TY containing 100 μg/mL ampicillin. Grow overnight at 37°C with shaking.
9. Extract plasmid DNA from each bacterial overnight culture using QIAprep Spin Miniprep Kit according to manufacturer's instruction. Elute DNA with 50 μL of elution buffer.
10. To check for the identity of yeast scFv-VP16 plasmid, digest 5 μL of DNA with *Sfi*I and *Not*I enzyme and fractionate on a 1.5% agarose gel. The 800- to 900-bp scFv fragment and the 7.5-kb vector band should be seen after digestion.
11. A quick comparison of the various scFv-VP16 clones can be achieved by a gel-DNA fingerprinting method of the product on

agarose gels. Definitive comparison is obtained from DNA sequence comparison of the scFv.

Retesting the scFv-VP16 plasmid using the yeast antibody–antigen interaction assay

Final verification of the intracellular binding of scFv with antigen requires retransfection of individual scFv-VP16 clones with the original antigen bait and assaying histidine-independent growth and β-galactosidase activation. Thus, retesting is similar to the initial yeast in vivo screen, except that it is done on a smaller scale and with individual isolated yeast scFv-VP16 plasmid instead of a library mixture.

1. Prepare competent yeast strain L40.
2. Mix 200 ng of bait plasmid DNA, (this must be a fish bait DNA in this re-test) 100 ng of the isolated yeast scFv-VP16 plasmid DNA, and 100 ng of salmon sperm carrier DNA (final volume should not exceed 10 μL). Add the DNA mixture into 100 μL of competent yeast and mix well.
3. Add 600 μL of freshly made PEG-LiAc solution and vortex mix. Incubate at 30°C for 30 min with shaking.
4. Add 70 μL of DMSO and mix gently by inversion. Do not vortex mix.
5. Heat shock at 42°C for 15 min. Place on ice for 2 min.
6. Pellet the cells by spinning in a benchtop microfuge at 11,600*g* for 10 s. Resuspend cells in 1X TE buffer.
7. Plate the yeast cells onto YC(-WLHUK) agar and incubate at 30°C for 3 to 4 d. Histidine-independent colonies should grow to a size of 1 to 2 mm in diameter. These should be tested for β-galactosidase activity using the β-galactosidase filter assay.

Characterization of the antigen-specific intracellular scFv

The various scFv-VP16 clones can be compared by using DNA fingerprinting after digestion with a frequent cutting restriction enzyme such as *Bst*NI (CCA/TGG). This comprises PCR amplification of insert scFv DNA followed by *Bst*NI digestion and analysis of the product on agarose gel. The isolated intracellular scFv can also be characterized by DNA sequencing of the yeast scFv-VP16 clone.

The primers for PCR and sequence analysis of yeast scFv-VP16 are:

Forward primer: 5'-TTG TTT CTT TTT CTG CAC AAT-3'

Back primer: 5'-CAA CAT GTC CAG ATC GAA-3'

1. Set up a 20-μL PCR with 2 μL of 10X PCR buffer, 2 μL of 25 m*M* $MgCl_2$, 2 μL of 2 m*M* dNTP, 1 μL of each primer (100 ng/mL), 1 mL of 50 ng/mL yeast scFv-VP16 plasmid, 10.8 μL of H_2O, and 1 μL of AmpliTaq polymerase. Overlay the reaction mixture with a drop of mineral oil.
2. Put onto a thermal cycler and heat up to 95°C for 10 min. Denature at 95°C for 1 min, anneal at 55°C for 1 min, and extend at 72°C for 1.5 min, for a total of 25 cycles.
3. Take 5 μL of the PCR and analyze on a 1.4% agarose gel. An approx 900-bp band should be seen.
4. Add to the remaining PCR product 17.5 μL of H_2O, 2 μL of NEB buffer 2, and 0.5 μL of *Bst*NI restriction enzyme. Mix well and incubate at 60°C for 3 h.
5. Fractionate 20 μL of the digested PCR product on a 4% NuSieve 3:1 agarose gel. Run at 75 V for about 1 h. Different clones should give rise to different patterns depending on the occurrence of the *Bst*NI sites.

Binding of scFv to antigen by immunodetection assay (Western blot)

The methods described above facilitate the isolation of scFv antibody fragments that can effectively bind to antigen in vivo for potential applications involving modification of function and antigen-specific cell killing. The soluble scFv fragment can be prepared for use in immunoblots (Western blots) by recloning in phage display vectors such as pHEN2 and production of periplasmic protein in *E. coli*.

1. Set up a digestion of 5 μg yeast scFv-VP16 plasmid DNA with 20 U of *Not*I enzyme in a 50-μL of reaction with 1X NEB buffer 2. Incubate at 37°C for 3 h.
2. Add 40 U of *Sfi*I enzyme to the *Not*I-digested DNA and mix. Cover with a drop of mineral oil and incubate overnight at 50°C. Run the digested sample on a 1.5% agarose gel at 50 V for about 45 min, and cut out the gel containing the scFv DNA fragment of around 800–900 bp. Extract the DNA using QIAEX II Gel Extraction Kit according to manufacturer's instructions. Elute DNA with 20 μL of elution buffer. Digest 1 μg of pHEN2 vector with *Not*I and *Sfi*I enzymes and purify the linearized DNA.
3. Set up a 10-μL ligation reaction with 1 μL of the purified *Sfi*I/*Not*I linear pHEN2 vector, 1 μL of the purified scFv DNA *Sfi*I/*Not*I fragment. 1 μL of T4 DNA ligase, and 1 μL of 10X ligase buffer. Incubate overnight at 15°C.

4. Transform DH5α bacteria with the ligation reaction using the Hanahan method. Identify bacteria containing the ligated plasmid (i.e., pHEN2-scFv) by screening with the scFv fragment or making DNA miniprep and restriction enzyme digestion.
5. For periplasmic protein production, transform HB2151 bacteria with pHEN2-scFv using the Hanahan method. Select colonies on TYE plates with 100 μg/mL ampicillin.
6. Inoculate a single colony of HB2151 containing pHEN2-scFv into 5 mL of 2X TY with 100 μg/mL ampicillin and 1% glucose. Incubate overnight at 30°C with shaking.
7. Put 0.5 mL of the overnight culture into 50 mL of 2X TY with 100 μg/mL ampicillin. Incubate at 30°C with shaking for about 2 to 3 h until OD_{600} is 0.5–0.6.
8. Add 100 μL of 500 m*M* isopropyl-β-Δ-thiogalactopyranoside (IPTG) to the culture to induce scFv protein, and continue incubation at 30°C with shaking for another 3.5 h.
9. Pellet the cells by centrifuging at 2000*g* for 20 min and resuspend in 400 μL of ice-cold 1X TES buffer. Transfer to a 2-mL Eppendorf tube.
10. Add 600 μL of 15 TES buffer (ice-cold) and mix gently by inversion. Place on ice for 30 min.
11. Spin at 4°C in a benchtop microfuge at 11,600*g* for 15 min. Keep the supernatant, which contains the periplasmic protein containing the scFv.
12. Use the periplasmic protein fresh for Western blots. Dilute periplasmic protein to 1:50 for immunodetection for Western blot. Incubate with the antigen immobilized on a nitrocellulose membrane overnight at 4°C with shaking.
13. Use 9E10 anti-myc tag mouse monoclonal antibody as the secondary antibody, because the scFv in pHEN2 is fused in-frame to a myc-tag.

Notes

1. Phage antibody libraries, displaying scFv, consisting of greater than 5 × 10^9 members should be used. In order to select the highest diversity of antigen-specific scFv binders, one round of in vitro panning of the library using antigen-coated immunotubes is used. This selection is carried out under low stringency (high concentration of antigen and short washes) to retain rare binders, resulting in a mixture of specific and nonspecific binding scFv-

phage. Typically, one would expect 10^4–10^5 in vitro binders from a library of about 5 × 10^9 initial complexity.

2. The eluted phage can be kept for about 1 wk at 4°C, but the titer will decrease with time.
3. The number of colonies obtained can be estimated by multiplying the number of colonies counted from to the dilution factor. Typically, 10^4 to 10^5 colonies should be obtained. Between 10^5 to 5 × 10^6 colonies can usually be plated onto one large square plate.
4. When plating the bacteria on agar plates, the surface of the agar should be flat to avoid uneven spreading of bacteria. The surface should also be dried, by leaving the agar plate with the lid off inside a sterile hood for 30 min.
5. The *Sfi*I restriction site (GGCCNNNNNGGCC) is variable. The *Sfi*I site (GGCCCAGCCG GCC) of the yeast VP16 vector described in this protocol is compatible with that found in phage libraries cloned in pHEN/pHEN2.
6. Amplification of the first round phage library and of the yeast scFv-VP16 library should preferably be done by plating on TYE agar plates containing ampicillin. Growing in liquid culture medium may result in loss or preferential enrichment of some clones.
7. For the salmon sperm carrier DNA used in yeast transformation, boiling for 5 min and then cooling to room temperature just before use can increase the transformation efficiency substantially.
8. If using other yeast expression vectors for the scFv-VP16 fusion, make sure that the scFv is cloned in-frame with the VP16. Also, we found that cloning scFv DNA at the 3' end of the VP16 sequence (i.e., fusing scFv at the carboxy terminus of the VP16 molecule instead of the amino terminus) can affect the interaction of the scFv with the antigen to a certain extent. Also a greater efficiency can be achieved by using a yeast bait strain stably expressing the hex A-DBD-Ag to transfect with the scFv-vp16 sub library.
9. The yeast colonies, which grow in the absence of histidine, do so, in theory, due to the interaction between the antigen and the scFv antibody fragment. However, in practice, we find a number of false-positive colonies on HIS-plating. Therefore it is prudent to reassess the colonies by testing their ability to activate the *LacZ* gene and to isolate the scFv-VP16 clone from a selected colony

and retransform this either alone or with the original bait plasmid preparation (the latter is essential, as the bait itself can mutate and give a false positive).

Nonradioactive Labeling and Detection of mRNAs Hybridized onto Nucleic Acid cDNA Arrays

Cellular gene expression changes during ontogenetic development of cell physiological activation–inhibtion and differentiation. Classical molecular assays like Southern, Northern, or Western blotting display only a few genes at once. The analysis of complex alterations of gene expression patterns, therefore, requires large quantities of biological materials, has significant experimental inter- and intravariability, and is quite time-consuming. Some of these problems became negligible with the advent of microarray techniques. cDNAs or oligonucleotids are immobilized on glass or membrane surfaces, cDNA or cRNA transcribed from cellular mRNA is hybridized, and signal detection is performed by radioactive or fluorescent techniques. The expression of up to several tens of thousands of genes can be made visible on small arrays, enabling investigators a rapid quantitative and qualitative analysis of pro- or eukaryotic gene expression patterns.

The biochemical conditions for the *reverse transcription* (RT) procedure of single genes (i.e., for detection in Northern blots) is optimized for each individual gene according to the primers used and the gene sequence. However, the optimal parameters vary between different genes, and it can be suggested that the RT procedure used for the generation of cDNA out of total mRNA might generate a quantitative bias of the mRNA mirror image of RT coupled with *polymerase chain reaction* (PCR)-generated cDNA used for hybridization onto high density gene arrays. This problem has been addressed recently comparing the gene expression patterns using cDNA and mRNA of human fibroblasts hybridized onto cDNA membrane arrays. Applying a novel nonradioactive mRNA labeling technique we have shown that a significant number of genes display a decreased expression rate using cDNA in comparison to hybridized mRNA, indicative of a quantitative bias of the RT step. A typical example of the Atlas Human cDNA Expression Array (receptors, cell surface antigens, and cell adhesion; Clontech) comparing nonradioactive mRNA hybridization using the digoxigenin (DIG) ChemLink labeling technique and the classical radioactive ^{32}P technique. The overall dot images applying both techniques are comparable. However, overexposure of dots and signal cross-talks between dots is a common problem with the ^{32}P technique (i.e., fibronectin receptor β-subunit; white arrow), which is not observed using the nonradioactive hybridization technique. On contrary, genes like IL2 receptor α chain, integrin α3, integrin-αL and β-catenin display a significant increased signal using the mRNA hybridization technique. We assume that this artifical bias is due to artefacts of the RT-step necessary for the generation of ^{32}P-labeled cDNA for array hybridization.

The novel nonradioactive mRNA labeling procedure described below is based on the noncovalent binding of a cisplatin derivative to mRNA. The cisplatin derivative is covalently modified either with biotin or DIG epitops, which are detected by streptavidin or monoclonal antibody binding. Recording of the cisplatin-labeled hybridized mRNA is performed either by luminescence or fluorescence labeling techniques. The advantages using this ChemLink labeling technique are several fold: hybridization of the primary biological reporter molecule mRNA, nonradioactive labeling and significant time-savings. A comparison of the conventional cDNA vs mRNA. This article describes the generation of ChemLink-modified mRNA and its hybridization onto cDNA membrane arrays. mRNAs of various sources have proved to be useful for this novel nonradioactive mRNA labeling and array hybridization technique. Although this technique is developed with fibroblastic cells, it will be very useful in lineage studies using embryonic stem cell cultures.

Materials

1. Phosphate-buffered saline (PBS).
2. Dulbecco's modified Eagle medium (DMEM) Medium.
3. Fetal calf serum (FCS), myoclone.
4. 0.25% (w/v) Trypsin.
5. RNeasy maxi kit.
6. mRNA isolation kit.
7. Phenol, pH 4.3.
8. Phenol:chloroform.
9. Phenol:chloroform:isoamylalcohol (25:λ24:λ1).
10. 8 *M* Lithium chloride (LiCl).
11. DIGChemlink Kit.
12. DIG wash and block buffer set.
13. DIG easy hyb solution.
14. Atlas membrane arrays.
15. CDP-Star ready to use substrate solution.

16. DNA, MB-grade.
17. Cot1 DNA.
18. Stripping buffer (50 m*M* NaOH, 0.. % sodium dodecyl sulfate [SDS]).
19. Stripping wash solutions 1: 2X standard saline citrate (SSC), 1% SDS.
20. Stripping wash solution 2: 0.1X SSC, 1% SDS.
21. Optional: Lumi-Imager.
22. Optional: DIG control teststrips plus DIG quantification teststrips.

Methods

Human fibroblast cell culture

1. Inoculate human fibroblasts in DMEM, 10% FCS in cell culture flasks and expand cell culture until subconfluency. A total of 5 × 10^7 cells are sufficient to extract mRNA for several experiments.
2. Wash the cells twice in culture flasks with PBS (37°C).
3. Add a sufficient volume of trypsin to detach cells from the cell culture flask for 5–10 min at 37°C.
4. Wash the cells twice with PBS (37°C). Discard supernatant to reduce the protein content of the medium. This decreases the risk of clogging the column.
5. Continue with RNA isolation as quickly as possible, as a delay may induce apoptotic or necrotic processes of cells.

Extraction of total RNA

The total RNA isolation described below is performed using the RNeasy Maxi Kit:

1. Resuspend up to 5 × 10^8 cells in 6 mL lysis buffer and shear it 5–6 times through a 18–20-gauge needle fitted to an RNase-free syringe. Do not apply extensive force, as this would result in air bubbles in the lysate.
2. Load the sample onto the column.
3. Wash columns with wash buffers (once with buffer RW1 and twice with buffer RPE) according to the protocol. Spin it down subsequently to drain it.
4. Elute RNA from the column twice in a suitable volume of RNase-free water into two separate elution tubes.
5. Measure the quantity and quality of total RNA using a spectrophotometer at wavelengths 260/280, avoiding absorption coefficients below 0.1 and above 0.6.

Extraction of mRNA

1. Dissolve total RNA in the kits' denaturation buffer, but not more than 2-fold.
2. Add biotin labeled oligo(dT)$_{20}$ primers. Allow mRNA oligonucleotide beads to attach onto streptavidin-coupled magnetic beads for 5 min at 37°C.
3. Wash beads twice in wash buffer.
4. Elute mRNA from beads in at least 100 μL RNase-free water twice for 2 min at 65°C.
5. Measure the quantity and quality of mRNA.
6. Optional: check quality of mRNA on Tris-borote EDTA (TBE) agarose gel. Quality can be checked inspecting residues of 23S and 18S RNA. Band intensities should be greatly reduced using an aliquot of total RNA as a standard.

Increase purity of mRNA

The quality of the mRNA is improved by additional phenol, phenol:chloroform, and phenol:chloroform:isoamylalcohol extraction steps.

1. Add 1 vol of phenol to isolated mRNA. Shake vigorously for 30 s and spin for 5 min in a microcentrifuge at maximum speed.
2. Remove the aqueous (upper) phase and transfer it to a clean tube, carefully avoiding contamination with the lower phase (organic).
3. Repeat steps 1 and 2 with phenol:chloroform and phenol:chloroform: isoamylalcohol.
4. Add 1/10 vol of 8 *M* LiCl to aqueous phase and mix thoroughly.
5. Add 2.5 vol 100% ethanol and mix carefully.
6. Incubate for 30 min at –70°C or overnight at –20°C (formation of precipitates).
7. Recover RNA by centrifugation for 30 min at maximum speed in table-top centrifuge at room temperature.
8. Carefully rinse mRNA pellet in 1 mL 70% ethanol (pellet should be brownish for good quality of RNA, if pellet is white due to high salt precipitation, rinse again in 70% ethanol).
9. Allow pellet to air-dry (do not use a vacuum device, as pellets might get lost) and dissolve in an appropriate amount of RNase-free water to obtain a mRNA concentration of at least 100 ng/μL.
10. Determine the quantity of mRNA.

11. Store aliquots of mRNA at –70°C as frequent freeze-thaw steps decrease the quality of mRNA. Do not store at temperatures above –70°C to decrease hydrolytic activity of water.

Labeling of mRNA with DIGChem link

This is performed using a DIGChemLink kit (components DIGChemLink, stop solution, and anti-digoxingen-AP, Fab fragments).

1. Set up the labeling reaction. Mix 0.5 up to 2 μg mRNA, 0.5 up to 2 μg DIGChemLink, and 20 μL diethyl pyrocarbonate (DEPC) H_2O. For the fibroblasts described here, an amount of 0.5 to 1 μg mRNA was optimal.
2. Vortex mix and spin down for 4 s at 1000*g* force at room temperature.
3. Incubate for 30 min at 85°C (this results in the best labeling efficacy and slightly fragmented mRNAs, which tend to have a better hybridization performance).
4. Spin down for 4 s at 1000*g* at room temperature.
5. Stop reaction by adding 5 μL of stop solution. Short-term storage at 4°C or –20°C (longtime storage not recommended).
6. Check labeling efficiency with DIG control teststrips plus DIG quantification teststrips (labeling efficiency should be at least 0.3 pg/μL).

cDNA array hybridization

1. Dilute 5 mg DNA MB-grade (heat-denatured for 5 min at 95°C) in 50 mL of DIG easy hyb and heat up to 50°C (prehybridization mixture).
2. Place cDNA array in a roller bottle without overlapping the membrane parts. Pour prehybridization mixture into roller bottle.
3. Incubate for 1 h at 50°C. Discard supernatant.
4. Dissolve labeled mRNA to a final concentration of 250–500 ng/mL in DIG easy hyb (8 mL in a roller bottle of 150 mm height). Incubate 5 μg of human Cot1 DNA in 100 μL RNase-free H_2O for 5 min at 95°C, mix well, spin down for 5 s at 1000*g* at room temperature, and add to DIG easy hyb.
5. Transfer hybridization solution to the array.
6. Hybridize overnight at 50°C.
7. Heat wash solutions up to 50°C. (stripping wash solution 1: 2X SSC, 1% SDS; and stripping wash solution 2: 0.1X SSC, 1% SDS).

8. Wash membrane with wash buffer 1 twice for 5 min at 50°C, and, subsequently, 2 times with wash buffer 2 for 15 min at 50°C, and proceed with DIG detection.

DIG detection and luminescence recording

1. Wash membrane array in DIG wash buffer for 5 min at room temperature.
2. Incubate for 30 min in DIG blocking buffer and discard supernatant.
3. Incubate for 30 min in anti-DIG antibody solution and discard supernatant.
4. Wash membrane array twice in DIG wash buffer for 15 min each at room temperature.
5. Incubate in DIG detection buffer for 5 min at room temperature and discard supernatant.
6. The detection of DIG-labeled RNA is performed by luminescence technique using a Lumi-Imager instrument for signal recording. Use CDP-Star as a substrate for peroxidase enzyme coupled to the DIG antibody. For this purpose, put membrane on a solid plastic sheet (do not use saran wrap) and drop 2 mL of the substrate close to the membrane (do not drop it on the membrane as this might result in an irregular signal intensity on the membrane). Take a second plastic sheet and allow it to glide onto the membrane preparing an even fluidic film, avoiding air bubbles on the membrane.
7. Incubate for 5 min at 37°C in dark.
8. Put the cDNA membrane in a Lumi-Imager Device using suitable software, including nearest neighbor background subtraction algorithms.

Reuse of cDNA array membranes

Nucleic acid arrays can be reused limited times performing a stripping and reprobing procedure. The following protocol is designed for usage on mRNA-cDNA hybrids, which display a higher affinity compared to cDNA-cDNA hybrids.

1. Incubate cDNA membranes in RNase-free water for 1 min to wash off substrate.
2. Gently agitate membrane in stripping buffer for 5 min at 37°C to remove mRNA from the membranes and discard supernatant. Repeat procedure a second time.
3. Rinse in 2X SSC, 1% SDS for 5 min at room temperature and subsequently in 0.1X SSC, 1% SDS.

4. Array membrane can be stored for a short time (up to some days) semi-dry at 4°C. For long-term storage, membrane must be stored at –20°C.

Notes

1. Although culture of human fibroblast cell culture is described, any cell type in addition to the human fibroblasts will be applicable to the DIGChemLink mRNA labeling technique.
2. The quality of RNA used for the procedure is of utmost importance, and, for this reason, it is of utmost importance to follow the extraction protocol of the RNeasy maxi kit as strictly as possible. The general guidelines working with RNA must be exercised.
3. For optimal extraction of total RNA, do not exceed the capacity of the column, for this would result in a lower performance and extraction quality.
4. Residual alcohol in the column reduces the quantity and quality of total RNA. Extraction of mRNA is performed in a 2-step procedure, as this increases stability and purity of mRNA.
5. The mRNA isolation kit is applied to extract mRNA out of total RNA. To extract mRNA post-translational polyadenylation is used, which is a common feature of the biogenesis of most eukaryotic mRNAs. The poly(A) tail allows mRNA extraction from total RNA using biotin-labeled oligo$(dT)_{20}$ primers bound to streptavidin-coupled magnetic beads. Do not allow the magnetic particles used for the mRNA extraction to dry out, as this may result in a non-reversible binding of mRNA to the beads.
6. The yield of mRNA extracted from total RNA should result in a yield of smaller than 5% of total RNA depending on the cells used.
7. Enrichment of mRNA from total RNA can be checked with a TBE gel comparing residual bands of ribosomal RNA in mRNA compared to total RNA.
8. For direct labeling of mRNA, the ChemLink molecule is utilized in the DIGChemLink kit. It consists of a digoxigenin-modified cisplatin, which binds noncovalently to mRNA. The pool of labeled mRNAs is hybridized onto cDNA membranes, and the binding of DIGChemLink to mRNA is identified by reporter molecule-labeled anti-DIG antibodies.
9. For prehybridization–hybridization, Cot 1 DNA and MB-grade DNA serve as competitor nucleic acids to reduce probe affinity to repetitive elements.

10. For hybridization, keep membranes wet and never let them dry out. Dry areas result in strong background signals. Use of roller bottles gives lower background heterogeneity on membranes compared to hybridization bags in a water bath.

Phage-Displayed Antibodies to Detect Cell Markers

Immunologic reagents such as antibodies can be valuable for identifying and isolating cells with particular characteristics. While antibodies derived from immunized animals are the most common immunologic reagents used for this purpose, "*antibodies*" selected by phage display have also proven to be useful. The advantage of this technique is that one can relatively rapidly clone specific antibodies that will be available in unlimited supply. Additionally, phage display may allow for the cloning of antibodies to antigens to which an animal's immune system would not normally respond due to tolerance. Finally, strategies have been developed to select for some antibodies and deselect for other antibodies using phage display approaches. This can enable investigators to select for antibodies that, for instance, bind to antigens on the surface of one cell type but do not bind to antigens on the surface of another cell type.

In antibody phage display, a portion of an antibody is expressed on a filamentous bacteriophage such as M13 or f1. A fusion gene encoding an antibody fused to the pIII bacteriophage surface protein is cloned in a phagemid. Bacteria harboring the phagemid are infected with helper phage. The bacteria produce phage particles that display the antibody attached to the pIII protein on their surface and contain the phagemid DNA. Phage particles displaying antibodies that bind to a target antigen can be selected, and the antibody gene can be isolated from the phagemid in the phage particles.

Antibody phage display is still a developing technique, and several problems may be encountered. Most of these problems are related to the fact that the antibodies isolated from phage display libraries are not usually complete antibody molecules, but instead are Fabs or single-chain fusion proteins of variable chains (scFvs). Unlike full-length immunoglobulin proteins that must be glycosylated, Fabs and scFvs can usually be expressed in bacteria. However, Fabs, and especially scFvs, may not always fold in the desired conformation and may form aggregates that may precipitate in some solutions. Furthermore, enzyme and fluorochrome-conjugated anti-IgG antibodies do not recognize scFvs and Fabs. Thus, if a scFv or Fab is to be visualized (for staining cells, for example), it must be tagged with an epitope that is recognized by a secondary antibody.

Prior to cloning an antibody, one first has to make or obtain a library. One can use a naive library constructed from germ line variable genes, or one can use a library made from the immune system of a previously immunized animal or person.

There are two potential advantages with using a library made from a previously immunized animal. First, the antibody of interest will be enriched. This advantage is minor, since in vitro selection methods are quite efficient, and this advantage might be nonexistent if the animal fails to respond to the antigen. Probably more important, however, is the fact that the in vivo immune response involves selection and mutation (affinity maturation) of antibodies. Hence, some antibodies that would not be present in a library made from the germ line genes might be present in a library made from DNA (or cDNA) from lymphocytes of immunized animals.

There are also advantages with using a naive library. Good naive libraries contain antibodies that would normally be deleted from an animal's immune system by tolerance mechanisms. Furthermore, some good naive libraries have already been constructed by investigators who are usually willing to share the resource for noncommercial purposes. This allows one to bypass substantial work.

Given these considerations, we recommend initially cloning antibodies from a naive library. Two groups have made phage display libraries of high complexity that have been widely used. The methods we describe here focus on using the library from Winter's group known as the Griffin.1 library, but most of these methods should apply to using other antibody phage display libraries.

Once the library is obtained, one selects for phage that bind to the antigen of interest and amplify those phage in bacteria. As a control, one can select for phage that are selected in the absence of the antigen of interest. The process is repeated approximately 3-6 times. After each selection, one can calculate the ratio of the number of phage selected in the presence of antigen to the number of phage selected in the absence of antigen. The higher this ratio, the higher the proportion of selected phage that display antibodies that bind to the target antigen. In addition, one can also determine whether the process is succeeding by using an *enzyme-linked immunosorbent assay* (ELISA) or similar assay. This is not usually worthwhile after the first selection and amplification, because the clone(s) of interest are likely to be very dilute at this stage.

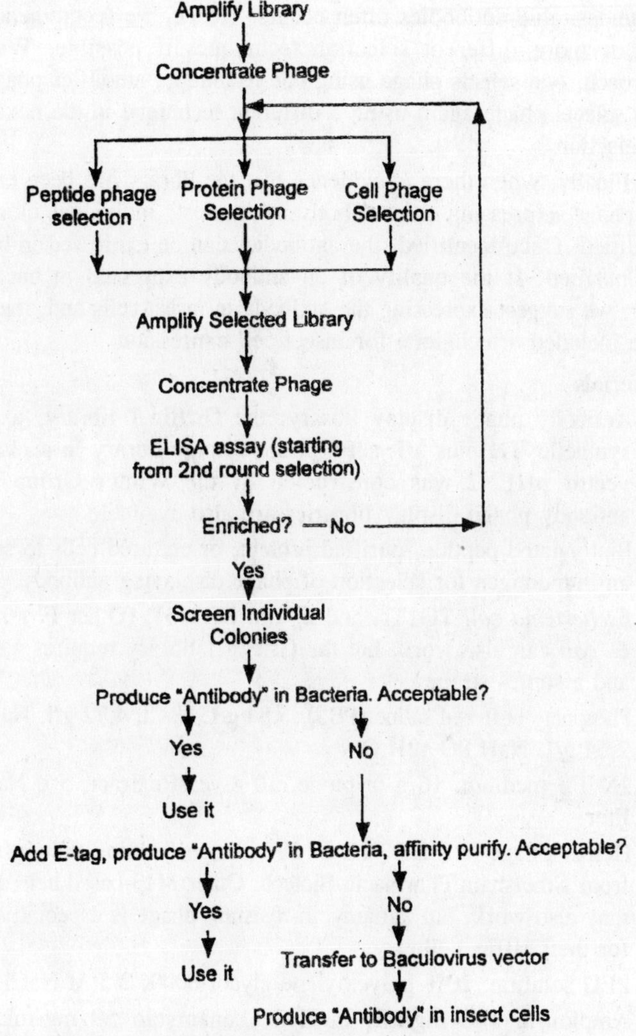

Fig. 12.1. Overview of antibody phase selection strategy.

Investigators may desire to make antibodies to cloned purified antigens, moderately pure antigens, or whole cells. In this chapter, we describe a cell selection technique, a protein selection technique, and a peptide selection technique. Undesirable antibodies that bind to contaminating antigens may be selected with any technique. The purer the antigen preparation, the lower the risk of selecting for undesirable antibodies. However, no selection technique is perfect, and selection for undesirable antibodies often occurs. Hence, we recommend using two or more different selection techniques if possible. With this approach, one selects phage using one technique, amplifies phage, and then selects phage again using a different technique in the next round of selection.

Finally, when there is evidence that the library has been enriched for phage expressing the antibody of interest, individual clones are identified. Once identified, the antibodies can be expressed in bacteria and purified. If the quality of an antibody expressed in bacteria is poor, we suggest expressing the antibody in insect cells and, therefore, have included a technique for insect cell expression.

Materials

1. Antibody phage display library: the Griffin.1 library, a human synthetic VH plus VL scFv phage display library in a phagemid vector pHEN2 was constructed by the Winter Group. Other antibody phage display libraries are also available.
2. Biotinylated peptide, purified protein, or cultured cells to serve as an immunogen for selection of phage displaying antibody.
3. *Escherichia coli* TG1Tr, and *E. coli* HB2151. (Other F' strains of *E. coli* can also work, but the Griffin.1 library requires a supE+ and a supE– strain.)
4. Phosphate-buffered saline (PBS): 5.84 g/L NaCl, 4.72 g/L Na_2HPO_4, 2.64 g/L $NaH_2PO_4 \bullet 2H_2O$
5. 2X TY medium: 16 g tryptone, 10 g yeast extract, 5 g NaCl per liter.
6. Helper phage M13KO7 (1×10^{12} plaque forming unit [pfu]/mL) from Amersham Pharmacia Biotech. Other M13-based helper phage may also work, but kanamycin-resistant phage is especially useful for the Griffin.1 library.
7. PEG solution: 20% polyethylene glycol 6000, 2.5 *M* NaCl.
8. Ampicillin 100 mg/mL in H_2O; kanamycin 50 mg/mL H_2O; isopropyl-β-D-thiogalactopyramoside (IPTG) 100 m*M* in H_2O. Each should be filter-sterilized and stored at –20°C.
9. Anti-M13 Horseradish peroxidase (HRP) conjugate, anti-myc, anti-mouse HRP conjugate.
10. ABTS: 2,2'-Azino-bis(3-ethylbenzothiazoline-6-sulfonic acid) diammonium salt: 2.2 mg in 10 mL sterile 0.05 *M* citric acid (pH 4.0), store at 4°C. The solution is stable for at least 2 wk.
11. H_2O_2 30%.

12. HRP-conjugated anti-E tag, which recognizes the synthetic E-tag amino acid sequence GAPVPYPDPLEPR.
13. Streptavidin magnetic beads.
14. Triethylamine.
15. BaculoGold Starter Package.
16. RPAS Purification Module.
17. Treated microtiter plates.
18. Magnetic particle concentrator.
19. TYE: 15 g bacto-agar, 8 g NaCl, 10 g tryptone, 5 g yeast extract per liter.

Methods

Phage display library amplification

This protocol was developed for the Griffin.1 library, but should work for most libraries cloned into phagemids.

1. Grow *E. coli* bacteria harboring the antibody phage display library in 500 mL 2X TY media with 100 μg/mL ampicillin, 1% glucose, shaking at 37°C.
2. When the OD_{600} = 0.5, add M13 KO7 helper phage to an aliquot of 25 mL of bacteria at a ratio of 120 (bacteria:helper phage) and incubate without shaking at 37°C for 30 min to infect the bacteria.
3. Shake the remaining 475 mL bacteria at 37°C for another 2 h, then centrifuge at 3300*g* for 10 min at room temperature, and resuspend in 10 mL 2X TY containing 15% glycerol. Store at –80°C.
4. Centrifuge the helper phage-infected bacteria in step 3 at 3300*g* for 10 min at room temperature, resuspend in 500 mL of prewarmed 2X TY containing 100 μg/mL ampicillin and 25 μg/mL kanamycin, and shake overnight at 30°C.
5. Centrifuge the culture at 3300*g* for 10 min, and transfer the supernatant to a clean tube. Centrifuge the supernatant again at 10,800*g* for 10 min, and transfer the supernatant containing phage to a clean container. The solution is now ready for phage concentration.

Concentration of unselected phage

1. Mix supernatant containing phage with 1/5 volume PEG solution, chill at 4°C for at least 1 h.
2. Centrifuge at 10,800*g* for 30 min. Discard the supernatant.
3. Resuspend the phage pellet in 40 mL PBS. Centrifuge at 10,800g for 5 min.

4. Transfer the supernatant to a new tube, mix with 1/5 volume PEG, and incubate at 4°C for at least 20 min.
5. Centrifuge at 10,800*g* for 10 min at 4°C. Discard the supernatant.
6. Resuspend the phage pellet in 5 mL of PBS to form a concentrated phage solution. Start selection in no more than 2 h following resuspension.

Peptide bead phage selection

In this technique, phage-expressing antibodies that bind to a specific peptide are selected. This technique requires biotinylated antigen. We have used biotinylated peptide, which can be synthesized by one of a number of peptide synthesis services. Theoretically, a biotinylated protein could also be used, but there would be no significant advantage over coated protein selection. Peptide phage selection is a modified version of a technique described elsewhere. One disadvantage of this technique is that antibodies to streptavidin are also selected and phage binding to streptavidin can out-compete almost everything else in the library. This problem can be minimized by preclearing the library and/or alternating this selection technique with other technique(s) in different selection cycles.

An advantage of this technique is that one can simultaneously deselect for some antibodies and select for others by adding a competing nonbiotinylated antigen in excess. For example, one could select for antibodies that bind to CD34 and deselect those that bind to c-kit.

1. Separately block 1 mL concentrated phage and streptavidin magnetic beads in PBS-2% fat free milk powder at room temperature for 30 min.
2. Add 100 μL of streptavidin magnetic beads to the concentrated phage and rotate for 0.5 h on a rotator. This step removes phage that bind to irrelevant antigens such as streptavidin or other epitopes on the magnetic beads.
3. Recover streptavidin magnetic beads from the solution by putting the tube on a magnetic rack for 2 min. Transfer the supernatant to a new tube, and discard the magnetic beads.
4. Repeat steps 2 and 3 once. Transfer the supernatant to a new tube.
5. Add biotinylated peptide to the phage solution to a final concentration of 20 n*M* (first round) or 2 n*M* (subsequent rounds), and incubate the mixture for 1–2 h. Optionally, add 100 to 1000-fold excess of a competing nonbiotinylated protein or peptide for simultaneous negative selection.

6. Add 50 μL of the milk powder-blocked streptavidin magnetic beads (for the first round) or 5 μL (for the subsequent rounds) to the phage–peptide solution, and rotate the solution for 15 min.
7. Recover streptavidin magnetic beads from the solution by putting the tube on a magnetic rack for 2 min.
8. Wash the beads by resuspending the beads in PBS, 0.1% Tween-20. Recover the beads using a magnetic rack.
9. Repeat washing the beads 2 more times with PBS, 0.1% Tween-20.
10. Wash the beads 3 times as in step 8 with PBS, and resuspend the beads in 100 μL PBS.
11. Infect 1 mL of exponentially growing TG1 *E. coli.* bacteria with selected phage by adding 1/2 of the magnetic beads in PBS and incubate at 37°C without shaking for 30 min.
12. Spread the bacteria on 2× TY plates containing 1% glucose and 100 μg/mL ampicillin, and grow at 30°C for 12–16 h.

Coated protein phage selection

In this technique, phage-expressing antibody that bind to a specific protein are selected. This technique was described elsewhere previously. In the original descriptions, phage were selected on protein-coated immunotubes. We have found this method to work well in the same microtiter tray wells normally used for ELISAs.

1. Coat 3 microtiter plate wells with purified protein in PBS (10 μg/mL) for 12–18 h at room temperature.
2. Separately block microtiter wells and phage with PBS-2% fat free milk powder for at least 2 h at room temperature.
3. Incubate concentrated phage solution (concentration approx 10^{12}–10^{13} pfu/mL) in microtiter wells for at least 1 h at room temperature.
4. Discard the phage solution, and wash the microtiter wells ten times with PBS, 0.1% Tween-20.
5. Wash the tube ten times with PBS.
6. Elute selected phage by adding freshly diluted 100 m*M* triethylamine to the immunotubes, and incubate for 10 minutes at room temperature.
7. Transfer the elution solution to a clean tube and neutralize the elution solution with 1/2 vol 1 *M* Tris-HCl, pH 7.4, and store at 4°C.

8. Use 1/5 to 1/2 of the neutralized elution solution to infect exponentially growing TG1 *E. coli* bacteria by mixing the bacteria and the elution solution for 30 min at 37°C without shaking. Then spread the mixture onto 2× TY, 1% glucose, 100 μg/mL ampicillin plates and incubate at 30°C overnight.

Cell phage selection

In this technique, which has been previously described, phage-expressing antibodies, which bind to specific cultured cells, are selected. Other techniques using fixed cells have also been described, but we prefer this technique, in which live cells with native antigens are exposed.

1. Grow cells in a 24-cm^2 tissue culture flask until almost confluent.
2. Change media three times.
3. Incubate for 1 h at 37°C.
4. Remove media. Add 10^{11}–10^{13} pfu phage in 2 mL media with supplements such as serum.
5. Shake flask gently at room temperature for 2 h.
6. Remove media and rinse rapidly ten times with PBS at room temperature.
7. Add 2 mL elution buffer (0.1 *M* glycine, pH 2.2, 0.1% bovine serum albumin [BSA]) for 10 min at room temperature.
8. Neutralize by adding 0.375 mL 1 *M* Tris-HCl, pH 9.1. Phage are in neutralized buffer. Remove buffer containing phage from flask and use to infect bacteria.

Growth of selected libraries

After selecting library and infecting bacteria, the infected bacteria harboring the selected library are stored. An aliquot is grown, and phage are produced by the bacteria. These phage are concentrated and can be assayed by ELISA and/or can be subjected to further rounds of selection.

1. Add 2× TY, 15% glycerol to a plate of bacterial colonies. Add about 0.5–1 mL to a 100-mm Petri dish or 5 to 6 mL to a bioassay dish.
2. Spread 2X TY, 15% glycerol with glass spreader to loosen colonies.
3. Inoculate 50–100 μL of bacteria into 100 mL of in 2X TY, 1% glucose, and 100 μg/mL ampicillin. Freeze the rest of the bacteria at -70°C for storage.
4. Grow with shaking at 37°C for about 2 h until $OD_{600} \approx 0.5$.

5. Infect 10 mL of culture with M13K07 helper phage by adding phage at a ratio of 20:1 (helper phage:bacteria). Incubate at 37°C without shaking for 30 min.
6. Centrifuge at 3300*g* for 10 min, and resuspend the pellet in 50 mL 2X TY, 100 μg/mL ampicillin, and 25 μg/mL kanamycin. Shake at 30°C overnight.
7. Centrifuge 40 mL at 10,800*g* for 30 min and add 8 mL PEG-NaCl to the supernatant. Mix and leave at 4°C for at least 1 h for the phage to precipitate.
8. Centrifuge at 10,800*g* for 10 min or at 3300*g* for 30 min to remove the supernatant, respin briefly, and aspirate any remaining PEG-NaCl.
9. Resuspend pellet in 1 to 2 mL PBS, then spin at 11,600*g* for 10 min in a microcentrifuge.
10. Transfer supernatant to fresh tubes and store phage at 4°C.

ELISA using phagemid particles or single-chain antibody

ELISAs are performed at three stages in the process. First, an ELISA can be performed using the polyclonal selected library to help determine whether the selection is succeeding. Second, an ELISA can be performed on individual phage clones to identify the phage of interest. Third, an ELISA can be performed to test the quality of a fresh preparation of a previously cloned antibody. ELISAs are performed essentially as previously described.

1. Phage supernatant, concentrated phage and bacteria culture supernatant containing singlechain antibody or purified single-chain antibody can be used in ELISA.
2. Coat microtiter plate wells with 200 μL of the immunogen protein (1–10 μg/mL PBS) for 2–18 h. Alternatively, coat wells with cell extracts or other cell preparations that contain the target antigen.
3. Block the wells with 2% dry milk in PBS (blocking buffer). Block phage or single-chain antibody with 1% dry milk in PBS.
4. Incubate single-clone phage supernatant or concentrated selected phage library or single-chain antibody (1–10 μg/mL) in microtiter wells for 1 h, then wash wells six times with PBS, 0.05% Tween-20.
5. Add 200 μL HRP conjugated:
   (a) anti-M13 antisera (for phage ELISA) diluted 5000:1 in blocking buffer; OR

(b) anti-epitope tag antibody such as anti-myc (clone 9E10) for the Griffin.1 library, or anti-E-tag antibody diluted 1000:1 in blocking buffer.

6. Incubate for at least 1 h.
7. Wash the wells six times with PBS, 0.05% Tween-20.
8. Mix ABTS substrate solution with H_2O_2 (21 mL ABTS with 36 μL H_2O_2). Then add 200 μL of the ABTS-H_2O_2 mixture to the wells for color development at room temperature or at 37°C. Measure optical density (415 nm) in a microplate reader 30 min after substrate is added.

Screening individual clones

The number of clones that should be picked depends on the degree of enrichment of the library, which is determined by the ratio of the clones selected in the presence of an antigen to the clones selected in the absence of an antigen. From poorly enriched libraries, one may want to test 96 clones; from highly enriched clones, one could test far fewer clones.

1. Pick individual colonies of TG1 *E. coli* harboring phagemids into 100 μL of 2X TY media with 1% glucose and 100 μg/mL ampicillin in a 96-well microtiter plate, and grow overnight at 37°C with vigorous shaking.
2. Transfer 2 μL of these cultures into a second microtiter plate containing 200 μL fresh 2X TY media with 1% glucose and 100 μg/mL ampicillin per well, and shake vigorously for 1 h.
3. Make glycerol stocks of the original 96-well plate by adding glycerol to a final concentration of 15%, and then store the plate at -80°C.
4. To each well of the second plate, add 25 μL 2X TY containing 100 μg/mL ampicillin and 10^9 pfu M13-K07 helper phage.
5. Stand for 30 min at 37°C, then shake vigorously for 1 h at 37°C.
6. Spin the plate at 1800*g* for 10 min at room temperature, and then aspirate off the supernatant.
7. Add 200 μL 2X TY containing 100 μg/mL ampicillin and 50 μg/mL kanamycin to each well to resuspend the bacteria pellet. Shake overnight at 30°C.
8. Spin at 1800*g* for 10 min, and use the supernatant from these cultures in phage ELISA assays.

Expressing single-chain antibody in E. coli strain HB2151

The Griffin.1 library vector contains an amber stop codon between the scFv gene and the region encoding the pIII phage protein. This amber stop codon is not recognized as a stop codon in TG1 bacteria (supE strain), resulting in expression of the fusion protein. In HB2151, the codon is recognized as a stop codon, and scFvs are produced. For other libraries, it may be necessary to subclone the antibody gene into a bacterial expression vector to express the antibody not fused to the pIII protein.

When expressed in bacteria, the cloned antibody may be secreted into the culture media or may accumulate in the bacterial periplasmic space. The location of antibody expression of each clone must be determined empirically as follows:

1. Infect 200 μL of exponentially growing HB2151 bacteria by adding 10 μL of supernatant and incubating 37°C for 30 min.
2. Plate about 1 μL of infected bacteria diluted in 100 μL of 2X TY on a TYE plate, 100 μg/mL ampicillin, 1% glucose, and incubate overnight at 37°C.
3. Pick colonies and grow in 2X TY with 100 μg/mL ampicillin, 1% glucose overnight.
4. For harvesting antibodies from supernatant
   (a) Transfer 100 μL of the overnight culture into 100 mL 2X TY with 1% glucose, 100 μg/mL ampicillin, and shake vigorously at 37°C until OD_{600} reaches 0.5–0.9.
   (b) Centrifuge at 3300*g* for 10 min at room temperature and discard the supernatant.
   (c) Resuspend the bacteria pellet in an equal vol of 2X TY containing 100 μg/mL ampicillin and 1 m*M* IPTG. Shake the culture at 30°C for 12–16 h.
   (d) Centrifuge at 3300*g* for 10 min. Transfer the supernatant to a clean tube and centrifuge again at 10,000*g* for 10 min at room temperature.
   (e) Transfer the supernatant to a clean container and neutralize with HCl (final pH 7.0–8.0).
   (f) If desired, test supernatant in an ELISA.
   (g) Purify antibody.
5. For harvesting antibodies from periplasmic space:
   (a) Innoculate 50 mL 2X TY media with 2% glucose, 100 μg/mL ampicillin.
   (b) When the OD_{600} reaches 0.7–1.0, centrifuge culture at 3000*g* for 10 min at room temperature. Resuspend in 50 mL 2X TY with 100 μg/mL ampicillin, and 1 m*M* IPTG.

(c) Grow at 30°C for 3–6 h. Centrifuge at 3000*g* at 4°C for 15 min. Resuspend in cold PBS, 1 *M* NaCl, 1 m*M* EDTA. Keep on ice for 15 min.

(d) Centrifuge at 3000*g* at 4°C for 15 min. Transfer supernatant containing the periplasmic fraction to a new tube. Add $MgCl_2$ to final concentration of 1 to 2 m*M*.

(e) If desired, test supernatant in an ELISA.

(f) Purify antibody.

Expressing single-chain antibody in baculovirus system

In our experience, scFvs expressed and secreted from insect cells are usually in the correct conformation and not significantly degraded. To express an antibody in insect cells, the antibody gene must be subcloned into a baculovirus expression vector, which encodes a signal sequence for protein secretion that will be fused to the antibody. The antibody can be recovered from the culture media. One can start with the E-tagged construct or with the original cloned antibody.

1. Using custom designed primers based on the sequence of the antibody, amplify the antibody gene, keeping the *Nco*I site at the 5' end of the gene, and adding an *Eco*RI site to the 3' end of the gene. Digest the *polymerase chain reaction* (PCR)-amplified DNA with *Nco*I and *Eco*RI, and purify the digested DNA using standard molecular biology techniques. Ligate this DNA fragment into the *Nco*I and *Eco*RI sites of Baculovirus vector pAcGP67B resulting in a pAcGP67B-scFv construct.
2. Transfect the pAcGP67B-SvAb DNA into insect cell line Sf9, according to the BaculoGold Starter Package's instruction manual, and harvest the culture supernatant containing virus particles containing the antibody cDNA by centrifuging the culture supernatant at 3300*g* for 10 min, and store the supernatant at 4°C. The virus in the supernatant will be stable for 6 mo.
3. Amplify the virus following the BaculoGold Starter Package's instruction manual, so that the titer of the virus particles in the supernatant reaches 2×10^8.
4. Infect Sf9 cells with amplified baculovirus containing single-chain antibody cDNA at a ratio of 1:3–10 (Sf9 cells:λvirus).
5. After 3 d at 27°C, Sf9 culture supernatant should contain a sufficient concentration of antibody for most purposes.
6. Harvest Sf9 culture supernatant expressing antibody by centrifugation at 3300*g* for 10 min at room temperature.

7. Transfer the supernatant to a clean container and neutralize with 1/10 volume of 1 *M* pH 7.4 Tris-HCl.

Purifying antibodies

For some purposes, satisfactory purification can be achieved using a nickle column to purify a histidine-tagged antibody such as those cloned from the Griffin.1 library. In addition, some antibodies bind to protein A and/or protein L, and can be purified using protein A or protein L affinity columns. We have had success with immunoaffinity purification, which results in a purer preparation of functional antibody than nickle columns. To this end, we suggest adding an "E-tag" and purifying antibodies using the RPAS module from Amersham-Pharmacia Biotech. Using this procedure, the concentration of the purified antibody in the most concentrated fraction is usually between 200–400 μg/mL. The purified antibody can be used in a range of 1–10 μg/mL for ELISA and Western blot assays. Higher concentrations may be needed for staining cells.

1. Introduce an artificial E-tag to the 3' end of the antibody cDNA by PCR using selected positive single-chain antibody single-clone phagemid DNA as the template. The nucleotide sequence of E-tag is: 5'-GGT GCG CCG GTG CCG TAT CCG GAT CCG CTG GAA CCG CGT-3'.
2. After ligating the PCR product to the plasmid DNA, transform competent *E. coli* cells. Spread the bacteria onto a 2X TY 1% glucose 100 μg/mL ampicillin plate, and incubate at 30°C overnight. Pick a single colony in 3 mL 2X TY, 1% glucose, 100 μg/mL ampicillin, and grow overnight at 37°C.
3. Grow and harvest supernatant or periplasmic fraction or express the antibody in insect cells.
4. Filter the supernatant through a 0.45-μm syringe filter.
5. Use the anti-E-tag column provided in RPAS module to purify the antibody.
   (a) Wash the new column with 15 mL of elution buffer (provided with the RPAS module) by pushing through the elution buffer with a syringe at a rate of 5 mL/min.
   (b) Equilibrate by applying 25 mL of binding buffer (provided with the RPAS module).
   (c) Bind antibody by applying neutralized filtered extract or supernatant to the column.
   (d) Wash column with 25 mL of binding buffer.

   (e) Elute with 15 mL of elution buffer. Discard the first 4.5 mL. Collect the next five 1-mL fractions into tubes containing 100 μL of neutralizing buffer (provided with the RPAS module). The neutralizing buffer should be 1/10 the volume of the eluted antibody.
   (f) Re-equilibrate the column with 25 mL of binding buffer.
6. Store the single-chain antibody at 4°C.

Notes

1. A phagemid is a plasmid that contains a phage origin of replication and will be incorporated into phage particles. Although most antibodies displayed on pIII are fused to pIII, theoretically, other phage surface proteins such as pVIII could also be used.
2. Precautions need to be taken throughout the protocol to avoid any carryover of phage from flasks and centrifuge bottles, etc., as autoclaving alone is not sufficient enough to remove all phage contamination. All nondisposable plasticware can be soaked for 1 h in 2% (v/v) hypochlorite (>1000 ppm free chlorine), followed by extensive washing, and then autoclaving. Glassware should be baked at 200°C for at least 4 h. The use of polypropylene tubes is recommended as phage may absorb nonspecifically to other plastics.
3. Glucose is added when bacteria are grown to amplify the library. Glucose represses transcription of the Griffin.1 library or any other library whose transcription is driven by the lac promoter. This improves growth of the bacteria and removes selective advantages that some clones have over others during growth.
4. The Griffin.1 library confers ampicillin resistance to bacteria. If another phagemid library carries a different antibiotic resistance gene, then the corresponding antibiotic should be substituted for ampicillin.
5. Other helper phage such as VCSM13 (Stratagene) also work. It is advantageous to use a helper phage that carries a kanamycin resistance gene, so that the helper phage and the phagemid can be simultaneously selected with two different antibiotics.
6. In contrast to the other selection techniques, the phage do not need to be eluted from the solid support (beads). Phage can be eluted from the beads using acid, but in our experience this does not improve the results.
7. After the first round, one needs to plate many colonies. We recommend using a Nunc Bioassay dish or a plate with similar

dimensions or multiple smaller plates. With further rounds of selection, fewer colonies are needed, and 100-mm round Petri dishes are convenient.

8. To empty the wash solution, the microtiter plate is usually inverted and drained on a paper towel. Although the plate can be tapped onto the paper towel, tapping with too much force can disrupt binding of desirable phage.
9. By using more stringent conditions to elute phage in coated protein phage selection, higher affinity single-chain antibody can be obtained.
10. At an OD_{600} = 0.5, the bacterial concentration is about 4×10^8 bacteria/mL.
11. For individual clones and for moderately enriched polyclonal phage libraries, the reaction is easily identified by the blue-green color. The microplate reader provides a quantitative measurement that may be useful for choosing individual clones.
12. Expression of single-chain antibody in the HB2151 *E. coli* strain may not be consistent. Sometimes a significant proportion of antibody may be degraded or incorrectly folded. Rather than spending significant time troubleshooting, we recommend expressing the antibody in the baculovirus system if the bacterially-expressed antibody is of poor quality.
13. The antibody must be fused in frame with the signal sequence for protein secretion. Antibodies cloned in the *Nco*I site of pHEN2 will be in the correct reading frame using the strategy described. If not using an antibody cloned into the *Nco*I site of pHEN2, then consider using pAcGP67A or pAcGP67C, which are in different reading frames.
14. Instead of the Sf9 cell line, another insect cell line such as Sf21 may be used.
15. Since yeastolate, which is supplemented in some of the TNM-FH media, can cause severe precipitation of scFvs when neutralizing the media with 1*M* Tris-HCl, pH 7.4, Hink's TNM-FH Medium without yeastolate should be used. The antibody will be expressed equally well in this media.
16. BAC purity note: baculovirus culture media has fewer interfering substances than bacterial culture media, and antibody in the supernatant may not need to be purified.
17. For nickle column purification, please see protocols from Qiagen or Novagen.

18. The goal is to be able to pick individual colonies. The amount of infected bacteria can be adjusted.
19. The bacteria should be supE-, such as strain HB2151, if an amber stop codon needs to be recognized as a stop codon. This is true for antibodies in the Griffin.1 library.
20. BSA (15 mg/mL) can be added to stabilize the single chain antibody. Thymerosal 0.01% can be added as a preservative.
21. Antibodies and specifically scFvs will degrade over time. Each specific clone has a different shelf life, which usually varies between a week to a few months. The quality of the preparation also affects the shelf life, with preparations from older affinity columns producing preparations with shorter shelf lives. We recommend using the purified antibody shortly after purification.

INDEX